CAMBRIDGE LIBRARY COLLECTION

Books of enduring scholarly value

Darwin

Two hundred years after his birth and 150 years after the publication of 'On the Origin of Species', Charles Darwin and his theories are still the focus of worldwide attention. This series offers not only works by Darwin, but also the writings of his mentors in Cambridge and elsewhere, and a survey of the impassioned scientific, philosophical and theological debates sparked by his 'dangerous idea'.

The Variation of Animals and Plants under Domestication

Charles Darwin (1809–82) first published this work in 1868 in two volumes. The book began as an expansion of the first two chapters of *On the Origin of Species:* 'Variation under Domestication' and 'Variation under Nature' and it developed into one of his largest works; Darwin referred to it as his 'big book'. In volume 2, concerned with *how* species inherit particular characteristics, Darwin first published his 'provisional hypothesis' of pangenesis. This theory of 'gemmules' was not met with much acceptance and today is not valuable as scientific explanation, but it was important in laying down the key questions that needed to be answered regarding the processes of genetic inheritance. Darwin also used volume 2 to challenge the theories of evolution by design, expounded by the botanist Asa Gray. Darwin's arguments were some of the very first in a long debate that remains hot today.

Cambridge University Press has long been a pioneer in the reissuing of out-of-print titles from its own backlist, producing digital reprints of books that are still sought after by scholars and students but could not be reprinted economically using traditional technology. The Cambridge Library Collection extends this activity to a wider range of books which are still of importance to researchers and professionals, either for the source material they contain, or as landmarks in the history of their academic discipline.

Drawing from the world-renowned collections in the Cambridge University Library, and guided by the advice of experts in each subject area, Cambridge University Press is using state-of-the-art scanning machines in its own Printing House to capture the content of each book selected for inclusion. The files are processed to give a consistently clear, crisp image, and the books finished to the high quality standard for which the Press is recognised around the world. The latest print-on-demand technology ensures that the books will remain available indefinitely, and that orders for single or multiple copies can quickly be supplied.

The Cambridge Library Collection will bring back to life books of enduring scholarly value (including out-of-copyright works originally issued by other publishers) across a wide range of disciplines in the humanities and social sciences and in science and technology.

The Variation of Animals and Plants under Domestication

VOLUME 2

CHARLES DARWIN

CAMBRIDGE UNIVERSITY PRESS

Cambridge, New York, Melbourne, Madrid, Cape Town, Singapore,
São Paolo, Delhi, Dubai, Tokyo

Published in the United States of America by Cambridge University Press, New York

www.cambridge.org
Information on this title: www.cambridge.org/9781108014236

© in this compilation Cambridge University Press 2010

This edition first published 1868
This digitally printed version 2010

ISBN 978-1-108-01423-6 Paperback

THE VARIATION

OF

ANIMALS AND PLANTS

UNDER DOMESTICATION.

By CHARLES DARWIN, M.A., F.R.S., &c.

IN TWO VOLUMES.—Vol. II.

WITH ILLUSTRATIONS

LONDON:

JOHN MURRAY, ALBEMARLE STREET.

1868.

CONTENTS OF VOLUME II.

CHAPTER XII.

INHERITANCE.

CHAPTER XIII.

INHERITANCE *continued* — REVERSION OR ATAVISM.

CHAPTER XIV.

INHERITANCE *continued* — FIXEDNESS OF CHARACTER — PREPOTENCY — SEXUAL LIMITATION — CORRESPONDENCE OF AGE.

CHAPTER XV.

ON CROSSING.

CHAPTER XVI.

CAUSES WHICH INTERFERE WITH THE FREE CROSSING OF VARIETIES — INFLUENCE OF DOMESTICATION ON FERTILITY.

CHAPTER XVII.

ON THE GOOD EFFECTS OF CROSSING, AND ON THE EVIL EFFECTS OF CLOSE INTERBREEDING.

CHAPTER XVIII.

ON THE ADVANTAGES AND DISADVANTAGES OF CHANGED CONDITIONS OF LIFE: STERILITY FROM VARIOUS CAUSES.

CHAPTER XIX.

SUMMARY OF THE FOUR LAST CHAPTERS, WITH REMARKS ON HYBRIDISM.

CHAPTER XX.

SELECTION BY MAN.

CHAPTER XXI.

SELECTION—*continued.*

CHAPTER XXII.

CAUSES OF VARIABILITY.

CHAPTER XXIII.

DIRECT AND DEFINITE ACTION OF THE EXTERNAL CONDITIONS OF LIFE.

CHAPTER XXIV.

LAWS OF VARIATION — USE AND DISUSE, ETC.

CHAPTER XXV.

LAWS OF VARIATION, *continued* — CORRELATED VARIABILITY.

CHAPTER XXVI.

LAWS OF VARIATION, *continued* — SUMMARY.

CHAPTER XXVII.

PROVISIONAL HYPOTHESIS OF PANGENESIS.

CONTENTS OF VOL. II.

CHAPTER XXVIII.

CONCLUDING REMARKS.

ERRATA.

Vol. II., pp. 18, 232, 258, *for* Cratægus oxycantha, *read* oxyacantha.
  ,,   p. 98, 8 lines from top, *for* Dianthus armoria *read* armeria.
  ,,   ,, 156, 15 lines from bottom, *for* Casuarinus *read* Casuarius.
  ,,   ,, ,, 4 lines from bottom, *for* Grus cineria *read* cinerea.
  ,,   ,, 168, 11 lines from top, *for* Œsculus *read* Æsculus.
  ,,   ,, 300, 3 lines from top, *for* anastomising *read* anastomosing.
  ,,   ,, ,, foot-note, *for* Birckell *read* Brickell.

THE

VARIATION OF ANIMALS AND PLANTS

UNDER DOMESTICATION.

CHAPTER XII.

INHERITANCE.

WONDERFUL NATURE OF INHERITANCE — PEDIGREES OF OUR DOMESTICATED ANIMALS — INHERITANCE NOT DUE TO CHANCE — TRIFLING CHARACTERS INHERITED — DISEASES INHERITED — PECULIARITIES IN THE EYE INHERITED — DISEASES IN THE HORSE — LONGEVITY AND VIGOUR — ASYMMETRICAL DEVIATIONS OF STRUCTURE — POLYDACTYLISM AND REGROWTH OF SUPERNUMERARY DIGITS AFTER AMPU- TATION — CASES OF SEVERAL CHILDREN SIMILARLY AFFECTED FROM NON-AFFECTED PARENTS — WEAK AND FLUCTUATING INHERITANCE : IN WEEPING TREES, IN DWARFNESS, COLOUR OF FRUIT AND FLOWERS, COLOUR OF HORSES — NON- INHERITANCE IN CERTAIN CASES — INHERITANCE OF STRUCTURE AND HABITS OVERBORNE BY HOSTILE CONDITIONS OF LIFE, BY INCESSANTLY RECURRING VARIABILITY, AND BY REVERSION — CONCLUSION.

THE subject of inheritance is an immense one, and has been treated by many authors. One work alone, 'De l'Hérédité Naturelle,' by Dr. Prosper Lucas, runs to the length of 1562 pages. We must confine ourselves to certain points which have an important bearing on the general subject of variation, both with domestic and natural productions. It is obvious that a variation which is not inherited throws no light on the derivation of species, nor is of any service to man, except in the case of perennial plants, which can be propagated by buds.

If animals and plants had never been domesticated, and wild ones alone had been observed, we should probably never have heard the saying, that "like begets like." The proposition would have been as self-evident, as that all the buds on the same tree are alike, though neither proposition is strictly true. For, as has often been remarked, probably no two individuals are

B

identically the same. All wild animals recognise each other, which shows that there is some difference between them; and when the eye is well practised, the shepherd knows each sheep, and man can distinguish a fellow-man out of millions on millions of other men. Some authors have gone so far as to maintain that the production of slight differences is as much a necessary function of the powers of generation, as the production of offspring like their parents. This view, as we shall see in a future chapter, is not theoretically probable, though practically it holds good. The saying that "like begets like" has in fact arisen from the perfect confidence felt by breeders, that a superior or inferior animal will generally reproduce its kind; but this very superiority or inferiority shows that the individual in question has departed slightly from its type.

The whole subject of inheritance is wonderful. When a new character arises, whatever its nature may be, it generally tends to be inherited, at least in a temporary and sometimes in a most persistent manner. What can be more wonderful than that some trifling peculiarity, not primordially attached to the species, should be transmitted through the male or female sexual cells, which are so minute as not to be visible to the naked eye, and afterwards through the incessant changes of a long course of development, undergone either in the womb or in the egg, and ultimately appear in the offspring when mature, or even when quite old, as in the case of certain diseases? Or again, what can be more wonderful than the well-ascertained fact that the minute ovule of a good milking cow will produce a male, from whom a cell, in union with an ovule, will produce a female, and she, when mature, will have large mammary glands, yielding an abundant supply of milk, and even milk of a particular quality? Nevertheless, the real subject of surprise is, as Sir H. Holland has well remarked,[1] not that a character should be inherited, but that any should ever fail to be inherited. In a future chapter, devoted to an hypothesis which I have termed pangenesis, an attempt will be made to show the means by which characters of all kinds are transmitted from generation to generation.

---

[1] 'Medical Notes and Reflections,' 3rd edit., 1855, p. 267.

Some writers,[2] who have not attended to natural history, have attempted to show that the force of inheritance has been much exaggerated. The breeders of animals would smile at such simplicity; and if they condescended to make any answer, might ask what would be the chance of winning a prize if two inferior animals were paired together? They might ask whether the half-wild Arabs were led by theoretical notions to keep pedigrees of their horses? Why have pedigrees been scrupulously kept and published of the Shorthorn cattle, and more recently of the Hereford breed? Is it an illusion that these recently improved animals safely transmit their excellent qualities even when crossed with other breeds? have the Shorthorns, without good reason, been purchased at immense prices and exported to almost every quarter of the globe, a thousand guineas having been given for a bull? With greyhounds pedigrees have likewise been kept, and the names of such dogs, as Snowball, Major, &c., are as well known to coursers as those of Eclipse and Herod on the turf. Even with the Gamecock pedigrees of famous strains were formerly kept, and extended back for a century. With pigs, the Yorkshire and Cumberland breeders "preserve and print pedigrees;" and to show how such highly-bred animals are valued, I may mention that Mr. Brown, who won all the first prizes for small breeds at Birmingham in 1850, sold a young sow and boar of his breed to Lord Ducie for 43 guineas; the sow alone was afterwards sold to the Rev. F. Thursby for 65 guineas; who writes, "she paid me very well, having sold her produce for 300l., and having now four breeding sows from her."[3] Hard cash paid down, over and over again, is an excellent test of inherited superiority. In fact, the whole art of breeding, from which such great results have been attained during the present century, depends on the inheritance of each small

[2] Mr. Buckle, in his grand work on 'Civilisation,' expresses doubts on the subject owing to the want of statistics. See also Mr. Bowen, Professor of Moral Philosophy, in 'Proc. American Acad. of Sciences,' vol. v. p. 102.

[3] For greyhounds, see Low's 'Domest. Animals of the British Islands,' 1845, p. 721. For game-fowls, see 'The Poultry Book,' by Mr. Tegetmeier, 1866, p. 123. For pigs, see Mr. Sidney's edit. of 'Youatt on the Pig,' 1860, pp. 11, 22.

detail of structure. But inheritance is not certain; for if it were, the breeder's art[4] would be reduced to a certainty, and there would be little scope left for all that skill and perseverance shown by the men who have left an enduring monument of their success in the present state of our domesticated animals.

It is hardly possible, within a moderate compass, to impress on the mind of those who have not attended to the subject, the full conviction of the force of inheritance which is slowly acquired by rearing animals, by studying the many treatises which have been published on the various domestic animals, and by conversing with breeders. I will select a few facts of the kind, which, as far as I can judge, have most influenced my own mind. With man and the domestic animals, certain peculiarities have appeared in an individual, at rare intervals, or only once or twice in the history of the world, but have reappeared in several of the children and grandchildren. Thus Lambert, "the porcupine-man," whose skin was thickly covered with warty projections, which were periodically moulted, had all his six children and two grandsons similarly affected.[5] The face and body being covered with long hair, accompanied by deficient teeth (to which I shall hereafter refer), occurred in three successive generations in a Siamese family; but this case is not unique, as a woman[6] with a completely hairy face was exhibited in London in 1663, and another instance has recently occurred. Colonel Hallam[7] has described a race of two-legged pigs, "the hinder extremities being entirely wanting;" and this deficiency was transmitted through three generations. In fact, all races presenting any remarkable peculiarity, such as solid-hoofed swine, Mauchamp sheep, niata cattle, &c., are instances of the long-continued inheritance of rare deviations of structure.

When we reflect that certain extraordinary peculiarities have

---

[4] 'The Stud Farm,' by Cecil, p. 39.

[5] 'Philosophical Transactions,' 1755, p. 23. I have seen only second-hand accounts of the two grandsons. Mr. Sedgwick, in a paper to which I shall hereafter often refer, states that *four* generations were affected, and in each the males alone.

[6] Barbara Van Beck, figured, as I am informed by the Rev. W. D. Fox, in Woodburn's 'Gallery of Rare Portraits.' 1816, vol. ii.

[7] 'Proc. Zoolog. Soc.,' 1833, p. 16.

thus appeared in a single individual out of many millions, all exposed in the same country to the same general conditions of life, and, again, that the same extraordinary peculiarity has sometimes appeared in individuals living under widely different conditions of life, we are driven to conclude that such peculiarities are not directly due to the action of the surrounding conditions, but to unknown laws acting on the organisation or constitution of the individual;—that their production stands in hardly closer relation to the conditions than does life itself. If this be so, and the occurrence of the same unusual character in the child and parent cannot be attributed to both having been exposed to the same unusual conditions, then the following problem is worth consideration, as showing that the result cannot be due, as some authors have supposed, to mere coincidence, but must be consequent on the members of the same family inheriting something in common in their constitution. Let it be assumed that, in a large population, a particular affection occurs on an average in one out of a million, so that the à priori chance that an individual taken at random will be so affected is only one in a million. Let the population consist of sixty millions, composed, we will assume, of ten million families, each containing six members. On these data, Professor Stokes has calculated for me that the odds will be no less than 8333 millions to 1 that in the ten million families there will not be even a single family in which one parent and two children will be affected by the peculiarity in question. But numerous cases could be given, in which several children have been affected by the same rare peculiarity with one of their parents; and in this case, more especially if the grandchildren be included in the calculation, the odds against mere coincidence become something prodigious, almost beyond enumeration.

In some respects the evidence of inheritance is more striking when we consider the reappearance of trifling peculiarities. Dr. Hodgkin formerly told me of an English family in which, for many generations, some members had a single lock differently coloured from the rest of the hair. I knew an Irish gentleman, who, on the right side of his head, had a small white lock in the midst of his dark hair: he assured me that his grandmother had

a similar lock on the same side, and his mother on the opposite
side. But it is superfluous to give instances; every shade
of expression, which may often be seen alike in parents
and children, tells the same story. On what a curious combination of corporeal structure, mental character, and training,
must handwriting depend! yet every one must have noted the
occasional close similarity of the handwriting in father and son,
although the father had not taught his son. A great collector
of franks assured me that in his collection there were several
franks of father and son hardly distinguishable except by their
dates. Hofacker, in Germany, remarks on the inheritance of handwriting; and it has even been asserted that English boys when
taught to write in France naturally cling to their English manner
of writing.[8] Gait, gestures, voice, and general bearing are all
inherited, as the illustrious Hunter and Sir A. Carlisle have
insisted.[9] My father communicated to me two or three striking
instances, in one of which a man died during the early infancy
of his son, and my father, who did not see this son until grown
up and out of health, declared that it seemed to him as if his old
friend had risen from the grave, with all his highly peculiar
habits and manners. Peculiar manners pass into tricks, and
several instances could be given of their inheritance; as in the
case, often quoted, of the father who generally slept on his back,
with his right leg crossed over the left, and whose daughter,
whilst an infant in the cradle, followed exactly the same habit,
though an attempt was made to cure her.[10] I will give one
instance which has fallen under my own observation, and which
is curious from being a trick associated with a peculiar state of
mind, namely, pleasurable emotion. A boy had the singular
habit, when pleased, of rapidly moving his fingers parallel to each
other, and, when much excited, of raising both hands, with
the fingers still moving, to the sides of his face on a level with
the eyes; this boy, when almost an old man, could still hardly
resist this trick when much pleased, but from its absurdity
concealed it. He had eight children. Of these, a girl, when

---

[8] Hofacker, 'Ueber die Eigenschaften,' &c., 1828, s. 34. Report by Pariset in 'Comptes Rendus,' 1847, p. 592.

[9] Hunter, as quoted in Harlan's 'Med.

Researches,' p. 530. Sir A. Carlisle, 'Phil. Transact.,' 1814, p. 94.

[10] Girou de Buzareignues, 'De la Génération,' p. 282.

pleased, at the age of four and a half years, moved her fingers in exactly the same way, and what is still odder, when much excited, she raised both her hands, with her fingers still moving, to the sides of her face, in exactly the same manner as her father had done, and sometimes even still continued to do when alone. I never heard of any one excepting this one man and his little daughter who had this strange habit; and certainly imitation was in this instance out of the question.

Some writers have doubted whether those complex mental attributes, on which genius and talent depend, are inherited, even when both parents are thus endowed. But he who will read Mr. Galton's able paper[11] on hereditary talent will have his doubts allayed.

Unfortunately it matters not, as far as inheritance is concerned, how injurious a quality or structure may be if compatible with life. No one can read the many treatises[12] on hereditary disease and doubt this. The ancients were strongly of this opinion, or, as Ranchin expresses it, *Omnes Græci, Arabes, et Latini in eo consentiunt.* A long catalogue could be given of all sorts of inherited malformations and of predisposition to various diseases. With gout, fifty per cent. of the cases observed in hospital practice are, according to Dr. Garrod, inherited, and a greater percentage in private practice. Every one knows how often insanity runs in families, and some of the cases given by Mr. Sedgwick are awful,—as of a surgeon, whose brother, father, and four paternal uncles were all insane, the latter dying by suicide; of a Jew, whose father, mother, and six brothers and sisters were all mad; and in some other cases several members of the same family, during three or four successive generations, have committed suicide. Striking instances

[11] 'Macmillan's Magazine,' July and August, 1865.

[12] The works which I have read and found most useful are Dr. Prosper Lucas's great work, 'Traité de l'Hérédité Naturelle,' 1847. Mr. W. Sedgwick, in 'British and Foreign Medico-Chirurg. Review,' April and July, 1861, and April and July, 1863 : Dr. Garrod on Gout is quoted in these articles. Sir Henry Holland, 'Medical Notes and Reflections,' 3rd edit., 1855. Piorry, 'De l'Hérédité dans les Maladies,' 1840. Adams, 'A Philosophical Treatise on Hereditary Peculiarities,' 2nd edit., 1815. Essay on 'Hereditary Diseases,' by Dr. J. Steinan, 1843. *See* Paget, in 'Medical Times,' 1857, p. 192, on the Inheritance of Cancer ; Dr. Gould, in 'Proc. of American Acad. of Sciences,' Nov. 8, 1853, gives a curious case of hereditary bleeding in four generations. Harlan, 'Medical Researches,' p. 593.

have been recorded of epilepsy, consumption, asthma, stone in the bladder, cancer, profuse bleeding from the slightest injuries, of the mother not giving milk, and of bad parturition being inherited. In this latter respect I may mention an odd case given by a good observer,[13] in which the fault lay in the offspring, and not in the mother: in a part of Yorkshire the farmers continued to select cattle with large hind-quarters, until they made a strain called "Dutch-buttocked," and "the monstrous size of the buttocks of the calf was frequently fatal to the cow, and numbers of cows were annually lost in calving."

Instead of giving numerous details on various inherited malformations and diseases, I will confine myself to one organ, that which is the most complex, delicate, and probably best-known in the human frame, namely, the eye, with its accessory parts. To begin with the latter: I have heard of a family in which parents and children were affected by drooping eyelids, in so peculiar a manner, that they could not see without throwing their heads backwards; and Sir A. Carlisle[14] specifies a pendulous fold to the eyelids as inherited. "In a family," says Sir H. Holland,[15] "where the father had a singular elongation of the upper eyelid, seven or eight children were born with the same deformity; two or three other children having it not." Many persons, as I hear from Mr. Paget, have two or three of the hairs in their eyebrows (apparently corresponding with the vibrissæ of the lower animals) much longer than the others; and even so trifling a peculiarity as this certainly runs in families.

With respect to the eye itself, the highest authority in England, Mr. Bowman, has been so kind as to give me the following remarks on certain inherited imperfections. First, hypermetropia, or morbidly long sight: in this affection, the organ, instead of being spherical, is too flat from front to back, and is often altogether too small, so that the retina is brought too forward for the focus of the humours; consequently a convex glass is required for clear vision of near objects, and frequently even of distant ones. This state occurs congenitally, or at a very early age, often in several children of the same family, where one of the parents has presented it.[16] Secondly, myopia, or short-sight, in which the eye is egg-shaped, and too long from front to back; the retina in this case lies behind the focus, and is therefore fitted to see distinctly only very near objects. This condition is not commonly congenital, but comes on in youth, the liability to it being well known to be transmissible from parent to child. The change from the spherical to the ovoidal shape seems the immediate con-

---

[13] Marshall, quoted by Youatt in his work on Cattle, p. 284.

[14] 'Philosoph. Transact.,' 1814, p. 94.

[15] 'Medical Notes and Reflections,' 3rd edit., p. 33.

[16] This affection, as I hear from Mr. Bowman, has been ably described and spoken of as hereditary by Dr. Donders, of Utrecht, whose work was published in English by the Sydenham Society in 1864.

sequence of something like inflammation of the coats, under which they yield, and there is ground for believing that it may often originate in causes acting directly on the individual affected, and may thenceforward become transmissible. When both parents are myopic Mr. Bowman has observed the hereditary tendency in this direction to be heightened, and some of the children to be myopic at an earlier age or in a higher degree than their parents. Thirdly, squinting is a familiar example of hereditary transmission : it is frequently a result of such optical defects as have been above mentioned; but the more primary and uncomplicated forms of it are also sometimes in a marked degree transmitted in a family. Fourthly, *Cataract*, or opacity of the crystalline lens, is commonly observed in persons whose parents have been similarly affected, and often at an earlier age in the children than in the parents. Occasionally more than one child in a family is thus afflicted, one of whose parents or other relation presents the senile form of the complaint. When cataract affects several members of a family in the same generation, it is often seen to commence at about the same age in each ; *e.g.*, in one family several infants or young persons may suffer from it; in another, several persons of middle age. Mr. Bowman also informs me that he has occasionally seen, in several members of the same family, various defects in either the right or left eye; and Mr. White Cooper has often seen peculiarities of vision confined to one eye reappearing in the same eye in the offspring.[17]

The following cases are taken from an able paper by Mr. W. Sedgwick, and from Dr. Prosper Lucas.[18] Amaurosis, either congenital or coming on late in life, and causing total blindness, is often inherited ; it has been observed in three successive generations. Congenital absence of the iris has likewise been transmitted for three generations, a cleft-iris for four generations, being limited in this latter case to the males of the family. Opacity of the cornea and congenital smallness of the eyes have been inherited. Portal records a curious case, in which a father and two sons were rendered blind, whenever the head was bent downwards, apparently owing to the crystalline lens, with its capsule, slipping through an unusually large pupil into the anterior chamber of the eye. Day-blindness, or imperfect vision under a bright light, is inherited, as is night-blindness, or an incapacity to see except under a strong light : a case has been recorded, by M. Cunier, of this latter defect having affected eighty-five members of the same family during six generations. The singular incapacity of distinguishing colours, which has been called *Daltonism*, is notoriously hereditary, and has been traced through five generations, in which it was confined to the female sex.

With respect to the colour of the iris : deficiency of colouring matter is well known to be hereditary in albinoes. The iris of one eye being of a different colour from that of the other, and the iris being spotted, are cases which have been inherited. Mr. Sedgwick gives, in addition, on the

---

[17] Quoted by Mr. Herbert Spencer, ' Principles of Biology,' vol. i. p. 244.

[18] ' British and Foreign Medico-   Chirurg. Review,' April, 1861, p. 482-6; 'l'Héréd. Nat.,' tom. i. pp. 391-408.

authority of Dr. Osborne,[19] the following curious instance of strong inheritance: a family of sixteen sons and five daughters all had eyes
"resembling in miniature the markings on the back of a tortoiseshell cat."
The mother of this large family had three sisters and a brother all similarly
marked, and they derived this peculiarity from their mother, who belonged
to a family notorious for transmitting it to their posterity.

Finally, Dr. Lucas emphatically remarks that there is not one single
faculty of the eye which is not subject to anomalies; and not one which is
not subjected to the principle of inheritance. Mr. Bowman agrees with
the general truth of this proposition; which of course does not imply that
all malformations are necessarily inherited; this would not even follow
if both parents were affected by an anomaly which in most cases was
transmissible.

Even if no single fact had been known with respect to the
inheritance of disease and malformations by man, the evidence
would have been ample in the case of the horse. And this
might have been expected, as horses breed much quicker than
man, are matched with care, and are highly valued. I
have consulted many works, and the unanimity of the belief by
veterinaries of all nations in the transmission of various morbid
tendencies is surprising. Authors, who have had wide experience,
give in detail many singular cases, and assert that contracted
feet, with the numerous contingent evils, of ring-bones, curbs,
splints, spavin, founder and weakness of the front legs, roaring
or broken and thick wind, melanosis, specific ophthalmia, and
blindness (the great French veterinary Huzard going so far
as to say that a blind race could soon be formed), crib-biting,
jibbing, and ill-temper, are all plainly hereditary. Youatt sums
up by saying "there is scarcely a malady to which the horse
is subject which is not hereditary;" and M. Bernard adds that
the doctrine "that there is scarcely a disease which does not
run in the stock, is gaining new advocates every day."[20] So it

[19] Dr. Osborne, Pres. of Royal College
of Phys. in Ireland, published this case
in the 'Dublin Medical Journal' for
1835.

[20] These various statements are taken
from the following works and papers :—
Youatt on 'The Horse,' pp. 35, 220.
Lawrence, 'The Horse,' p. 30. Karkeek,
in an excellent paper in 'Gard. Chronicle,' 1853, p. 92. Mr. Burke, in
'Journal of R. Agricul. Soc. of Eng-

land,' vol. v. p. 511. 'Encyclop. of
Rural Sports,' p. 279. Girou de Buzareignues, 'Philosoph. Phys.,' p. 215.
See following papers in 'The Veterinary:' Roberts, in vol. ii. p. 144; M.
Marrimpoey, vol. ii. p. 387; Mr. Karkeek,
vol. iv. p. 5; Youatt on Goître in Dogs,
vol. v. p. 483; Youatt, in vol. vi. pp.
66, 348, 412; M. Bernard, vol. xi. p.
539; Dr. Samesreuther, on Cattle, in
vol. xii. p. 181; Percivall, in vol. xiii.

is in regard to cattle, with consumption, good and bad teeth, fine skin, &c. &c. But enough, and more than enough, has been said on disease. Andrew Knight, from his own experience, asserts that disease is hereditary with plants; and this assertion is endorsed by Lindley.[21]

Seeing how hereditary evil qualities are, it is fortunate that good health, vigour, and longevity are equally inherited. It was formerly a well-known practice, when annuities were purchased to be received during the lifetime of a nominee, to search out a person belonging to a family of which many members had lived to extreme old age. As to the inheritance of vigour and endurance, the English race-horse offers an excellent instance. Eclipse begot 334, and King Herod 497 winners. A "cock-tail" is a horse not purely bred, but with only one-eighth or one-sixteenth impure blood in his veins, yet very few instances have ever occurred of such horses having won a great race. They are sometimes as fleet for short distances as thoroughbreds, but as Mr. Robson, the great trainer, asserts, they are deficient in wind, and cannot keep up the pace. Mr. Lawrence also remarks, "perhaps no instance has ever occurred of a three-part-bred horse saving his 'distance' in running two miles with thoroughbred racers." It has been stated by Cecil, that when unknown horses, whose parents were not celebrated, have unexpectedly won great races, as in the case of Priam, they can always be proved to be descended on both sides, through many generations, from first-rate ancestors. On the Continent, Baron Cameronn challenges, in a German veterinary periodical, the opponents of the English race-horse, to name one good horse on the Continent which has not some English race-blood in his veins.[22]

With respect to the transmission of the many slight, but in-

p. 47. With respect to blindness in horses, see also a whole row of authorities in Dr. P. Lucas's great work, tom. i. p. 399. Mr. Baker, in 'The Veterinary,' vol. xiii. p. 721, gives a strong case of hereditary imperfect vision and of jibbing.

[21] Knight on 'The Culture of the Apple and Pear,' p. 34. Lindley's 'Horticulture,' p. 180.

[22] These statements are taken from the following works in order:—Youatt on 'The Horse,' p. 48; Mr. Darvill, in 'The Veterinary,' vol. viii. p. 50. With respect to Robson, see 'The Veterinary,' vol. iii. p. 580; Mr. Lawrence on 'The Horse,' 1829, p. 9; 'The Stud Farm,' by Cecil, 1851; Baron Cameronn, quoted in 'The Veterinary,' vol. x. p. 500.

finitely diversified characters, by which the domestic races of
animals and plants are distinguished, nothing need be said; for
the very existence of persistent races proclaims the power of
inheritance.

A few special cases, however, deserve some consideration. It
might have been anticipated, that deviations from the law of
symmetry would not have been inherited. But Anderson[23] states
that a rabbit produced in a litter a young animal having only
one ear; and from this animal a breed was formed which
steadily produced one-eared rabbits. He also mentions a bitch,
with a single leg deficient, and she produced several puppies
with the same deficiency. From Hofacker's account[24] it appears
that a one-horned stag was seen in 1781 in a forest in Germany,
in 1788 two, and afterwards, from year to year, many were
observed with only one horn on the right side of the head.
A cow lost a horn by suppuration,[25] and she produced three
calves which had on the same side of the head, instead of a
horn, a small bony lump attached merely to the skin; but
we here approach the doubtful subject of inherited mutilations.
A man who is left-handed, and a shell in which the spire turns
in the wrong direction, are departures from the normal though
a symmetrical condition, and they are well known to be inherited.

Polydactylism.—Supernumerary fingers and toes are eminently liable, as
various authors have insisted, to transmission, but they are noticed here
chiefly on account of their occasional regrowth after amputation. Polydactylism graduates[26] by multifarious steps from a mere cutaneous appendage, not including any bone, to a double hand. But an additional
digit, supported on a metacarpal bone, and furnished with all the proper
muscles, nerves, and vessels, is sometimes so perfect, that it escapes detection, unless the fingers are actually counted. Occasionally there are
several supernumerary digits; but usually only one, making the total
number six. This one may represent either a thumb or finger, being
attached to the inner or outer margin of the hand. Generally, through
the law of correlation, both hands and feet are similarly affected. I have
tabulated the cases recorded in various works or privately communicated

---

[23] ' Recreations in Agriculture and
Nat. Hist.,' vol. i. p. 68.

[24] ' Ueber die Eigenschaften,' &c.,
1828, s. 107.

[25] Bronn's ' Geschichte der Natur,'
band ii. s. 132.

[26] Vrolik has discussed this point at
full length in a work published in Dutch,
from which Mr. Paget has kindly translated for me passages. *See*, also, Isidore
Geoffroy St. Hilaire's 'Hist. des Anomalies,' 1832, tom. i. p. 684.

to me, of forty-six persons with extra digits on one or both hands and feet; if in each case all four extremities had been similarly affected, the table would have shown a total of ninety-two hands and ninety-two feet each with six digits. As it is, seventy-three hands and seventy-five feet were thus affected. This proves, in contradiction to the result arrived at by Dr. Struthers,[27] that the hands are not more frequently affected than the feet.

The presence of more than five digits is a great anomaly, for this number is not normally exceeded by any mammal, bird, or existing reptile.[28] Nevertheless, supernumerary digits are strongly inherited; they have been transmitted through five generations; and in some cases, after disappearing for one, two, or even three generations, have reappeared through reversion. These facts are rendered, as Professor Huxley has observed, more remarkable from its being known in most cases that the affected person had not married one similarly affected. In such cases the child of the fifth generation would have only 1-32nd part of the blood of his first sedigitated ancestor. Other cases are rendered remarkable by the affection gathering force, as Dr. Struthers has shown, in each generation, though in each the affected person had married one not affected; moreover such additional digits are often amputated soon after birth, and can seldom have been strengthened by use. Dr. Struthers gives the following instance: in the first generation an additional digit appeared on one hand; in the second, on both hands; in the third, three brothers had both hands, and one of the brothers a foot affected; and in the fourth generation all four limbs were affected. Yet we must not over-estimate the force of inheritance. Dr. Struthers asserts that cases of non-inheritance and of the first appearance of additional digits in unaffected families are much more frequent than cases of inheritance. Many other deviations of structure, of a nature almost as anomalous as supernumerary digits, such as deficient phalanges, thickened joints, crooked fingers, &c., are in like manner strongly inherited, and are equally subject to intermission with reversion, though in such cases there is no reason to suppose that both parents had been similarly affected.[29]

---

[27] 'Edinburgh New Phil. Journal,' July, 1863.

[28] Some great anatomists, as Cuvier and Meckel, believe that the tubercle on one side of the hinder foot of the tailless Batrachians represents a sixth digit. Certainly, when the hinder foot of a toad, as soon as it first sprouts from the tadpole, is dissected, the partially ossified cartilage of this tubercle resembles under the microscope, in a remarkable manner, a digit. But the highest authority on such subjects, Gegenbaur (Untersuchung. zur vergleich. anat. der Wirbelthiere : Carpus et Tarsus, 1864, s. 63), concludes that this resemblance is not real, only superficial.

[29] For these several statements, see Dr. Struthers, in work cited, especially on intermissions in the line of descent. Prof. Huxley, 'Lectures on our Knowledge of Organic Nature,' 1863, p. 97. With respect to inheritance, see Dr. Prosper Lucas, 'L'Hérédité Nat.,' tom. i. p. 325. Isid. Geoffroy, 'Anom.,' tom. i. p. 701. Sir A. Carlisle, in 'Phil. Transact.,' 1814, p. 94. A. Walker, on 'Intermarriage,' 1838, p. 140, gives a case of five generations; as does Mr. Sedgwick, in 'Brit. and Foreign Medico-Chirurg. Review,' April, 1863, p. 462.

Additional digits have been observed in negroes as well as in other races of man, and in several of the lower animals. Six toes have been described on the hind feet of the newt (*Salamandra cristata*), and, as it is said, of the frog. It deserves notice from what follows, that the six-toed newt, though adult, had preserved some of its larval characters; for part of the hyoidal apparatus, which is properly absorbed during the act of metamorphosis, was retained. In the dog, six toes on the hinder feet have been transmitted through three generations; and I have heard of a race of six-toed cats. In several breeds of the fowl the hinder toe is double, and is generally transmitted truly, as is well shown when Dorkings are crossed with common four-toed breeds.[30] With animals which have properly less than five digits, the number is sometimes increased to five, especially in the front legs, though rarely carried beyond that number; but this is due to the development of a digit already existing in a more or less rudimentary state. Thus the dog has properly four toes behind, but in the larger breeds a fifth toe is commonly, though not perfectly, developed. Horses, which properly have one toe alone fully developed with rudiments of the others, have been described with each foot bearing two or three small separate hoofs: analogous facts have been noticed with sheep, goats, and pigs.[31]

The most interesting point with respect to supernumerary digits is their occasional regrowth after amputation. Mr. White[32] describes a child, three years old, with a thumb double from the first joint. He removed the lesser thumb, which was furnished with a nail; but to his astonishment it grew again, and reproduced a nail. The child was then taken to an eminent London surgeon, and the newly-grown thumb was wholly removed by its socket-joint, but again it grew and reproduced a nail. Dr. Struthers mentions a case of partial regrowth of an additional thumb, amputated when the child was three months old; and the late Dr. Falconer communicated to me an analogous case which had fallen under his own observation. A gentleman, who first called my attention to this subject, has given me the following facts which occurred in his own family. He himself, two brothers, and a sister were born with an extra digit to each extremity. His parents were not affected, and there was no tradition in the family, or in the village in which the family had long resided, of any member having been thus affected. Whilst a child, both additional toes, which were attached by bones, were rudely cut off; but the stump of one grew again, and a second operation was performed in his thirty-third year.

On the inheritance of other anomalies in the extremities, *see* Dr. H. Dobell, in vol. xlvi. of 'Medico-Chirurg. Transactions,' 1863; also Mr. Sedgwick, in op. cit., April, 1863, p. 460. With respect to additional digits in the negro, *see* Prichard, 'Physical History of Mankind.' Dr. Dieffenbach ('Journ. Royal Geograph. Soc.,' 1841, p. 208) says this anomaly is not uncommon with the Polynesians of the Chatham Islands.

[30] 'The Poultry Chronicle,' 1854, p. 559.

[31] The statements in this paragraph are taken from Isidore Geoffroy St. Hilaire, 'Hist. des Anomalies,' tom. i. pp. 688-693.

[32] As quoted by Carpenter, 'Princ. of Comp. Physiology,' 1854, p. 480.

He has had fourteen children, of whom three have inherited additional digits; and one of them, when about six weeks old, was operated on by an eminent surgeon. The additional finger, which was attached by bone to the outer side of the hand, was removed at the joint; the wound healed, but immediately the digit began growing; and in about three months' time the stump was removed for the second time by the root. But it has since grown again, and is now fully a third of an inch in length, including a bone; so that it will for the third time have to be operated on.

Now the normal digits in adult man and other mammals, in birds, and, as I believe, in true reptiles, have no power of regrowth. The nearest approach to this power is exhibited by the occasional reappearance in man of imperfect nails on the stumps of his fingers after amputation.[33] But man in his embryonic condition has a considerable power of reproduction, for Sir J. Simpson[34] has several times observed arms which had been cut off in the womb by bands of false membrane, and which had grown again to a certain extent. In one instance, the extremity was "divided into three minute nodules, on two of which small points of nails could be detected;" so that these nodules clearly represented fingers in process of regrowth. When, however, we descend to the lower vertebrate classes, which are generally looked at as representing the higher classes in their embryonic condition, we find ample powers of regrowth. Spallanzani[35] cut off the legs and tail of a salamander six times, and Bonnet eight times, successively, and they were reproduced. An additional digit beyond the proper number was occasionally formed after Bonnet had cut off or had divided longitudinally the hand or foot, and in one instance three additional digits were thus formed.[36] These latter cases appear at first sight quite distinct from the congenital production of additional digits in the higher animals; but theoretically, as we shall see in a future chapter, they probably present no real difference. The larvæ or tadpoles of the tailless Batrachians, but not the adults,[37] are capable of reproducing lost members.[38] Lastly, as I have been informed by Mr. J. J. Briggs and Mr. F. Buckland, when portions of the pectoral and tail fins of various fresh-

---

[33] Müller's 'Phys.,' Eng. translat., vol. i. 1838, p. 407. A thrush, however, was exhibited before the British Association at Hull, in 1853, which had lost its tarsus, and this member, it was asserted, had been thrice reproduced : I presume it was lost each time by disease.

[34] 'Monthly Journal of Medical Science,' Edinburgh, 1848, new series, vol. ii. p. 890.

[35] 'An Essay on Animal Reproduction,' trans. by Dr. Maty, 1769, p. 79.

[36] Bonnet, ' Œuvres d'Hist. Nat.,' tom. v., part i., 4to. edit., 1781, pp. 343, 350, 353.

[37] So with insects, the larvæ reproduce lost limbs, but, except in one order, the mature insect has no such power. But the Myriapoda, which apparently represent the larvæ of true insects, have, as Newport has shown, this power until their last moult. See an excellent discussion on this whole subject by Dr. Carpenter in his 'Princ. Comp. Phys.,' 1854, p. 479.

[38] Dr. Günther, in Owen's ' Anatomy of Vertebrates,' vol. i., 1866, p. 567. Spallanzani has made similar observations.

water fish are cut off, they are perfectly reproduced in about six weeks'
time.

From these several facts we may infer that supernumerary
digits in man retain to a certain extent an embryonic condition, and that they resemble in this respect the normal digits
and limbs in the lower vertebrate classes. They also resemble
the digits of some of the lower animals in the number exceeding
five ; for no mammal, bird, existing reptile, or amphibian (unless
the tubercle on the hind feet of the toad and other tailless
Batrachians be viewed as a digit) has more than five ; whilst
fishes sometimes have in their pectoral fins as many as twenty
metacarpal and phalangeal bones, which, together with the bony
filaments, apparently represent our digits with their nails. So,
again, in certain extinct reptiles, namely, the Ichthyopterygia,
" the digits may be seven, eight, or nine in number, a significant
mark," says Professor Owen, " of piscine affinity." [39]

We encounter much difficulty in attempting to reduce these
various facts to any rule or law. The inconstant number of
the additional digits—their irregular attachment to either the
inner or outer margin of the hand—the gradation which can
be traced from a mere loose rudiment of a single digit to a
completely double hand—the occasional appearance of additional digits in the salamander after a limb has been amputated—these various facts appear to indicate mere fluctuating
monstrosity ; and this perhaps is all that can be safely said.
Nevertheless, as supernumerary digits in the higher animals, from
their power of regrowth and from the number thus acquired
exceeding five, partake of the nature of the digits in the lower
vertebrate animals ;—as they occur by no means rarely, and are
transmitted with remarkable strength, though perhaps not more
strongly than some other anomalies ;—and as with animals
which have fewer than five digits, when an additional one
appears it is generally due to the development of a visible
rudiment ;—we are led in all cases to suspect, that, although no
actual rudiment can be detected, yet that a latent tendency to
the formation of an additional digit exists in all mammals, including man. On this view, as we shall more plainly see in the

[39] 'On the Anatomy of Vertebrates,' 1866, p. 170 : with respect to the pectoral
fins of fishes, pp. 166-168.

next chapter when discussing latent tendencies, we should have to look at the whole case as one of reversion to an enormously remote, lowly-organised, and multidigitate progenitor.

I may here allude to a class of facts closely allied to, but somewhat different from, ordinary cases of inheritance. Sir H. Holland [40] states that brothers and sisters of the same family are frequently affected, often at about the same age, by the same peculiar disease, not known to have previously occurred in the family. He specifies the occurrence of diabetes in three brothers under ten years old; he also remarks that children of the same family often exhibit in common infantile diseases the same peculiar symptoms. My father mentioned to me the case of four brothers who died between the ages of sixty and seventy, in the same highly peculiar comatose state. An instance has been already given of supernumerary digits appearing in four children out of six in a previously unaffected family. Dr. Devay states [41] that two brothers married two sisters, their first-cousins, none of the four nor any relation being an albino; but the seven children produced from this double marriage were all perfect albinoes. Some of these cases, as Mr. Sedgwick [42] has shown, are probably the result of reversion to a remote ancestor, of whom no record had been preserved; and all these cases are so far directly connected with inheritance that no doubt the children inherited a similar constitution from their parents, and, from being exposed to nearly similar conditions of life, it is not surprising that they should be affected in the same manner and at the same period of life.

Most of the facts hitherto given have served to illustrate the force of inheritance, but we must now consider cases, grouped as well as the subject allows into classes, showing how feeble, capricious, or deficient the power of inheritance sometimes is. When a new peculiarity first appears, we can never predict whether it will be inherited. If both parents from their birth present

---

[40] 'Medical Notes and Reflections,' 1839, pp. 24, 34. See, also, Dr. P. Lucas, 'l'Héréd. Nat.,' tom. ii. p. 33.
[41] 'Du Danger des Mariages Con-

sanguins,' 2nd edit., 1862, p. 103.
[42] 'British and Foreign Medico-Chirurg. Review,' July, 1863, pp. 183, 189.

the same peculiarity, the probability is strong that it will be transmitted to at least some of their offspring. We have seen that variegation is transmitted much more feebly by seed from a branch which had become variegated through bud-variation, than from plants which were variegated as seedlings. With most plants the power of transmission notoriously depends on some innate capacity in the individual: thus Vilmorin[43] raised from a peculiarly coloured balsam some seedlings, which all resembled their parent; but of these seedlings some failed to transmit the new character, whilst others transmitted it to all their descendants during several successive generations. So again with a variety of the rose, two plants alone out of six were found by Vilmorin to be capable of transmitting the desired character.

The weeping or pendulous growth of trees is strongly inherited in some cases, and, without any assignable reason, feebly in other cases. I have selected this character as an instance of capricious inheritance, because it is certainly not proper to the parent-species, and because, both sexes being borne on the same tree, both tend to transmit the same character. Even supposing that there may have been in some instances crossing with adjoining trees of the same species, it is not probable that all the seedlings would have been thus affected. At Moccas Court there is a famous weeping oak; many of its branches "are 30 feet long, and no thicker in any part of this length than a common rope:" this tree transmits its weeping character, in a greater or less degree, to all its seedlings; some of the young oaks being so flexible that they have to be supported by props; others not showing the weeping tendency till about twenty years old.[44] Mr. Rivers fertilized, as he informs me, the flowers of a new Belgian weeping thorn (*Cratœgus oxycantha*) with pollen from a crimson not-weeping variety, and three young trees, "now six or seven years old, show a decided tendency to be pendulous, but as yet are not so much so as the mother-plant." According to Mr. MacNab,[45] seedlings from a magnificent weeping birch (*Betula alba*), in the Botanic Garden at Edinburgh, grew for the first ten or fifteen years upright, but then all became weepers like their parent. A peach with pendulous branches, like those of the weeping willow, has been found capable of propagation by seed.[46] Lastly, a weeping and almost prostrate yew (*Taxus baccata*) was found in a hedge in Shropshire; it was a male, but one branch bore female flowers, and produced berries; these,

---

[43] Verlot, 'La Production des Variétés,' 1865, p. 32.

[44] Loudon's 'Gard. Mag.,' vol. xii., 1836, p. 368.

[45] Verlot, ' La Product. des Variétés,' 1865, p. 94.

[46] Bronn's 'Geschichte der Natur,' b. ii. s. 121.

being sown, produced seventeen trees, all of which had exactly the same peculiar habit with the parent-tree.[47]

These facts, it might have been thought, would have been sufficient to render it probable that a pendulous habit would in all cases be strictly inherited. But let us look to the other side. Mr. MacNab[48] sowed seeds of the weeping beech (*Fagus sylvatica*), but succeeded in raising only common beeches. Mr. Rivers, at my request, raised a number of seedlings from three distinct varieties of weeping elm; and at least one of the parent-trees was so situated that it could not have been crossed by any other elm; but none of the young trees, now about a foot or two in height, show the least signs of weeping. Mr. Rivers formerly sowed above twenty thousand seeds of the weeping ash (*Fraxinus excelsior*), and not a single seedling was in the least degree pendulous: in Germany, M. Borchmeyer raised a thousand seedlings, with the same result. Nevertheless, Mr. Anderson, of the Chelsea Botanic Garden, by sowing seed from a weeping ash, which was found before the year 1780, in Cambridgeshire, raised several pendulous trees.[49] Professor Henslow also informs me that some seedlings from a female weeping ash in the Botanic Garden at Cambridge were at first a little pendulous, but afterwards became quite upright: it is probable that this latter tree, which transmits to a certain extent its pendulous habit, was derived by a bud from the same original Cambridgeshire stock; whilst other weeping ashes may have had a distinct origin. But the crowning case, communicated to me by Mr. Rivers, which shows how capricious is the inheritance of a pendulous habit, is that a variety of another species of ash (*F. lentiscifolia*) which was formerly pendulous, "now about twenty years old, has long lost this habit, every shoot being "remarkably erect; but seedlings formerly raised from it were perfectly "prostrate, the stems not rising more than two inches above the ground." Thus the weeping variety of the common ash, which has been extensively propagated by buds during a long period, did not, with Mr. Rivers, transmit its character to one seedling out of above twenty thousand; whereas the weeping variety of a second species of ash, which could not, whilst grown in the same garden, retain its own weeping character, transmitted to its seedlings the pendulous habit in excess!

Many analogous facts could be given, showing how apparently capricious is the principle of inheritance. All the seedlings from a variety of the Barberry (*B. vulgaris*) with red leaves inherited the same character; only about one-third of the seedlings of the copper Beech (*Fagus sylvatica*) had purple leaves. Not one out of a hundred seedlings of a variety of the *Cerasus padus*, with yellow fruit, bore yellow fruit: one-twelfth of the seedlings of the variety of *Cornus mascula*, with yellow fruit, came true:[50] and lastly, all the trees raised by my father from a yellow-berried holly (*Ilex aquifolium*),

[47] Rev. W. A. Leighton, 'Flora of Shropshire,' p. 497; and Charlesworth's 'Mag. of Nat. Hist.,' vol. i., 1837, p. 30.
[48] Verlot, op. cit., p. 93.
[49] For these several statements, see

Loudon's 'Gard. Magazine,' vol. x., 1834, pp. 408, 180; and vol. ix., 1833, p. 597.
[50] These statements are taken from Alph. De Candolle, 'Bot. Géograph.,' p. 1083.

found wild, produced yellow berries. Vilmorin[51] observed in a bed of *Saponaria calabrica* an extremely dwarf variety, and raised from it a large number of seedlings; some of these partially resembled their parent, and he selected their seed; but the grandchildren were not in the least dwarfed: on the other hand, he observed a stunted and bushy variety of *Tagetes signata* growing in the midst of the common varieties by which it was probably crossed; for most of the seedlings raised from this plant were intermediate in character, only two perfectly resembling their parent; but seed saved from these two plants reproduced the new variety so truly, that hardly any selection has since been necessary.

Flowers transmit their colour truly, or most capriciously. Many annuals come true: thus I purchased German seeds of thirty-four named sub-varieties of one *race* of ten-week stocks (*Matthiola annua*), and raised a hundred and forty plants, all of which, with the exception of a single plant, came true. In saying this, however, it must be understood that I could distinguish only twenty kinds out of the thirty-four named sub-varieties; nor did the colour of the flower always correspond with the name affixed to the packet; but I say that they came true, because in each of the thirty-six short rows every plant was absolutely alike, with the one single exception. Again, I procured packets of German seed of twenty-five named varieties of common and quilled asters, and raised a hundred and twenty-four plants; of these, all except ten were true in the above limited sense; and I considered even a wrong shade of colour as false.

It is a singular circumstance that white varieties generally transmit their colour much more truly than any other variety. This fact probably stands in close relation with one observed by Verlot,[52] namely, that flowers which are normally white rarely vary into any other colour. I have found that the white varieties of *Delphinium consolida* and of the Stock are the truest. It is, indeed, sufficient to look through a nurseryman's seed-list, to see the large number of white varieties which can be propagated by seed. The several coloured varieties of the sweet-pea (*Lathyrus odoratus*) are very true; but I hear from Mr. Masters, of Canterbury, who has particularly attended to this plant, that the white variety is the truest. The hyacinth, when propagated by seed, is extremely inconstant in colour, but "white hyacinths almost always give by seed white-flowered plants;"[53] and Mr. Masters informs me that the yellow varieties also reproduce their colour, but of different shades. On the other hand, pink and blue varieties, the latter being the natural colour, are not nearly so true: hence, as Mr. Masters has remarked to me, "we see that a garden variety may acquire a more permanent habit than a natural species;" but it should have been added, that this occurs under cultivation, and therefore under changed conditions.

With many flowers, especially perennials, nothing can be more fluctuating than the colour of the seedlings, as is notoriously the case with verbenas, carnations, dahlias, cinerarias, and others.[54] I sowed seed of twelve

[51] Verlot, op. cit., p. 38.
[52] Op. cit., p. 59.
[53] Alph. De Candolle, 'Géograph. Bot.,' p. 1082.

[54] *See* 'Cottage Gardener,' April 10, 1860, p. 18, and Sept. 10, 1861, p. 456; 'Gard. Chron.,' 1845, p. 102.

named varieties of Snapdragon (*Antirrhinum majus*), and utter confusion was the result. In most cases the extremely fluctuating colour of seedling plants is probably in chief part due to crosses between differently-coloured varieties during previous generations. It is almost certain that this is the case with the polyanthus and coloured primrose (*Primula veris* and *vulgaris*), from their reciprocally dimorphic structure;[55] and these are plants which florists speak of as never coming true by seed : but if care be taken to prevent crossing, neither species is by any means very inconstant in colour; thus I raised twenty-three plants from a purple primrose, fertilised by Mr. J. Scott with its own pollen, and eighteen came up purple of different shades, and only five reverted to the ordinary yellow colour : again, I raised twenty plants from a bright-red cowslip, similarly treated by Mr. Scott, and every one perfectly resembled its parent in colour, as likewise did, with the exception of a single plant, 73 grandchildren. Even with the most variable flowers, it is probable that each delicate shade of colour might be permanently fixed so as to be transmitted by seed, by cultivation in the same soil, by long-continued selection, and especially by the prevention of crosses. I infer this from certain annual larkspurs (*Delphinium consolida* and *ajacis*), of which common seedlings present a greater diversity of colour than any other plant known to me ; yet on procuring seed of five named German varieties of *D. consolida*, only nine plants out of ninety-four were false; and the seedlings of six varieties of *D. ajacis* were true in the same manner and degree as with the stocks above described. A distinguished botanist maintains that the annual species of Delphinium are always self-fertilised ; therefore I may mention that thirty-two flowers on a branch of *D. consolida*, enclosed in a net, yielded twenty-seven capsules, with an average of 17·2 seed in each; whilst five flowers, under the same net, which were artificially fertilised, in the same manner as must be effected by bees during their incessant visits, yielded five capsules with an average of 35·2 fine seed; and this shows that the agency of insects is necessary for the full fertility of this plant. Analogous facts could be given with respect to the crossing of many other flowers, such as carnations, &c., of which the varieties fluctuate much in colour.

As with flowers, so with our domesticated animals, no character is more variable than colour, and probably in no animal more so than with the horse. Yet with a little care in breeding, it appears that races of any colour might soon be formed. Hofacker gives the result of matching two hundred and sixteen mares of four different colours with like-coloured stallions, without regard to the colour of their ancestors; and of the two hundred and sixteen colts born, eleven alone failed to inherit the colour of their parents : Autenrieth and Ammon assert that, after two generations, colts of a uniform colour are produced with certainty.[56]

In a few rare cases peculiarities fail to be inherited, apparently from the force of inheritance being too strong. I have been assured by breeders of the canary-bird that to get a good jonquil-

[55] Darwin, in ' Journal of Proc. Linn. Soc. Bot.,' 1862, p. 94.
[56] Hofacker, ' Ueber die Eigenschaften,' &c., s. 10.

coloured bird it does not answer to pair two jonquils, as the colour then comes out too strong, or is even brown. So again, if two crested canaries are paired, the young birds rarely inherit this character : [57] for in crested birds a narrow space of bare skin is left on the back of the head, where the feathers are up-turned to form the crest, and, when both parents are thus characterised, the bareness becomes excessive, and the crest itself fails to be developed. Mr. Hewitt, speaking of Laced Sebright Bantams, says [58] that, " why this should be so, I know not, but I am confident that those that are best laced frequently produce offspring very far from perfect in their markings, whilst those exhibited by myself, which have so often proved successful, were bred from the union of heavily-laced birds with those that were scarcely sufficiently laced."

It is a singular fact that, although several deaf-mutes often occur in the same family, and though their cousins and other relations are often in the same condition, yet their parents are very rarely deaf-mutes. To give a single instance : not one scholar out of 148, who were at the same time in the London Institution, was the child of parents similarly afflicted. So again, when a male or a female deaf-mute marries a sound person, their children are most rarely affected : in Ireland out of 203 children thus produced one alone was mute. Even when both parents have been deaf-mutes, as in the case of forty-one marriages in the United States and of six in Ireland, only two deaf and dumb children were produced. Mr. Sedgwick,[59] in commenting on this remarkable and fortunate failure in the power of transmission in the direct line, remarks that it may possibly be owing to " excess having reversed the action of some natural law in development." But it is safer in the present state of our knowledge to look at the whole case as simply unintelligible.

With respect to the inheritance of structures mutilated by injuries or altered by disease it is difficult to come to any

[57] Bechstein, 'Naturgesch. Deutschlands,' b. iv. s. 462. Mr. Brent, a great breeder of canaries, informs me that he believes that these statements are correct.

[58] 'The Poultry Book,' by W. B. Tegetmeier, 1866, p. 245.

[59] 'British and Foreign Med.-Chirurg. Review,' July, 1861, pp. 200-204. Mr. Sedgwick has given such full details on this subject, with ample references, that I need refer to no other authorities.

definite conclusion. In some cases mutilations have been practised for a vast number of generations without any inherited result.
Godron has remarked[60] that different races of man have from
time immemorial knocked out their upper incisors, cut off joints
of their fingers, made holes of immense size through the lobes of
their ears or through their nostrils, made deep gashes in various
parts of their bodies, and there is no reason whatever to suppose
that these mutilations have ever been inherited. Adhesions
due to inflammation and pits from the small-pox (and formerly
many consecutive generations must have been thus pitted)
are not inherited. With respect to Jews, I have been assured
by three medical men of the Jewish faith that circumcision,
which has been practised for so many ages, has produced no
inherited effect; Blumenbach, on the other hand, asserts[61] that
in Germany Jews are often born in a condition rendering circumcision difficult, so that a name is there applied to them
signifying " born circumcised." The oak and other trees must
have borne galls from primeval times, yet they do not produce inherited excrescences; many other such facts could be
adduced.

On the other hand, various cases have been recorded of cats,
dogs, and horses, which have had their tails, legs, &c., amputated
or injured, producing offspring with the same parts ill-formed;
but as it is not at all rare for similar malformations to appear
spontaneously, all such cases may be due to mere coincidence.
Nevertheless, Dr. Prosper Lucas has given, on good authorities,
such a long list of inherited injuries, that it is difficult not to
believe in them. Thus, a cow that had lost a horn from an
accident with consequent suppuration, produced three calves
which were hornless on the same side of the head. With the
horse, there seems hardly a doubt that bony exostoses on the
legs, caused by too much travelling on hard roads, are inherited.
Blumenbach records the case of a man who had his little finger
on the right hand almost cut off, and which in consequence
grew crooked, and his sons had the same finger on the same
hand similarly crooked. A soldier, fifteen years before his
marriage, lost his left eye from purulent ophthalmia, and his

[60] ' De l'Espèce,' tom. ii., 1859, p. 299.
[61] ' Philosoph. Magazine,' vol. iv., 1799, p. 5.

two sons were microphthalmic on the same side.[62] In all such
cases, if truthfully reported, in which the parent has had an
organ injured on one side, and more than one child has been
born with the same organ affected on the same side, the chances
against mere coincidence are enormous. But perhaps the most
remarkable and trustworthy fact is that given by Dr. Brown-
Séquard,[63] namely, that many young guinea-pigs inherited an
epileptic tendency from parents which had been subjected to
a particular operation, inducing in the course of a few weeks
a convulsive disease like epilepsy: and it should be especially
noted that this eminent physiologist bred a large number of
guinea-pigs from animals which had not been operated on, and
not one of these manifested the epileptic tendency. On the
whole, we can hardly avoid admitting, that injuries and mutilations, especially when followed by disease, or perhaps exclusively when thus followed, are occasionally inherited.

Although many congenital monstrosities are inherited, of
which examples have already been given, and to which may
be added the lately recorded case of the transmission during
a century of hare-lip with a cleft-palate in the writer's own
family,[64] yet other malformations are rarely or never inherited.
Of these latter cases, many are probably due to injuries in the
womb or egg, and would come under the head of non-inherited
injuries or mutilations. With plants, a long catalogue of inherited monstrosities of the most serious and diversified nature
could easily be given; and with plants, there is no reason
to suppose that monstrosities are caused by direct injuries to
the seed or embryo.

Causes of Non-inheritance.

A large number of cases of non-inheritance are intelligible on
the principle, that a strong tendency to inheritance does exist, but

---

[62] This last case is quoted by Mr.
Sedgwick in 'British and Foreign
Medico-Chirurg. Review,' April, 1861,
p. 484. For Blumenbach, see above-
cited paper. See, also, Dr. P. Lucas,
'Traité de l'Héréd. Nat.,' tom. ii. p.
492. Also 'Transact. Linn. Soc.,' vol.
ix. p. 323. Some curious cases are given
by Mr. Baker in 'The Veterinary,' vol.
xiii. p. 723. Another curious case is
given in the 'Annales des Scienc. Nat.,'
1st series, tom. xi. p. 324.

[63] 'Proc. Royal Soc.,' vol. x. p.
297.

[64] Mr. Sproule, in 'British Medical
Journal,' April 18, 1863.

that it is overborne by hostile or unfavourable conditions of life. No one would expect that our improved pigs, if forced during several generations to travel about and root in the ground for their own subsistence, would transmit, as truly as they now do, their tendency to fatten, and their short muzzles and legs. Dray-horses assuredly would not long transmit their great size and massive limbs, if compelled to live on a cold, damp mountainous region; we have indeed evidence of such deterioration in the horses which have run wild on the Falkland Islands. European dogs in India often fail to transmit their true character. Our sheep in tropical countries lose their wool in a few generations. There seems also to be a close relation between certain peculiar pastures and the inheritance of an enlarged tail in fat-tailed sheep, which form one of the most ancient breeds in the world. With plants, we have seen that the American varieties of maize lose their proper character in the course of two or three generations, when cultivated in Europe. Our cabbages, which here come so true by seed, cannot form heads in hot countries. Under changed circumstances, periodical habits of life soon fail to be transmitted, as the period of maturity in summer and winter wheat, barley, and vetches. So it is with animals; for instance, a person whose statement I can trust, procured eggs of Aylesbury ducks from that town, where they are kept in houses and are reared as early as possible for the London market; the ducks bred from these eggs in a distant part of England, hatched their first brood on January 24th, whilst common ducks, kept in the same yard and treated in the same manner, did not hatch till the end of March; and this shows that the period of hatching was inherited. But the grandchildren of these Aylesbury ducks completely lost their early habit of incubation, and hatched their eggs at the same time with the common ducks of the same place.

Many cases of non-inheritance apparently result from the con-ditions of life continually inducing fresh variability. We have seen that when the seeds of pears, plums, apples, &c., are sown, the seedlings generally inherit some degree of family likeness from the parent-variety. Mingled with these seedlings, a few, and sometimes many, worthless, wild-looking plants commonly appear; and their appearance may be attributed to the prin-ciple of reversion. But scarcely a single seedling will be found perfectly to resemble the parent-form; and this, I believe, may be accounted for by constantly recurring variability induced by the conditions of life. I believe in this, because it has been observed that certain fruit-trees truly propagate their kind whilst growing on their own roots, but when grafted on other stocks, and by this process their natural state is manifestly affected, they produce seedlings which vary greatly, departing from the parental type in many characters.[65] Metzger, as stated in the ninth chapter, found that certain kinds of wheat brought from Spain and cultivated in Germany, failed during many years to reproduce themselves truly; but that at last, when accustomed to their new conditions, they ceased to be variable,—that is, they became amenable to the power of inheritance. Nearly all the plants which cannot be propagated with any approach to certainty by seed, are kinds which have long been propagated by buds, cuttings, offsets, tubers, &c., and have in consequence been frequently exposed during their individual lives to widely diversified conditions of life. Plants thus propagated become so variable, that they are subject, as we have seen in the last chapter, even to bud-variation. Our domesticated animals, on the other hand, are not exposed during their individual lives to such extremely diversified conditions, and are not liable to such extreme variability; therefore they do not lose the power of transmitting most of their characteristic features. In the foregoing remarks on non-inheritance, crossed breeds are of course excluded, as their diversity mainly depends on the unequal development of characters derived from either parent, modified by the principles of reversion and prepotency.

Conclusion.

It has, I think, been shown in the early part of this chapter how strongly new characters of the most diversified nature, whether normal or abnormal, injurious or beneficial, whether affecting organs of the highest or most trifling importance, are inherited. Contrary to the common opinion, it is often sufficient for the inheritance of some peculiar character, that one parent alone should possess it, as in most cases in which the rarer

---

[65] Downing, 'Fruits of America,' p. 5; Sageret, 'Pom. Phys.,' pp. 43, 72.

anomalies have been transmitted.   But the power of transmission is extremely variable : in a number of individuals descended from the same parents, and treated in the same manner, some display this power in a perfect manner, and in some it is quite deficient ; and for this difference no reason can be assigned. In some cases the effects of injuries or mutilations apparently are inherited ; and we shall see in a future chapter that the effects of the long-continued use and disuse of parts are certainly inherited.   Even those characters which are considered the most fluctuating, such as colour, are with rare exceptions transmitted much more forcibly than is generally supposed.   The wonder, indeed, in all cases is not that any character should be transmitted, but that the power of inheritance should ever fail.   The checks to inheritance, as far as we know them, are, firstly, circumstances hostile to the particular character in question ; secondly, conditions of life incessantly inducing fresh variability ; and lastly, the crossing of distinct varieties during some previous generation, together with reversion or atavism— that is, the tendency in the child to resemble its grand-parents or more remote ancestors instead of its immediate parents. This latter subject will be fully discussed in the following chapter.

CHAPTER XIII.

INHERITANCE *continued* — REVERSION OR ATAVISM.

DIFFERENT FORMS OF REVERSION — IN PURE OR UNCROSSED BREEDS, AS IN PIGEONS, FOWLS, HORNLESS CATTLE AND SHEEP, IN CULTIVATED PLANTS — REVERSION IN FERAL ANIMALS AND PLANTS — REVERSION IN CROSSED VARIETIES AND SPECIES — REVERSION THROUGH BUD-PROPAGATION, AND BY SEGMENTS IN THE SAME FLOWER OR FRUIT — IN DIFFERENT PARTS OF THE BODY IN THE SAME ANIMAL — THE ACT OF CROSSING A DIRECT CAUSE OF REVERSION, VARIOUS CASES OF, WITH INSTINCTS — OTHER PROXIMATE CAUSES OF REVERSION — LATENT CHARACTERS — SECONDARY SEXUAL CHARACTERS — UNEQUAL DEVELOPMENT OF THE TWO SIDES OF THE BODY — APPEARANCE WITH ADVANCING AGE OF CHARACTERS DERIVED FROM A CROSS — THE GERM WITH ALL ITS LATENT CHARACTERS A WONDERFUL OBJECT — MONSTROSITIES — PELORIC FLOWERS DUE IN SOME CASES TO REVERSION.

THE great principle of inheritance to be discussed in this chapter has been recognised by agriculturists and authors of various nations, as shown by the scientific term *Atavism*, derived from atavus, an ancestor; by the English terms of *Reversion*, or *Throwing back;* by the French *Pas-en-arrière;* and by the German *Rück-schlag*, or *Rück-schritt*. When the child resembles either grandparent more closely than its immediate parents, our attention is not much arrested, though in truth the fact is highly remarkable; but when the child resembles some remote ancestor, or some distant member in a collateral line,—and we must attribute the latter case to the descent of all the members from a common progenitor,—we feel a just degree of astonishment. When one parent alone displays some newly-acquired and generally inheritable character, and the offspring do not inherit it, the cause may lie in the other parent having the power of prepotent transmission. But when both parents are similarly characterised, and the child does not, whatever the cause may be, inherit the character in question, but resembles its grandparents, we have one of the simplest cases of reversion. We continually see another and even more simple case of atavism, though not generally included under this head, namely, when

the son more closely resembles his maternal than his paternal
grandsire in some male attribute, as in any peculiarity in the
beard of man, the horns of the bull, the hackles or comb of
the cock, or, as in certain diseases necessarily confined to the
male sex; for the mother cannot possess or exhibit such male
attributes, yet the child has inherited them, through her blood,
from his maternal grandsire.

The cases of reversion may be divided into two main classes,
which, however, in some instances, blend into each other;
namely, first, those occurring in a variety or race which has
not been crossed, but has lost by variation some character
that it formerly possessed, and which afterwards reappears.
The second class includes all cases in which a distinguishable
individual, sub-variety, race, or species, has at some former
period been crossed with a distinct form, and a character derived
from this cross, after having disappeared during one or several
generations, suddenly reappears. A third class, differing only
in the manner of reproduction, might be formed to include
all cases of reversion effected by means of buds, and therefore
independent of true or seminal generation. Perhaps even a
fourth class might be instituted, to include reversions by segments in the same individual flower or fruit, and in different
parts of the body in the same individual animal as it grows old.
But the two first main classes will be sufficient for our purpose.

Reversion to lost Characters by pure or uncrossed forms.—
Striking instances of this first class of cases were given in the
sixth chapter, namely, of the occasional reappearance, in variously-
coloured pure breeds of the pigeon, of blue birds with all the
marks which characterise the wild *Columba livia.* Similar cases
were given in the case of the fowl. With the common ass, as we
now know that the legs of the wild progenitor are striped, we
may feel assured that the occasional appearance of such stripes
in the domestic animal is a case of simple reversion. But I
shall be compelled to refer again to these cases, and therefore
will here pass them over.

The aboriginal species from which our domesticated cattle and
sheep are descended, no doubt possessed horns; but several hornless breeds are now well established. Yet in these—for instance,

in Southdown sheep—"it is not unusual to find among the male lambs some with small horns." The horns, which thus occasionally reappear in other polled breeds, either "grow to the "full size, or are curiously attached to the skin alone and hang "loosely down, or drop off." [1] The Galloways and Suffolk cattle have been hornless for the last 100 or 150 years, but a horned calf, with the horn often loosely attached, is occasionally born.[2]

There is reason to believe that sheep in their early domesticated condition were "brown or dingy black;" but even in the time of David certain flocks were spoken of as white as snow. During the classical period the sheep of Spain are described by several ancient authors as being black, red, or tawny.[3] At the present day, notwithstanding the great care which is taken to prevent it, particoloured lambs and some entirely black are occasionally dropped by our most highly improved and valued breeds, such as the Southdowns. Since the time of the famous Bakewell, during the last century, the Leicester sheep have been bred with the most scrupulous care; yet occasionally grey-faced, or black-spotted, or wholly black lambs appear.[4] This occurs still more frequently with the less improved breeds, such as the Norfolks.[5] As bearing on this tendency in sheep to revert to dark colours, I may state (though in doing so I trench on the reversion of crossed breeds, and likewise on the subject of prepotency) that the Rev. W. D. Fox was informed that seven white Southdown ewes were put to a so-called Spanish ram, which had two small black spots on his sides, and they produced thirteen lambs, all perfectly black. Mr. Fox believes that this ram belonged to a breed which he has himself kept, and which is always spotted with black and white; and he finds that Leicester sheep crossed by rams of this breed always produce black lambs: he has gone on recrossing these crossed sheep with pure white Leicesters during three successive gene-

---

[1] Youatt on Sheep, pp. 20, 234. The same fact of loose horns occasionally appearing in hornless breeds has been observed in Germany : Bechstein, 'Naturgesch. Deutschlands,' b. i. s. 362.

[2] Youatt on Cattle, pp. 155, 174.

[3] Youatt on Sheep, 1838, pp. 17, 145.

[4] I have been informed of this fact through the Rev. W. D. Fox, on the excellent authority of Mr. Wilmot : see, also, remarks on this subject in an original article in the 'Quarterly Review,' 1849, p. 395.

[5] Youatt, pp. 19, 234.

rations, but always with the same result. Mr. Fox was also told by the friend from whom the spotted breed was procured, that he likewise had gone on for six or seven generations crossing with white sheep, but still black lambs were invariably produced.

Similar facts could be given with respect to tailless breeds of various animals. For instance, Mr. Hewitt[6] states that chickens bred from some Rumpless fowls, which were reckoned so good that they won a prize at an exhibition, " in a considerable number of instances were furnished with fully developed tail-feathers." On inquiry, the original breeder of these fowls stated that, from the time when he had first kept them, they had often produced fowls furnished with tails; but that these latter would again reproduce rumpless chickens.

Analogous cases of reversion occur in the vegetable kingdom; thus " from seeds gathered from the finest cultivated varieties of Heartsease (*Viola tricolor*), plants perfectly wild both in their foliage and their flowers are frequently produced;"[7] but the reversion in this instance is not to a very ancient period, for the best existing varieties of the heartsease are of comparatively modern origin. With most of our cultivated vegetables there is some tendency to reversion to what is known to be, or may be presumed to be, their aboriginal state; and this would be more evident if gardeners did not generally look over their beds of seedlings, and pull up the false plants or " rogues " as they are called. It has already been remarked, that some few seedling apples and pears generally resemble, but apparently are not identical with, the wild trees from which they are descended. In our turnip[8] and carrot-beds a few plants often " break "— that is, flower too soon; and their roots are generally found to be hard and stringy, as in the parent-species. By the aid of a little selection, carried on during a few generations, most of our cultivated plants could probably be brought back, without any great change in their conditions of life, to a wild or nearly wild condition: Mr. Buckman has effected this with the parsnip;[9]

[6] ' The Poultry Book,' by Mr. Tegetmeier,' 1866, p. 231.

[7] Loudon's ' Gard. Mag.,' vol. x., 1834, p. 396 : a nurseryman, with much experience on this subject, has likewise assured me that this sometimes occurs.

[8] ' Gardener's Chron.,' 1855, p. 777.

[9] Ibid., 1862, p. 721.

and Mr. Hewett C. Watson, as he informs me, selected, during three generations, "the most diverging plants of Scotch kail, perhaps one of the least modified varieties of the cabbage; and in the third generation some of the plants came very close to the forms now established in England about old castle-walls, and called indigenous."

Reversion in Animals and Plants which have run wild.—In the cases hitherto considered, the reverting animals and plants have not been exposed to any great or abrupt change in their conditions of life which could have induced this tendency; but it is very different with animals and plants which have become feral or run wild. It has been repeatedly asserted in the most positive manner by various authors, that feral animals and plants invariably return to their primitive specific type. It is curious on what little evidence this belief rests. Many of our domesticated animals could not subsist in a wild state; thus, the more highly improved breeds of the pigeon will not "field" or search for their own food. Sheep have never become feral, and would be destroyed by almost every beast of prey. In several cases we do not know the aboriginal parent-species, and cannot possibly tell whether or not there has been any close degree of reversion. It is not known in any instance what variety was first turned out; several varieties have probably in some cases run wild, and their crossing alone would tend to obliterate their proper character. Our domesticated animals and plants, when they run wild, must always be exposed to new conditions of life, for, as Mr. Wallace[10] has well remarked, they have to obtain their own food, and are exposed to competition with the native productions. Under these circumstances, if our domesticated animals did not undergo change of some kind, the result would be quite opposed to the conclusions arrived at in this work. Nevertheless, I do not doubt that the simple fact of animals and plants becoming feral, does cause some tendency to reversion to the primitive state; though this tendency has been much exaggerated by some authors.

[10] *See* some excellent remarks on this subject by Mr. Wallace, 'Journal Proc. Linn. Soc.,' 1858, vol. iii. p. 60.

I will briefly run through the recorded cases. With neither horses nor cattle is the primitive stock known; and it has been shown in former chapters that they have assumed different colours in different countries. Thus the horses which have run wild in South America are generally brownish-bay, and in the East dun-coloured; their heads have become larger and coarser, and this may be due to reversion. No careful description has been given of the feral goat. Dogs which have run wild in various countries have hardly anywhere assumed a uniform character; but they are probably descended from several domestic races, and aboriginally from several distinct species. Feral cats, both in Europe and La Plata, are regularly striped; in some cases they have grown to an unusually large size, but do not differ from the domestic animal in any other character. When variously-coloured tame rabbits are turned out in Europe, they generally reacquire the colouring of the wild animal; there can be no doubt that this does really occur, but we should remember that oddly-coloured and conspicuous animals would suffer much from beasts of prey and from being easily shot; this at least was the opinion of a gentleman who tried to stock his woods with a nearly white variety; and when thus destroyed, they would in truth be supplanted by, instead of being transformed into, the common rabbit. We have seen that the feral rabbits of Jamaica, and especially of Porto Santo, have assumed new colours and other new characters. The best known case of reversion, and that on which the widely-spread belief in its universality apparently rests, is that of pigs. These animals have run wild in the West Indies, South America, and the Falkland Islands, and have everywhere acquired the dark colour, the thick bristles, and great tusks of the wild boar; and the young have reacquired longitudinal stripes. But even in the case of the pig, Roulin describes the half-wild animals in different parts of South America as differing in several respects. In Louisiana the pig[11] has run wild, and is said to differ a little in form, and much in colour, from the domestic animal, yet does not closely resemble the wild boar of Europe. With pigeons and fowls,[12] it is not known what variety was first turned out, nor what character the feral birds have assumed. The guinea-fowl in the West Indies, when feral, seems to vary more than in the domesticated state.

With respect to plants run wild, Dr. Hooker[13] has strongly insisted on what slight evidence the common belief in their power of reversion rests. Godron[14] describes wild turnips, carrots, and celery; but these plants in their cultivated state hardly differ from their wild prototypes, except in the

---

[11] Dureau de la Malle, in 'Comptes Rendus,' tom. xli., 1855, p. 807. From the statements above given, the author concludes that the wild pigs of Louisiana are not descended from the European *Sus scrofa*.

[12] Capt. W. Allen, in his 'Expedition to the Niger,' states that fowls have run wild on the island of Annobon, and have become modified in form and voice. The account is so meagre and vague

that it did not appear to me worth copying; but I now find that Dureau de la Malle ('Comptes Rendus,' tom. xli., 1855, p. 690) advances this as a good instance of reversion to the primitive stock, and as confirmatory of a still more vague statement in classical times by Varro.

[13] 'Flora of Australia,' 1859, Introduct., p. ix.

[14] 'De l'Espèce,' tom. ii. pp. 54, 58, 60

succulency and enlargement of certain parts,—characters which would be surely lost by plants growing in a poor soil and struggling with other plants. No cultivated plant has run wild on so enormous a scale as the cardoon (*Cynara cardunculus*) in La Plata. Every botanist who has seen it growing there, in vast beds, as high as a horse's back, has been struck with its peculiar appearance; but whether it differs in any important point from the cultivated Spanish form, which is said not to be prickly like its American descendant, or whether it differs from the wild Mediterranean species, which is said not to be social, I do not know.

Reversion to Characters derived from a Cross, in the case of Sub-varieties, Races, and Species.—When an individual having some recognizable peculiarity unites with another of the same sub-variety, not having the peculiarity in question, it often reappears in the descendants after an interval of several generations. Every one must have noticed, or heard from old people of children closely resembling in appearance or mental disposition, or in so small and complex a character as expression, one of their grandparents, or some more distant collateral relation. Very many anomalies of structure and diseases,[15] of which instances have been given in the last chapter, have come into a family from one parent, and have reappeared in the progeny after passing over two or three generations. The following case has been communicated to me on good authority, and may, I believe, be fully trusted: a pointer-bitch produced seven puppies; four were marked with blue and white, which is so unusual a colour with pointers that she was thought to have played false with one of the greyhounds, and the whole litter was condemned; but the gamekeeper was permitted to save one as a curiosity. Two years afterwards a friend of the owner saw the young dog, and declared that he was the image of his old pointer-bitch Sappho, the only blue and white pointer of pure descent which he had ever seen. This led to close inquiry, and it was proved that he was the great-great-grandson of Sappho; so that, according to the common expression, he had only 1-16th of her blood in his veins. Here it can hardly be doubted that a character derived from a cross with an individual of the same variety reappeared after passing over three generations.

[15] Mr. Sedgwick gives many instances in the 'British and Foreign Med.-Chirurg. Review,' April and July, 1863, pp. 448, 188.

When two distinct races are crossed, it is notorious that the tendency in the offspring to revert to one or both parent-forms is strong, and endures for many generations. I have myself seen the clearest evidence of this in crossed pigeons and with various plants. Mr. Sidney [16] states that, in a litter of Essex pigs, two young ones appeared which were the image of the Berkshire boar that had been used twenty-eight years before in giving size and constitution to the breed. I observed in the farmyard at Betley Hall some fowls showing a strong likeness to the Malay breed, and was told by Mr. Tollet that he had forty years before crossed his birds with Malays; and that, though he had at first attempted to get rid of this strain, he had subsequently given up the attempt in despair, as the Malay character would reappear.

This strong tendency in crossed breeds to revert has given rise to endless discussions in how many generations after a single cross, either with a distinct breed or merely with an inferior animal, the breed may be considered as pure, and free from all danger of reversion. No one supposes that less than three generations suffices, and most breeders think that six, seven, or eight are necessary, and some go to still greater lengths. [17] But neither in the case of a breed which has been contaminated by a single cross, nor when, in the attempt to form an intermediate breed, half-bred animals have been matched together during many generations, can any rule be laid down how soon the tendency to reversion will be obliterated. It depends on the difference in the strength or prepotency of transmission in the two parent-forms, on their actual amount of difference, and on the nature of the conditions of life to which the crossed offspring are exposed. But we must be careful not to confound these cases of reversion to characters gained from a cross, with those given under the first class, in which characters originally common to *both* parents, but lost at some former period, re-appear; for such characters may recur after an almost indefinite number of generations.

---

[16] In his edit. of 'Youatt on the Pig,' 1860, p. 27.

[17] Dr. P. Lucas, ' Héréd. Nat.,' tom. ii pp. 314, 892 : *see* a good practical article on this subject in ' Gard. Chronicle,' 1856, p. 620. I could add a vast number of references, but they would be superfluous.

The law of reversion is equally powerful with hybrids, when they are sufficiently fertile to breed together, or when they are repeatedly crossed with either pure parent-form, as with mongrels. It is not necessary to give instances, for in the case of plants almost every one who has worked on this subject from the time of Kölreuter to the present day has insisted on this tendency. Gärtner has recorded some good instances; but no one has given more striking cases than Naudin.[18] The tendency differs in degree or strength in different groups, and partly depends, as we shall presently see, on the fact of the parent-plants having been long cultivated. Although the tendency to reversion is extremely general with nearly all mongrels and hybrids, it cannot be considered as invariably characteristic of them; there is, also, reason to believe that it may be mastered by long-continued selection; but these subjects will more properly be discussed in a future chapter on Crossing. From what we see of the power and scope of reversion, both in pure races and when varieties or species are crossed, we may infer that characters of almost every kind are capable of reappearance after having been lost for a great length of time. But it does not follow from this that in each particular case certain characters will reappear: for instance, this will not occur when a race is crossed with another endowed with prepotency of transmission. In some few cases the power of reversion wholly fails, without our being able to assign any cause for the failure: thus it has been stated that in a French family in which 85 out of above 600 members, during six generations, had been subject to night-blindness, "there has not been a single example of this affection in the children of parents who were themselves free from it." [19]

Reversion through Bud-propagation—Partial Reversion, by segments in the same flower or fruit, or in different parts of the

[18] Kölreuter gives cases in his 'Dritte Fortsetzung,' 1766, s. 53, 59; and in his well-known 'Memoirs on Lavatera and Jalapa.' Gärtner, 'Bastarderzeugung,' s. 437, 441, &c. Naudin, in his 'Recherches sur l'Hybridité, Nouvelles Archives du Muséum,' tom. i. p. 25.

[19] Quoted by Mr. Sedgwick in 'Med.-Chirurg. Review,' April, 1861, p. 485. Dr. H. Dobell, in 'Med.-Chirurg. Transactions,' vol. xlvi., gives an analogous case, in which, in a large family, fingers with thickened joints were transmitted to several members during five generations; but when the blemish once disappeared it never reappeared.

body in the same individual animal.—In the eleventh chapter, many cases of reversion by buds, independently of seminal generation, were given—as when a leaf-bud on a variegated, curled, or laciniated variety suddenly reassumes its proper character; or as when a Provence-rose appears on a moss-rose, or a peach on a nectarine-tree. In some of these cases only half the flower or fruit, or a smaller segment, or mere stripes, reassumed their former character; and here we have with buds reversion by segments. Vilmorin[20] has also recorded several cases with plants derived from seed, of flowers reverting by stripes or blotches to their primitive colours : he states that in all such cases a white or pale-coloured variety must first be formed, and, when this is propagated for a length of time by seed, striped seedlings occasionally make their appearance; and these can afterwards by care be multiplied by seed.

The stripes and segments just referred to are not due, as far as is known, to reversion to characters derived from a cross, but to characters lost by variation. These cases, however, as Naudin[21] insists in his discussion on disjunction of character, are closely analogous with those given in the eleventh chapter, in which crossed plants are known to have produced half-and-half or striped flowers and fruit, or distinct kinds of flowers on the same root resembling the two parent-forms. Many piebald animals probably come under this same head. Such cases, as we shall see in the chapter on Crossing, apparently result from certain characters not readily blending together, and, as a consequence of this incapacity for fusion, the offspring either perfectly resemble one of their two parents, or resemble one parent in one part and the other parent in another part; or whilst young are intermediate in character, but with advancing age revert wholly or by segments to either parent-form, or to both. Thus young trees of the *Cytisus adami* are intermediate in foliage and flowers between the two parent-forms; but when older the buds continually revert either partially or wholly to both forms. The cases given in the eleventh chapter on the changes which occurred during growth

[20] Verlot, ' Des Variétés,' 1865, p. 63.
[21] ' Nouvelles Archives du Muséum,' tom. i. p. 25. Alex. Braun (in his ' Rejuvenescence,' Ray Soc., 1853, p. 315) apparently holds a similar opinion.

in crossed plants of Tropæolum, Cereus, Datura, and Lathyrus
are all analogous. As however these plants are hybrids of
the first generation, and as their buds after a time come to
resemble their parents and not their grandparents, these cases
do not at first appear to come under the law of reversion in
the ordinary sense of the word; nevertheless, as the change is
effected through a succession of bud-generations on the same
plant, they may be thus included.

Analogous facts have been observed in the animal kingdom,
and are more remarkable, as they occur strictly in the same
individual, and not as with plants through a succession of
bud-generations. With animals the act of reversion, if it can
be so designated, does not pass over a true generation, but
merely over the early stages of growth in the same individual.
For instance, I crossed several white hens with a black cock, and
many of the chickens were during the first year perfectly white,
but acquired during the second year black feathers; on the
other hand, some of the chickens which were at first black
became during the second year piebald with white. A great
breeder [22] says, that a Pencilled Brahma hen which has any of
the blood of the Light Brahma in her, will " occasionally produce a pullet well pencilled during the first year, but she will
most likely moult brown on the shoulders and become quite
unlike her original colours in the second year." The same thing
occurs with Light Brahmas if of impure blood. I have observed
exactly similar cases with the crossed offspring from differently
coloured pigeons. But here is a more remarkable fact: I
crossed a turbit, which has a frill formed by the feathers being
reversed on its breast, with a trumpeter; and one of the young
pigeons thus raised showed at first not a trace of the frill, but,
after moulting thrice, a small yet unmistakably distinct frill
appeared on its breast. According to Girou,[23] calves produced
from a red cow by a black bull, or from a black cow by a
red bull, are not rarely born red, and subsequently become
black.

In the foregoing cases, the characters which appear with
advancing age are the result of a cross in the previous or some

---

[22] Mr. Teebay, in ' The Poultry Book,' by Mr. Tegetmeier,' 1866, p. 72.
[23] Quoted by Hofacker, ' Ueber die Eigenschaften,' &c., s. 98.

former generation; but in the following cases, the characters which thus reappear formerly appertained to the species, and were lost at a more or less remote epoch. Thus, according to Azara,[24] the calves of a hornless race of cattle which originated in Corrientes, though at first quite hornless, as they become adult sometimes acquire small, crooked, and loose horns; and these in succeeding years occasionally become attached to the skull. White and black bantams, both of which generally breed true, sometimes assume as they grow old a saffron or red plumage. For instance, a first-rate black bantam has been described, which during three seasons was perfectly black, but then annually became more and more red; and it deserves notice that this tendency to change, whenever it occurs in a bantam, "is almost certain to prove hereditary." [25] The cuckoo or blue-mottled Dorking cock, when old, is liable to acquire yellow or orange hackles in place of his proper bluish-grey hackles.[26] Now, as *Gallus bankiva* is coloured red and orange, and as Dorking fowls and both kinds of bantams are descended from this species, we can hardly doubt that the change which occasionally occurs in the plumage of these birds as their age advances, results from a tendency in the individual to revert to the primitive type.

Crossing as a direct cause of Reversion.—It has long been notorious that hybrids and mongrels often revert to both or to one of their parent-forms, after an interval of from two to seven or eight, or according to some authorities even a greater number of generations. But that the act of crossing in itself gives an impulse towards reversion, as shown by the reappearance of long-lost characters, has never, I believe, been hitherto proved. The proof lies in certain peculiarities, which do not characterise the immediate parents, and therefore cannot have been derived from them, frequently appearing in the offspring of two breeds when crossed, which peculiarities never appear, or appear with extreme rarity, in these same breeds, as long as they are pre-

[24] 'Essais Hist. Nat. du Paraguay,' tom. ii. 1801, p. 372.
[25] These facts are given on the high authority of Mr. Hewitt, in 'The Poultry Book,' by Mr. Tegetmeier, 1866, p. 248.
[26] 'The Poultry Book,' by Tegetmeier, 1866, p. 97.

cluded from crossing. As this conclusion seems to me highly curious and novel, I will give the evidence in detail.

My attention was first called to this subject, and I was led to make numerous experiments, by MM. Boitard and Corbié having stated that, when they crossed certain breeds, pigeons coloured like the wild *C. livia*, or the common dovecot, namely, slaty-blue, with double black wing-bars, sometimes chequered with black, white loins, the tail barred with black, with the outer feathers edged with white, were almost invariably produced. The breeds which I crossed, and the remarkable results attained, have been fully described in the sixth chapter. I selected pigeons, belonging to true and ancient breeds, which had not a trace of blue or any of the above specified marks; but when crossed, and their mongrels recrossed, young birds were continually produced, more or less plainly coloured slaty-blue, with some or all of the proper characteristic marks. I may recall to the reader's memory one case, namely, that of a pigeon, hardly distinguishable from the wild Shetland species, the grandchild of a red-spot, white fantail, and two black barbs, from any of which, when purely-bred, the production of a pigeon coloured like the wild *C. livia* would have been almost a prodigy.

I was thus led to make the experiments, recorded in the seventh chapter, on fowls. I selected long-established, pure breeds, in which there was not a trace of red, yet in several of the mongrels feathers of this colour appeared; and one magnificent bird, the offspring of a black Spanish cock and white Silk hen, was coloured almost exactly like the wild *Gallus bankiva*. All who know anything of the breeding of poultry will admit that tens of thousands of pure Spanish and of pure white Silk fowls might have been reared without the appearance of a red feather. The fact, given on the authority of Mr. Tegetmeier, of the frequent appearance, in mongrel fowls, of pencilled or transversely-barred feathers, like those common to many gallinaceous birds, is likewise apparently a case of reversion to a character formerly possessed by some ancient progenitor of the family. I owe to the kindness of this same excellent observer the inspection of some neck-hackles and tail-feathers from a hybrid between the common fowl and a very distinct species, the *Gallus varius*; and these feathers are transversely striped in a conspicuous manner with dark metallic blue and grey, a character which could not have been derived from either immediate parent.

I have been informed by Mr. B. P. Brent, that he crossed a white Aylesbury drake and a black so-called Labrador duck, both of which are true breeds, and he obtained a young drake closely like the mallard (*A. boschas*). Of the musk-duck (*A. moschata*, Linn.) there are two sub-breeds, namely, white and slate-coloured; and these I am informed breed true, or nearly true. But the Rev. W. D. Fox tells me that, by putting a white drake to a slate-coloured duck, black birds, pied with white, like the wild musk-duck, were always produced.

We have seen in the fourth chapter, that the so-called Himalayan rabbit, with its snow-white body, black ears, nose, tail, and feet, breeds perfectly true. This race is known to have been formed by the union of two varieties of silver-grey rabbits. Now, when a Himalayan doe was crossed by a sandy-coloured buck, a silver-grey rabbit was produced; and this is evidently a case of reversion to one of the parent varieties. The young of the Himalayan rabbit are born snow-white, and the dark marks do not appear until some time subsequently; but occasionally young Himalayan rabbits are born of a light silver-grey, which colour soon disappears; so that here we have a trace of reversion, during an early period of life, to the parent-varieties, independently of any recent cross.

In the third chapter it was shown that at an ancient period some breeds of cattle in the wilder parts of Britain were white with dark ears, and that the cattle now kept half wild in certain parks, and those which have run quite wild in two distant parts of the world, are likewise thus coloured. Now, an experienced breeder, Mr. J. Beasley, of Northamptonshire,[27] crossed some carefully selected West Highland cows with purely-bred shorthorn bulls. The bulls were red, red and white, or dark roan; and the Highland cows were all of a red colour, inclining to a light or yellow shade. But a considerable number of the offspring—and Mr. Beasley calls attention to this as a remarkable fact—were white, or white with red ears. Bearing in mind that none of the parents were white, and that they were purely-bred animals, it is highly probable that here the offspring reverted, in consequence of the cross, to the colour either of the aboriginal parent-species or of some ancient and half-wild parent-breed. The following case, perhaps, comes under the same head: cows in their natural state have their udders but little developed, and do not yield nearly so much milk as our domesticated animals. Now there is some reason to believe[28] that cross-bred animals between two kinds, both of which are good milkers, such as Alderneys and Shorthorns, often turn out worthless in this respect.

In the chapter on the Horse reasons were assigned for believing that the primitive stock was striped and dun-coloured; and details were given, showing that in all parts of the world stripes of a dark colour frequently appear along the spine, across the legs, and on the shoulders, where they are occasionally double or treble, and even sometimes on the face and body of horses of all breeds and of all colours. But the stripes appear most frequently on the various kinds of duns. They may sometimes plainly be seen on foals, and subsequently disappear. The dun-colour and the stripes are strongly transmitted when a horse thus characterised is crossed with any other; but I was not able to prove that striped duns are generally produced from the crossing of two distinct breeds, neither of which are duns, though this does sometimes occur.

The legs of the ass are often striped, and this may be considered as a reversion to the wild parent-form, the *Asinus tæniopus* of Abyssinia,[29] which is thus striped. In the domestic animal the stripes on the shoulder are occasionally double, or forked at the extremity, as in certain zebrine

---

[27] 'Gardener's Chron. and Agricultural Gazette,' 1866, p. 528.
[28] Ibid., 1860, p. 343.   [29] Sclater, in 'Proc. Zoolog. Soc.,' 1862, p. 163.

species. There is reason to believe that the foal is frequently more plainly striped on the legs than the adult animal. As with the horse, I have not acquired any distinct evidence that the crossing of differently-coloured varieties of the ass brings out the stripes.

But now let us turn to the result of crossing the horse and ass. Although mules are not nearly so numerous in England as asses, I have seen a much greater number with striped legs, and with the stripes far more conspicuous than in either parent-form. Such mules are generally light-coloured, and might be called fallow-duns. The shoulder-stripe in one instance was deeply forked at the extremity, and in another instance was double, though united in the middle. Mr. Martin gives a figure of a Spanish mule with strong zebra-like marks on its legs,[30] and remarks, that mules are particularly liable to be thus striped on their legs. In South America, according to Roulin,[31] such stripes are more frequent and conspicuous in the mule than in the ass. In the United States, Mr. Gosse,[32] speaking of these animals, says, " that in a great number, perhaps in nine out of every ten, the legs are banded with transverse dark stripes."

Many years ago I saw in the Zoological Gardens a curious triple hybrid, from a bay mare, by a hybrid from a male ass and female zebra. This animal when old had hardly any stripes; but I was assured by the superintendent, that when young it had shoulder-stripes, and faint stripes on its flanks and legs. I mention this case more especially as an instance of the stripes being much plainer during youth than in old age.

As the zebra has such conspicuously striped legs, it might have been expected that the hybrids from this animal and the common ass would have had their legs in some degree striped; but it appears from the figures given in Dr. Gray's ' Knowsley Gleanings,' and still more plainly from that given by Geoffroy and F. Cuvier,[33] that the legs are much more conspicuously striped than the rest of the body; and this fact is intelligible only on the belief that the ass aids in giving, through the power of reversion, this character to its hybrid offspring.

The quagga is banded over the whole front part of its body like a zebra, but has no stripes on its legs, or mere traces of them. But in the famous hybrid bred by Lord Morton,[34] from a chesnut, nearly purely-bred, Arabian mare, by a male quagga, the stripes were " more strongly defined and darker than those on the legs of the quagga." The mare was subsequently put to a black Arabian horse, and bore two colts, both of which, as formerly stated, were plainly striped on the legs, and one of them likewise had stripes on the neck and body.

The *Asinus Indicus*[35] is characterised by a spinal stripe, without shoulder

---

[30] ' History of the Horse,' p. 212.

[31] ' Mém. présentés par divers Savans à l'Acad. Royale,' tom. vi. 1835, p. 338.

[32] 'Letters from Alabama,' 1859, p. 280.

[33] ' Hist. Nat. des Mammifères,' 1820, tom. i.

[34] ' Philosoph. Transact.,' 1821, p. 20.

[35] Sclater, in 'Proc. Zoolog. Soc.,' 1862, p. 163 : this species is the Ghor-Khur of N.W. India, and has often been called the Hemionus of Pallas. *See*, also, Mr. Blyth's excellent paper in ' Journ. of Asiatic Soc. of Bengal,' vol. xxviii., 1860, p. 229.

or leg stripes; but traces of these latter stripes may occasionally be seen even in the adult;[36] and Colonel S. Poole, who has had ample opportunities for observation, informs me that in the foal, when first born, the head and legs are often striped, but the shoulder-stripe is not so distinct as in the domestic ass; all these stripes, excepting that along the spine, soon disappear. Now a hybrid, raised at Knowsley[37] from a female of this species by a male domestic ass, had all four legs transversely and conspicuously striped, had three short stripes on each shoulder, and had even some zebra-like stripes on its face! Dr. Gray informs me that he has seen a second hybrid of the same parentage similarly striped.

From these facts we see that the crossing of the several equine species tends in a marked manner to cause stripes to appear on various parts of the body, especially on the legs. As we do not know whether the primordial parent of the genus was striped, the appearance of the stripes can only hypothetically be attributed to reversion. But most persons, after considering the many undoubted cases of variously coloured marks reappearing by reversion in crossed pigeons, fowls, ducks, &c., will come to the same conclusion with respect to the horse-genus; and in this case we must admit that the progenitor of the group was striped on the legs, shoulders, face, and probably over the whole body, like a zebra. If we reject this view, the frequent and almost regular appearance of stripes in the several foregoing hybrids is left without any explanation.

It would appear that with crossed animals a similar tendency to the recovery of lost characters holds good even with instincts. There are some breeds of fowls which are called "everlasting layers," because they have lost the instinct of incubation; and so rare is it for them to incubate that I have seen notices published in works on poultry, when hens of such breeds have taken to sit.[38] Yet the aboriginal species was of course a good incubator; for with birds in a state of nature hardly any

[36] Another species of wild ass, the true *A. hemionus* or *Kiang*, which ordinarily has no shoulder-stripes, is said occasionally to have them; and these, as with the horse and ass, are sometimes double: see Mr. Blyth, in the paper just quoted, and in 'Indian Sporting Review,' 1856, p. 320; and Col. Hamilton Smith. in 'Nat. Library, Horses,' p. 318; and 'Dict. Class. d'Hist. Nat.,' tom. iii. p. 563.

[37] Figured in the 'Gleanings from the Knowsley Menageries,' by Dr. J. E. Gray.

[38] Cases of both Spanish and Polish hens sitting are given in the 'Poultry Chronicle,' 1855, vol. iii. p. 477.

instinct is so strong as this. Now, so many cases have been recorded of the crossed offspring from two races, neither of which are incubators, becoming first-rate sitters, that the re-appearance of this instinct must be attributed to reversion from crossing. One author goes so far as to say, "that a cross between two non-sitting varieties almost invariably produces a mongrel that becomes broody, and sits with remarkable steadi-ness." [39]   Another author, after giving a striking example, remarks that the fact can be explained only on the principle that "two negatives make a positive." It cannot, however, be maintained that hens produced from a cross between two non-sitting breeds invariably recover their lost instinct, any more than that crossed fowls or pigeons invariably recover the red or blue plumage of their prototypes. I raised several chickens from a Polish hen by a Spanish cock,—breeds which do not incubate,—and none of the young hens at first recovered their instinct, and this appeared to afford a well-marked exception to the foregoing rule; but one of these hens, the only one which was preserved, in the third year sat well on her eggs and reared a brood of chickens. So that here we have the appearance with advancing age of a primitive instinct, in the same manner as we have seen that the red plumage of the *Gallus bankiva* is sometimes reacquired by crossed and purely-bred fowls of various kinds as they grow old.

The parents of all our domesticated animals were of course aboriginally wild in disposition; and when a domesticated species is crossed with a distinct species, whether this is a domesticated or only tamed animal, the hybrids are often wild

---

[39] 'The Poultry Book,' by Mr. Teget-meier, 1866, pp. 119, 163. The author, who remarks on the two negatives ('Journ. of Hort.,' 1862, p. 325), states that two broods were raised from a Spanish cock and Silver-pencilled Ham-burgh hen, neither of which are incu-bators, and no less than seven out of eight hens in these two broods "showed a perfect obstinacy in sitting." The Rev. E. S. Dixon ('Ornamental Poultry,' 1848, p. 200) says that chickens reared from a cross between Golden and Black Polish fowls, are "good and steady birds to sit." Mr. B. P. Brent informs me that he raised some good sitting hens by crossing Pencilled Hamburgh and Polish breeds. A cross-bred bird from a Spanish non-incubating cock and Cochin incubating hen is mentioned in the 'Poultry Chronicle,' vol. iii. p. 13, as an "exemplary mother." On the other hand, an exceptional case is given in the 'Cottage Gardener,' 1860, p. 388, of a hen raised from a Spanish cock and black Polish hen which did not incubate.

to such a degree, that the fact is intelligible only on the principle that the cross has caused a partial return to the primitive disposition.

The Earl of Powis formerly imported some thoroughly domesticated humped cattle from India, and crossed them with English breeds, which belong to a distinct species; and his agent remarked to me, without any question having been asked, how oddly wild the cross-bred animals were. The European wild boar and the Chinese domesticated pig are almost certainly specifically distinct: Sir F. Darwin crossed a sow of the latter breed with a wild Alpine boar which had become extremely tame, but the young, though having half-domesticated blood in their veins, were "extremely wild in confinement, and would not eat swill like common English pigs." Mr. Hewitt, who has had great experience in crossing tame cock-pheasants with fowls belonging to five breeds, gives as the character of all " extraordinary wildness;"[40] but I have myself seen one exception to this rule. Mr. S. J. Salter,[41] who raised a large number of hybrids from a bantam-hen by *Gallus Sonneratii*, states that " all were exceedingly wild." Mr. Waterton[42] bred some wild ducks from eggs hatched under a common duck, and the young were allowed to cross freely both amongst themselves and with the tame ducks; they were " half wild and half tame; they came to the windows to be fed, but still they had a wariness about them quite remarkable."

On the other hand, mules from the horse and ass are certainly not in the least wild, yet they are notorious for obstinacy and vice. Mr. Brent, who has crossed canary-birds with many kinds of finches, has not observed, as he informs me, that the hybrids were in any way remarkably wild. Hybrids are often raised between the common and musk duck, and I have been assured by three persons, who have kept these crossed birds, that they were not wild; but Mr. Garnett[43] observed that his female hybrids exhibited "migratory propensities," of which there is not a vestige in the common or musk duck. No case is

---

[40] 'The Poultry Book,' by Tegetmeier,' 1866, pp. 165, 167.

[41] '.Natural History Review,' 1863, April, p. 277.

[42] ' Essays on Natural History,' p. 197.

[43] As stated by Mr. Orton, in his ' Physiology of Breeding,' p. 12.

known of this latter bird having escaped and become wild in Europe or Asia, except, according to Pallas, on the Caspian Sea; and the common domestic duck only occasionally becomes wild in districts where large lakes and fens abound. Nevertheless, a large number of cases have been recorded[44] of hybrids from these two ducks, although so few are reared in comparison with purely-bred birds of either species, having been shot in a completely wild state. It is improbable that any of these hybrids could have acquired their wildness from the musk-duck having paired with a truly wild duck; and this is known not to be the case in North America; hence we must infer that they have reacquired, through reversion, their wildness, as well as renewed powers of flight.

These latter facts remind us of the statements, so frequently made by travellers in all parts of the world, on the degraded state and savage disposition of crossed races of man. That many excellent and kind-hearted mulattos have existed no one will dispute ; and a more mild and gentle set of men could hardly be found than the inhabitants of the island of Chiloe, who consist of Indians commingled with Spaniards in various proportions. On the other hand, many years ago, long before I had thought of the present subject, I was struck with the fact that, in South America, men of complicated descent between Negroes, Indians, and Spaniards, seldom had, whatever the cause might be, a good expression.[45] Livingstone,—and a more unimpeachable authority cannot be quoted,—after speaking of a half-caste man on the Zambesi, described by the Portuguese as a rare monster of inhumanity, remarks, " It is unaccountable why half-castes, such as he, are so much more cruel than the Portuguese, but such is undoubtedly the case." An inhabitant remarked to Livingstone, " God made white men, and God made black men, but the Devil made half-castes."[46] When two races, both

---

[44] M. E. de Selys-Longchamps refers (' Bulletin Acad. Roy. de Bruxelles,' tom. xii. No. 10) to more than seven of these hybrids shot in Switzerland and France. M. Deby asserts (' Zoologist,' vol. v., 1845-46, p. 1254) that several have been shot in various parts of Belgium and Northern France. Audubon

(' Ornitholog. Biography,' vol. iii. p. 168), speaking of these hybrids, says that, in North America, they " now and then wander off and become quite wild."

[45] 'Journal of Researches,' 1845, p. 71.

[46] ' Expedition to the Zambesi,' 1865, pp. 25, 150.

low in the scale, are crossed, the progeny seems to be eminently bad. Thus the noble-hearted Humboldt, who felt none
of that prejudice against the inferior races now so current in
England, speaks in strong terms of the bad and savage disposition of Zambos, or half-castes between Indians and Negroes;
and this conclusion has been arrived at by various observers.[47]
From these facts we may perhaps infer that the degraded state
of so many half-castes is in part due to reversion to a primitive
and savage condition, induced by the act of crossing, as well as
to the unfavourable moral conditions under which they generally
exist.

Summary on the proximate causes leading to Reversion.—When
purely-bred animals or plants reassume long-lost characters,—
when the common ass, for instance, is born with striped legs,
when a pure race of black or white pigeons throws a slaty-
blue bird, or when a cultivated heartsease with large and
rounded flowers produces a seedling with small and elongated
flowers,—we are quite unable to assign any proximate cause.
When animals run wild, the tendency to reversion, which,
though it has been greatly exaggerated, no doubt exists, is
sometimes to a certain extent intelligible. Thus, with feral pigs,
exposure to the weather will probably favour the growth of the
bristles, as is known to be the case with the hair of other domesticated animals, and through correlation the tusks will tend to
be redeveloped. But the reappearance of coloured longitudinal
stripes on young feral pigs cannot be attributed to the direct
action of external conditions. In this case, and in many others,
we can only say that changed habits of life apparently have
favoured a tendency, inherent or latent in the species, to return
to the primitive state.

It will be shown in a future chapter that the position of
flowers on the summit of the axis, and the position of seeds
within the capsule, sometimes determine a tendency towards
reversion; and this apparently depends on the amount of sap or
nutriment which the flower-buds and seeds receive. The position, also, of buds, either on branches or on roots, sometimes
determines, as was formerly shown, the transmission of the

[47] Dr. P. Broca, on 'Hybridity in the Genus Homo,' Eng. translat., 1864, p. 39.

proper character of the variety, or its reversion to a former state.

We have seen in the last section that when two races or species are crossed there is the strongest tendency to the re-appearance in the offspring of long-lost characters, possessed by neither parent nor immediate progenitor. When two white, or red, or black pigeons, of well-established breeds, are united, the offspring are almost sure to inherit the same colours ; but when differently-coloured birds are crossed, the opposed forces of inheritance apparently counteract each other, and the tendency which is inherent in both parents to produce slaty-blue offspring becomes predominant. So it is in several other cases. But when, for instance, the ass is crossed with *A. Indicus* or with the horse,—animals which have not striped legs,—and the hybrids have conspicuous stripes on their legs and even on their faces, all that can be said is, that an inherent tendency to reversion is evolved through some disturbance in the organisation caused by the act of crossing.

Another form of reversion is far commoner, indeed is almost universal with the offspring from a cross, namely, to the cha- racters proper to either pure parent-form. As a general rule, crossed offspring in the first generation are nearly intermediate between their parents, but the grandchildren and succeeding generations continually revert, in a greater or lesser degree, to one or both of their progenitors. Several authors have main- tained that hybrids and mongrels include all the characters of both parents, not fused together, but merely mingled in different proportions in different parts of the body ; or, as Naudin [48] has expressed it, a hybrid is a living mosaic-work, in which the eye cannot distinguish the discordant elements, so completely are they intermingled. We can hardly doubt that, in a certain sense, this is true, as when we behold in a hybrid the elements of both species segregating themselves into segments in the same flower or fruit, by a process of self-attraction or self-affinity ; this segregation taking place either by seminal or by bud-propagation. Naudin further believes that the segregation of the two specific elements or essences is eminently liable to occur in the male and female reproductive matter ; and he thus explains the almost

[48] ' Nouvelles Archives du Muséum,' tom. i. p. 151.

universal tendency to reversion in successive hybrid generations. For this would be the natural result of the union of pollen and ovules, in both of which the elements of the same species had been segregated by self-affinity. If, on the other hand, pollen which included the elements of one species happened to unite with ovules including the elements of the other species, the intermediate or hybrid state would still be retained, and there would be no reversion. But it would, as I suspect, be more correct to say that the elements of both parent-species exist in every hybrid in a double state, namely, blended together and completely separate. How this is possible, and what the term specific essence or element may be supposed to express, I shall attempt to show in the hypothetical chapter on pangenesis.

But Naudin's view, as propounded by him, is not applicable to the reappearance of characters lost long ago by variation; and it is hardly applicable to races or species which, after having been crossed at some former period with a distinct form, and having since lost all traces of the cross, nevertheless occasionally yield an individual which reverts (as in the case of the great-great-grandchild of the pointer Sappho) to the crossing form. The most simple case of reversion, namely, of a hybrid or mongrel to its grandparents, is connected by an almost perfect series with the extreme case of a purely-bred race recovering characters which had been lost during many ages; and we are thus led to infer that all the cases must be related by some common bond.

Gärtner believed that only those hybrid plants which are highly sterile exhibit any tendency to reversion to their parent-forms. It is rash to doubt so good an observer, but this conclusion must I think be an error; and it may perhaps be accounted for by the nature of the genera observed by him, for he admits that the tendency differs in different genera. The statement is also directly contradicted by Naudin's observations, and by the notorious fact that perfectly fertile mongrels exhibit the tendency in a high degree,—even in a higher degree, according to Gärtner himself, than hybrids.[49]

Gärtner further states that reversions rarely occur with

[49] 'Bastarderzeugung,' s. 582. 438, &c.

hybrid plants raised from species which have not been cultivated, whilst, with those which have been long cultivated, they are of frequent occurrence. This conclusion explains a curious discrepancy: Max Wichura,[50] who worked exclusively on willows, which had not been subjected to culture, never saw an instance of reversion; and he goes so far as to suspect that the careful Gärtner had not sufficiently protected his hybrids from the pollen of the parent-species: Naudin, on the other hand, who chiefly experimented on cucurbitaceous and other cultivated plants, insists more strenuously than any other author on the tendency to reversion in all hybrids. The conclusion that the condition of the parent-species, as affected by culture, is one of the proximate causes leading to reversion, agrees fairly well with the converse case of domesticated animals and cultivated plants being liable to reversion when they become feral; for in both cases the organisation or constitution must be disturbed, though in a very different way.

Finally, we have seen that characters often reappear in purely-bred races without our being able to assign any proximate cause; but when they become feral this is either indirectly or directly induced by the change in their conditions of life. With crossed breeds, the act of crossing in itself certainly leads to the recovery of long-lost characters, as well as of those derived from either parent-form. Changed conditions, consequent on cultivation, and the relative position of buds, flowers, and seeds on the plant, all apparently aid in giving this same tendency. Reversion may occur either through seminal or bud generation, generally at birth, but sometimes only with an advance of age. Segments or portions of the individual may alone be thus affected. That a being should be born resembling in certain characters an ancestor removed by two or three, and in some cases by hundreds or even thousands of generations, is assuredly a wonderful fact. In these cases the child is commonly said to inherit such characters directly from its grandparents or more remote ancestors. But this view is hardly conceivable. If, however, we suppose that every character is derived exclu-

[50] 'Die Bastardbefruchtung .... der Weiden,' 1865, s. 23. For Gärtner's remarks on this head, see 'Bastarderzeugung,' s. 474, 582.

sively from the father or mother, but that many characters lie latent in both parents during a long succession of generations, the foregoing facts are intelligible. In what manner characters may be conceived to lie latent, will be considered in a future chapter to which I have lately alluded.

Latent Characters. — But I must explain what is meant by characters lying latent. The most obvious illustration is afforded by secondary sexual characters. In every female all the secondary male characters, and in every male all the secondary female characters, apparently exist in a latent state, ready to be evolved under certain conditions. It is well known that a large number of female birds, such as fowls, various pheasants, partridges, peahens, ducks, &c., when old or diseased, or when operated on, partly assume the secondary male characters of their species. In the case of the hen-pheasant this has been observed to occur far more frequently during certain seasons than during others.[51] A duck ten years old has been known to assume both the perfect winter and summer plumage of the drake.[52] Waterton[53] gives a curious case of a hen which had ceased laying, and had assumed the plumage, voice, spurs, and warlike disposition of the cock; when opposed to an enemy she would erect her hackles and show fight. Thus every character, even to the instinct and manner of fighting, must have lain dormant in this hen as long as her ovaria continued to act. The females of two kinds of deer, when old, have been known to acquire horns; and, as Hunter has remarked, we see something of an analogous nature in the human species.

On the other hand, with male animals, it is notorious that the secondary sexual characters are more or less completely lost when they are subjected to castration. Thus, if the operation be performed on a young cock, he never, as Yarrell states, crows

[51] Yarrell, ' Phil. Transact.,' 1827, p. 268; Dr. Hamilton, in ' Proc. Zoolog. Soc.,' 1862, p. 23.

[52] 'Archiv. Skand. Beiträge zur Naturgesch,' viii. s. 397-413.

[53] In his ' Essays on Nat. Hist.,' 1838. Mr. Hewitt gives analogous cases with hen-pheasants in ' Journal of Horticulture,' July 12, 1864, p. 37. Isidore Geoffroy Saint Hilaire, in his ' Essais de Zoolog. Géu.' (suites à Buffon, 1842, pp. 496-513), has collected such cases in ten different kinds of birds. It appears that Aristotle was well aware of the change in mental disposition in old hens. The case of the female deer acquiring horns is given at p. 513.

again; the comb, wattles, and spurs do not grow to their full size,
and the hackles assume an intermediate appearance between
true hackles and the feathers of the hen. Cases are recorded
of confinement alone causing analogous results. But characters
properly confined to the female are likewise acquired; the capon
takes to sitting on eggs, and will bring up chickens; and what
is more curious, the utterly sterile male hybrids from the pheasant and the fowl act in the same manner, "their delight being
to watch when the hens leave their nests, and to take on
themselves the office of a sitter." [54] That admirable observer
Réaumur [55] asserts that a cock, by being long confined in solitude
and darkness, can be taught to take charge of young chickens;
he then utters a peculiar cry, and retains during his whole life
this newly acquired maternal instinct. The many well-ascertained cases of various male mammals giving milk, show that
their rudimentary mammary glands retain this capacity in a
latent condition.

We thus see that in many, probably in all cases, the secondary
characters of each sex lie dormant or latent in the opposite sex,
ready to be evolved under peculiar circumstances. We can
thus understand how, for instance, it is possible for a good
milking cow to transmit her good qualities through her male
offspring to future generations; for we may confidently believe
that these qualities are present, though latent, in the males
of each generation. So it is with the game-cock, who can transmit his superiority in courage and vigour through his female to
his male offspring; and with man it is known [56] that diseases,
such as hydrocele, necessarily confined to the male sex, can be
transmitted through the female to the grandson. Such cases
as these offer, as was remarked at the commencement of this
chapter, the simplest possible examples of reversion; and they
are intelligible on the belief that characters common to the
grandparent and grandchild of the same sex are present, though
latent, in the intermediate parent of the opposite sex.

The subject of latent characters is so important, as we shall
see in a future chapter, that I will give another illustration.

[54] 'Cottage Gardener,' 1860, p. 379.
[55] 'Art de faire Eclorre,' &c., 1749, tom. ii. p. 8.
[56] Sir H. Holland, 'Medical Notes and Reflections,' 3rd edit., 1855, p. 31.

Many animals have the right and left sides of their body unequally developed: this is well known to be the case with flat-fish, in which the one side differs in thickness and colour, and in the shape of the fins, from the other; and during the growth of the young fish one eye actually travels, as shown by Steenstrup, from the lower to the upper surface.[57] In most flat-fishes the left is the blind side, but in some it is the right; though in both cases " wrong fishes," which are developed in a reversed manner to what is usual, occasionally occur, and in *Platessa flesus* the right or left side is indifferently developed, the one as often as the other. With gasteropods or shell-fish, the right and left sides are extremely unequal; the far greater number of species are dextral, with rare and occasional reversals of development, and some few are normally sinistral; but certain species of Bulimus, and many Achatinellæ,[58] are as often sinistral as dextral. I will give an analogous case in the great Articulate kingdom: the two sides of Verruca[59] are so wonderfully unlike, that without careful dissection it is extremely difficult to recognise the corresponding parts on the opposite sides of the body; yet it is apparently a mere matter of chance whether it be the right or the left side that undergoes so singular an amount of change. One plant is known to me[60] in which the flower, according as it stands on the one or other side of the spike, is unequally developed. In all the foregoing cases the two sides of the animal are perfectly symmetrical at an early period of growth. Now, whenever a species is as liable to be unequally developed on the one as on the other side, we may infer that the capacity for such development is present, though latent, in the undeveloped side. And as a reversal of development occasionally occurs in animals of many kinds, this latent capacity is probably very common.

The best yet simplest instances of characters lying dormant are, perhaps, those previously given, in which chickens and

---

[57] Prof. Thomson on Steenstrup's Views on the Obliquity of Flounders: 'Annals and Mag. of Nat. Hist.,' May, 1865, p. 361.

[58] Dr. E. von Martens, in 'Annals and Mag. of Nat. Hist.,' March, 1866, p. 209.

[59] Darwin, 'Balanidæ,' Ray Soc., 1854, p. 499: *see* also the appended remarks on the apparently capricious development of the thoracic limbs on the right and left sides in the higher crustaceans.

[60] Mormodes ignea: Darwin, 'Fertilization of Orchids,' 1862, p. 251.

young pigeons, raised from a cross between differently coloured birds, are at first of one colour, but in a year or two acquire feathers of the colour of the other parent; for in this case the tendency to a change of plumage is clearly latent in the young bird. So it is with hornless breeds of cattle, some of which acquire, as they grow old, small horns. Purely bred black and white bantams, and some other fowls, occasionally assume, with advancing years, the red feathers of the parent-species. I will here add a somewhat different case, as it connects in a striking manner latent characters of two classes. Mr. Hewitt[61] possessed an excellent Sebright gold-laced hen bantam, which, as she became old, grew diseased in her ovaria, and assumed male characters. In this breed the males resemble the females in all respects except in their combs, wattles, spurs, and instincts; hence it might have been expected that the diseased hen would have assumed only those masculine characters which are proper to the breed, but she acquired, in addition, well-arched tail sickle-feathers quite a foot in length, saddle-feathers on the loins, and hackles on the neck,—ornaments which, as Mr. Hewitt remarks, "would be held as abominable in this breed." The Sebright bantam is known[62] to have originated about the year 1800 from a cross between a common bantam and a Polish fowl, recrossed by a hen-tailed bantam, and carefully selected; hence there can hardly be a doubt that the sickle-feathers and hackles which appeared in the old hen were derived from the Polish fowl or common bantam; and we thus see that not only certain masculine characters proper to the Sebright bantam, but other masculine characters derived from the first progenitors of the breed, removed by a period of above sixty years, were lying latent in this hen-bird, ready to be evolved as soon as her ovaria became diseased.

From these several facts it must be admitted that certain characters, capacities, and instincts may lie latent in an individual, and even in a succession of individuals, without our being able to detect the least signs of their presence. We have

[61] 'Journal of Horticulture,' July, 1864, p. 38. I have had the opportunity of examining these remarkable feathers through the kindness of Mr. Tegetmeier.

[62] 'The Poultry Book,' by Mr. Tegetmeier, 1866, p. 241.

already seen that the transmission of a character from the grandparent to the grandchild, with its apparent omission in the intermediate parent of the opposite sex, becomes simple on this view. When fowls, pigeons, or cattle of different colours are crossed, and their offspring change colour as they grow old, or when the crossed turbit acquired the characteristic frill after its third moult, or when purely-bred bantams partially assume the red plumage of their prototype, we cannot doubt that these qualities were from the first present, though latent, in the individual animal, like the characters of a moth in the cater-pillar. Now, if these animals had produced offspring before they had acquired with advancing age their new characters, nothing is more probable than that they would have transmitted them to some of their offspring, which in this case would in appearance have received such characters from their grand-parents or more distant progenitors. We should then have had a case of reversion, that is, of the reappearance in the child of an ancestral character, actually present, though during youth completely latent, in the parent; and this we may safely con-clude is what occurs with reversions of all kinds to progenitors however remote.

This view of the latency in each generation of all the cha-racters which appear through reversion, is also supported by their actual presence in some cases during early youth alone, or by their more frequent appearance and greater distinctness at this age than during maturity. We have seen that this is often the case with the stripes on the legs and faces of the several species of the horse-genus. The Himalayan rabbit, when crossed, sometimes produces offspring which revert to the parent silver-grey breed, and we have seen that in purely bred animals pale-grey fur occasionally reappears during early youth. Black cats, we may feel assured, would occasionally produce by reversion tabbies; and on young black kittens, with a pedi-gree [63] known to have been long pure, faint traces of stripes may almost always be seen which afterwards disappear. Horn-less Suffolk cattle occasionally produce by reversion horned animals; and Youatt [64] asserts that even in hornless individuals

---

[63] Carl Vogt, 'Lectures on Man,' Eng. translat., 1864, p. 411.
[64] On Cattle, p. 174.

" the rudiment of a horn may be often felt at an early age."

No doubt it appears at first sight in the highest degree improbable that in every horse of every generation there should be a latent capacity and tendency to produce stripes, though these may not appear once in a thousand generations; that in every white, black, or other coloured pigeon, which may have transmitted its proper colour during centuries, there should be a latent capacity in the plumage to become blue and to be marked with certain characteristic bars; that in every child in a six-fingered family there should be the capacity for the production of an additional digit; and so in other cases. Nevertheless there is no more inherent improbability in this being the case than in a useless and rudimentary organ, or even in only a tendency to the production of a rudimentary organ, being inherited during millions of generations, as is well known to occur with a multitude of organic beings. There is no more inherent improbability in each domestic pig, during a thousand generations, retaining the capacity and tendency to develop great tusks under fitting conditions, than in the young calf having retained for an indefinite number of generations rudimentary incisor teeth, which never protrude through the gums.

I shall give at the end of the next chapter a summary of the three preceding chapters; but as isolated and striking cases of reversion have here been chiefly insisted on, I wish to guard the reader against supposing that reversion is due to some rare or accidental combination of circumstances. When a character, lost during hundreds of generations, suddenly reappears, no doubt some such combination must occur; but reversions may be constantly observed, at least to the immediately preceding generations, in the offspring of most unions. This has been universally recognised in the case of hybrids and mongrels, but it has been recognised simply from the difference between the united forms rendering the resemblance of the offspring to their grandparents or more remote progenitors of easy detection. Reversion is likewise almost invariably the rule, as Mr. Sedgwick has shown, with certain diseases. Hence we must conclude that a tendency to this peculiar form of transmission is an integral part of the general law of inheritance.

Monstrosities.—A large number of monstrous growths and of lesser anomalies are admitted by every one to be due to an arrest of development, that is to the persistence of an embryonic condition. If every horse or ass had striped legs whilst young, the stripes which occasionally appear on these animals when adult would have to be considered as due to the anomalous retention of an early character, and not as due to reversion. Now, the leg-stripes in the horse-genus, and some other characters in analogous cases, are apt to occur during early youth and then to disappear; thus the persistence of early characters and reversion are brought into close connexion.

But many monstrosities can hardly be considered as the result of an arrest of development; for parts of which no trace can be detected in the embryo, but which occur in other members of the same class of animals or plants, occasionally appear, and these may probably with truth be attributed to reversion. For instance: supernumerary mammæ, capable of secreting milk, are not extremely rare in women; and as many as five have been observed. When four are developed, they are generally arranged symmetrically on each side of the chest; and in one instance a woman (the daughter of another with supernumerary mammæ) had one mamma, which yielded milk, developed in the inguinal region. This latter case, when we remember the position of the mammæ in some of the lower animals on both the chest and inguinal region, is highly remarkable, and leads to the belief that in all cases the additional mammæ in woman are due to reversion. The facts given in the last chapter on the tendency in supernumerary digits to regrowth after amputation, indicate their relation to the digits of the lower vertebrate animals, and lead to the suspicion that their appearance may in some manner be connected with reversion. But I shall have to recur, in the chapter on pangenesis, to the abnormal multiplication of organs, and likewise to their occasional transposition. The occasional development in man of the coccygeal vertebræ into a short and free tail, though it thus becomes in one sense more perfectly developed, may at the same time be considered as an arrest of development, and as a case of reversion. The greater frequency of a monstrous kind of proboscis in the pig than in any other mammal, considering the position of the pig in the mammalian series, has likewise been attributed, perhaps truly, to reversion.[65]

When flowers which are properly irregular in structure become regular or peloric, the change is generally looked at by botanists as a return to the primitive state. But Dr. Maxwell Masters,[66] who has ably discussed this subject, remarks that when, for instance, all the sepals of a Tropæolum become green and of the same shape, instead of being coloured with one alone prolonged into a spur, or when all the petals of a Linaria become simple and regular, such cases may be due merely to an arrest of development; for in these flowers all the organs during their earliest condition are symmetrical, and, if arrested at this stage of growth, they would not become irregular. If, moreover, the arrest were to take place at a still earlier period of development, the result would be a simple tuft of green leaves; and no one probably would call this a case of reversion. Dr. Masters designates the cases first alluded to as regular peloria; and others, in which all the corresponding parts assume a similar form of irregularity, as when all the petals in a Linaria become spurred, as irregular peloria. We have no right to attribute these latter cases to reversion, until it can be shown to be probable that the parent-form, for instance, of the genus Linaria had had all its petals spurred; for a change of this nature might result from the spreading of an anomalous structure, in accordance with the law, to be discussed in a future chapter, of homologous parts tending to vary in the same manner. But as both forms of peloria frequently occur on the same individual plant of the Linaria,[67] they probably stand in some close relation to each other. On the doctrine that peloria is simply the result of an arrest of development, it is difficult to understand how an organ arrested at a very early period of growth should acquire its full functional perfection;—how a petal, supposed to be thus arrested, should acquire its brilliant colours, and serve as an envelope to the flower, or a stamen produce efficient pollen; yet this occurs with many peloric flowers. That pelorism is not due to mere chance variability, but either to an arrest of development or to reversion, we may infer from an observation made by Ch. Morren,[68] namely, that families which have irregular flowers often " return by these monstrous growths to their regular form; whilst we never see a regular flower realise the structure of an irregular one."

Some flowers have almost certainly become more or less completely peloric through reversion. *Corydalis tuberosa* properly has one of its two nectaries colourless, destitute of nectar, only half the size of the other, and

---

[65] Isid. Geoffroy St. Hilaire, 'Des Anomalies,' tom. iii. p. 353. With respect to the mammæ in women, see tom. i. p. 710.

[66] 'Natural Hist. Review,' April, 1863, p. 258. See also his Lecture, Royal Institution, March 16, 1860. On same subject, see Moquin-Tandon, 'Eléments

[67] Verlot, 'Des Variétés,' 1865, p. 89; Naudin, 'Nouvelles Archives du Muséum,' tom. i. p. 137.

[68] In his discussion on some curious peloric calceolarias, quoted in 'Journal of Horticulture,' Feb. 24, 1863, p. 152.

de Tératologie,' 1841. pp. 184, 352.

therefore, to a certain extent, in a rudimentary state; the pistil is curved towards the perfect nectary, and the hood, formed of the inner petals, slips off the pistil and stamens in one direction alone, so that, when a bee sucks the perfect nectary, the stigma and stamens are exposed and rubbed against the insect's body. In several closely allied genera, as in Dielytra, &c., there are two perfect nectaries, the pistil is straight, and the hood slips off on either side, according as the bee sucks either nectary. Now, I have examined several flowers of *Corydalis tuberosa*, in which both nectaries were equally developed and contained nectar; in this we see only the redevelopment of a partially aborted organ; but with this redevelopment the pistil becomes straight, and the hood slips off in either direction; so that these flowers have acquired the perfect structure, so well adapted for insect agency, of Dielytra and its allies. We cannot attribute these coadapted modifications to chance, or to correlated variability; we must attribute them to reversion to a primordial condition of the species.

The peloric flowers of Pelargonium have their five petals in all respects alike, and there is no nectary; so that they resemble the symmetrical flowers of the closely allied Geranium-genus; but the alternate stamens are also sometimes destitute of anthers, the shortened filaments being left as rudiments, and in this respect they resemble the symmetrical flowers of the closely allied genus, Erodium. Hence we are led to look at the peloric flowers of Pelargonium as having probably reverted to the state of some primordial form, the progenitor of the three closely related genera of Pelargonium, Geranium, and Erodium.

In the peloric form of *Antirrhinum majus*, appropriately called the "*Wonder*," the tubular and elongated flowers differ wonderfully from those of the common snapdragon; the calyx and the mouth of the corolla consist of six equal lobes, and include six equal instead of four unequal stamens. One of the two additional stamens is manifestly formed by the development of a microscopically minute papilla, which may be found at the base of the upper lip of the flower in all common snapdragons, at least in nineteen plants examined by me. That this papilla is a rudiment of a stamen was well shown by its various degrees of development in crossed plants between the common and peloric Antirrhinum. Again, a peloric *Galeobdolon luteum*, growing in my garden, had five equal petals, all striped like the ordinary lower lip, and included five equal instead of four unequal stamens; but Mr. R. Keeley, who sent me this plant, informs me that the flowers vary greatly, having from four to six lobes to the corolla, and from three to six stamens.[69] Now, as the members of the two great families to which the Antirrhinum and Galeobdolon belong are properly pentamerous, with some of the parts confluent and others suppressed, we ought not to look at the sixth stamen and the sixth lobe to the corolla in either case as due to reversion, any more than the additional petals in double flowers in these same two families. But the case is different with the fifth stamen in the peloric Antirrhinum, which

[69] For other cases of six divisions in peloric flowers of the Labiatæ and Scrophulariaceæ, *see* Moquin-Tandon, 'Tératologie,' p. 192.

is produced by the redevelopment of a rudiment always present, and which probably reveals to us the state of the flower, as far as the stamens are concerned, at some ancient epoch. It is also difficult to believe that the other four stamens and the petals, after an arrest of development at a very early embryonic age, would have come to full perfection in colour, structure, and function, unless these organs had at some former period normally passed through a similar course of growth. Hence it appears to me probable that the progenitor of the genus Antirrhinum must at some remote epoch have included five stamens and borne flowers in some degree resembling those now produced by the peloric form.

Lastly, I may add that many instances have been recorded of flowers, not generally ranked as peloric, in which certain organs, normally few in number, have been abnormally augmented. As such an increase of parts cannot be looked at as an arrest of development, nor as due to the redevelopment of rudiments, for no rudiments are present, and as these additional parts bring the plant into closer relationship with its natural allies, they ought probably to be viewed as reversions to a primordial condition.

These several facts show us in an interesting manner how intimately certain abnormal states are connected together; namely, arrests of development causing parts to become rudimentary or to be wholly suppressed,—the redevelopment of parts at present in a more or less rudimentary condition,—the reappearance of organs of which not a vestige can now be detected,—and to these may be added, in the case of animals, the presence during youth, and subsequent disappearance, of certain characters which occasionally are retained throughout life. Some naturalists look at all such abnormal structures as a return to the ideal state of the group to which the affected being belongs; but it is difficult to conceive what is meant to be conveyed by this expression. Other naturalists maintain, with greater probability and distinctness of view, that the common bond of connection between the several foregoing cases is an actual, though partial, return to the structure of the ancient progenitor of the group. If this view be correct, we must believe that a vast number of characters, capable of evolution, lie hidden in every organic being. But it would be a mistake to suppose that the number is equally great in all beings. We know, for instance, that plants of many orders occasionally become peloric; but many more cases have been observed in the Labiatæ and Scrophulariaceæ than in any other order; and in one genus of the Scrophulariaceæ, namely Linaria, no less

than thirteen species have been described in a peloric condition.[70]
On this view of the nature of peloric flowers, and bearing in
mind what has been said with respect to certain monstrosities
in the animal kingdom, we must conclude that the progenitors
of most plants and animals, though widely different in structure, have left an impression capable of redevelopment on the
germs of their descendants.

The fertilised germ of one of the higher animals, subjected as
it is to so vast a series of changes from the germinal cell to old
age,—incessantly agitated by what Quatrefages well calls the
tourbillon vital,—is perhaps the most wonderful object in nature.
It is probable that hardly a change of any kind affects
either parent, without some mark being left on the germ.   But
on the doctrine of reversion, as given in this chapter, the germ
becomes a far more marvellous object, for, besides the visible
changes to which it is subjected, we must believe that it is
crowded with invisible characters, proper to both sexes, to both
the right and left side of the body, and to a long line of male
and female ancestors separated by hundreds or even thousands
of generations from the present time; and these characters,
like those written on paper with invisible ink, all lie ready to
be evolved under certain known or unknown conditions.

[70] Moquin-Tandon, ' Tératologie,' p. 186.

CHAPTER XIV.

In the two last chapters the nature and force of Inheritance, the
circumstances which interfere with its power, and the tendency
to Reversion, with its many remarkable contingencies, were discussed.   In the present chapter some other related phenomena
will be treated of, as fully as my materials permit.

Fixedness of Character.

It is a general belief amongst breeders that the longer any
character has been transmitted by a breed, the more firmly it
will continue to be transmitted. I do not wish to dispute the
truth of the proposition, that inheritance gains strength simply
through long continuance, but I doubt whether it can be proved.
In one sense the proposition is little better than a truism; if
any character has remained constant during many generations,
it will obviously be little likely, the conditions of life remaining
the same, to vary during the next generation. So, again, in
improving a breed, if care be taken for a length of time to
exclude all inferior individuals, the breed will obviously tend
to become truer, as it will not have been crossed during many
generations by an inferior animal. We have previously seen,

but without being able to assign any cause, that, when a new
character appears, it is occasionally from the first well fixed,
or fluctuates much, or wholly fails to be transmitted. So it is
with the aggregate of slight differences which characterise a new
variety, for some propagate their kind from the first much truer
than others.   Even with plants multiplied by bulbs, layers, &c.,
which may in one sense be said to form parts of the same individual,
it is well known that certain varieties retain and transmit through
successive bud-generations their newly-acquired characters more
truly than others.   In none of these, nor in the following cases,
does there appear to be any relation between the force with
which a character is transmissible and the length of time during
which it has already been transmitted.   Some varieties, such as
white and yellow hyacinths and white sweet-peas, transmit their
colours more faithfully than do the varieties which have retained
their natural colour.   In the Irish family, mentioned in the
twelfth chapter, the peculiar tortoiseshell-like colouring of the
eyes was transmitted far more faithfully than any ordinary colour.
Ancon and Mauchamp sheep and niata cattle, which are all comparatively modern breeds, exhibit remarkably strong powers of
inheritance. Many similar cases could be adduced.

As all domesticated animals and cultivated plants have varied,
and yet are descended from aboriginally wild forms, which no
doubt had retained the same character from an immensely
remote epoch, we see that scarcely any degree of antiquity
ensures a character being transmitted perfectly true.   In this
case, however, it may be said that changed conditions of life
induce certain modifications, and not that the power of inheritance fails; but in every case of failure, some cause, either
internal or external, must interfere.   It will generally be found
that the parts in our domesticated productions which have
varied, or which still continue to vary,—that is, which fail to
retain their primordial state,—are the same with the parts which
differ in the natural species of the same genus. As, on the theory
of descent with modification, the species of the same genus
have been modified since they branched off from a common
progenitor, it follows that the characters by which they differ
from each other have varied whilst other parts of the organisation have remained unchanged ; and it might be argued that

these same characters now vary under domestication, or fail
to be inherited, owing to their lesser antiquity. But we must
believe structures, which have already varied, would be more
liable to go on varying, rather than structures which during
an immense lapse of time have remained unaltered; and this
variation is probably the result of certain relations between the
conditions of life and the organisation, quite independently of
the greater or less antiquity of each particular character.

Fixedness of character, or the strength of inheritance, has
often been judged of by the preponderance of certain characters
in the crossed offspring between distinct races; but prepotency
of transmission here comes into play, and this, as we shall immediately see, is a very different consideration from the strength
or weakness of inheritance. It has often been observed[1] that
breeds of animals inhabiting wild and mountainous countries
cannot be permanently modified by our improved breeds; and
as these latter are of modern origin, it has been thought that
the greater antiquity of the wilder breeds has been the cause
of their resistance to improvement by crossing; but it is
more probably due to their structure and constitution being
better adapted to the surrounding conditions. When plants
are first subjected to culture, it has been found that, during
several generations, they transmit their characters truly, that is,
do not vary, and this has been attributed to ancient characters
being strongly inherited; but it may with equal or greater
probability be consequent on changed conditions of life requiring
a long time for their accumulative action. Notwithstanding these
considerations, it would perhaps be rash to deny that characters
become more strongly fixed the longer they are transmitted;
but I believe that the proposition resolves itself into this,
—that all characters of all kinds, whether new or old, tend
to be inherited, and that those which have already withstood
all counteracting influences and been truly transmitted, will, as
a general rule, continue to withstand them, and consequently be
faithfully inherited.

[1] See Youatt on Cattle, pp. 92, 69, 78, 88, 163: also Youatt on Sheep, p. 325.
Also Dr. Lucas, 'L'Héréd. Nat.,' tom. ii. p. 310.

Prepotency in the Transmission of Character.

When individuals distinct enough to be recognised, but of the same family, or when two well-marked races, or two species, are crossed, the usual result, as stated in the previous chapter, is, that the offspring in the first generation are intermediate between their parents, or resemble one parent in one part and the other parent in another part. But this is by no means the invariable rule; for in many cases it is found that certain individuals, races, and species are prepotent in transmitting their likeness. This subject has been ably discussed by Prosper Lucas,[2] but is rendered extremely complicated by the prepotency sometimes running equally in both sexes, and sometimes more strongly in one sex than in the other; it is likewise complicated by the presence of secondary sexual characters, which render the comparison of mongrels with their parent-breeds difficult.

It would appear that in certain families some one ancestor, and after him others in the same family, must have had great power in transmitting their likeness through the male line; for we cannot otherwise understand how the same features should so often be transmitted after marriages with various females, as has been the case with the Austrian Emperors, and as, according to Niebuhr, formerly occurred in certain Roman families with their mental qualities.[3] The famous bull Favourite is believed[4] to have had a prepotent influence on the short-horn race. It has also been observed[5] with English race-horses that certain mares have generally transmitted their own character, whilst other mares of equally pure blood have allowed the character of the sire to prevail.

The truth of the principle of prepotency comes out more clearly when certain races are crossed. The improved Shorthorns, notwithstanding that the breed is comparatively modern, are generally acknowledged to possess great power in impressing their likeness on all other breeds; and it is chiefly in consequence of this power that they are so highly valued

[2] 'Héréd. Nat.,' tom. ii. pp. 112-120.

[3] Sir H. Holland, 'Chapters on Mental Physiology,' 1852, p. 234.

[4] 'Gardener's Chronicle,' 1860, p. 270.

[5] Mr. N. H. Smith, Observations on Breeding, quoted in ' Encyclop. of Rural Sports,' p. 278.

for exportation.[6] Godine has given a curious case of a ram of a goat-like breed of sheep from the Cape of Good Hope, which produced offspring hardly to be distinguished from himself, when crossed with ewes of twelve other breeds. But two of these half-bred ewes, when put to a merino ram, produced lambs closely resembling the merino breed. Girou de Buzareingues[7] found that of two races of French sheep the ewes of one, when crossed during successive generations with merino rams, yielded up their character far sooner than the ewes of the other race. Sturm and Girou have given analogous cases with other breeds of sheep and with cattle, the prepotency running in these cases through the male side; but I was assured on good authority in South America, that when niata cattle are crossed with common cattle, though the niata breed is pre-potent whether males or females are used, yet that the prepotency is strongest through the female line. The Manx cat is tailless and has long hind legs; Dr. Wilson crossed a male Manx with common cats, and, out of twenty-three kittens, seventeen were destitute of tails; but when the female Manx was crossed by common male cats all the kittens had tails, though they were generally short and imperfect.[8]

In making reciprocal crosses between pouter and fantail pigeons, the pouter-race seemed to be prepotent through both sexes over the fantail. But this is probably due to weak power in the fantail rather than to any unusually strong power in the pouter, for I have observed that barbs also preponderated over fantails. This weakness of transmission in the fantail, though the breed is an ancient one, is said[9] to be general; but I have observed one exception to the rule, namely, in a cross between a fantail and laugher. The most curious instance known to me of weak power in both sexes is in the trumpeter pigeon. This breed has been well known for at least 130 years: it breeds perfectly true, as I have been assured by those who have long kept many birds: it is characterised by a peculiar tuft of feathers over the beak, by a crest on the head, by a most peculiar coo quite unlike that of any other breed, and by much-feathered feet. I have crossed both sexes with turbits of two sub-breeds, with almond tumblers, spots, and runts, and reared many mongrels and recrossed them; and though the crest on the head and feathered feet were inherited (as is generally the case with most breeds), I have never seen a vestige of the tuft over the beak or heard the peculiar coo. Boitard and Corbié[10] assert that this is the invariable result of crossing trumpeters with any other breed: Neumeister,[11] however, states that in Germany mongrels have been obtained, though very rarely, which were furnished with the tuft and would trumpet: but a pair of these mongrels with a tuft, which I imported, never trumpeted. Mr. Brent states[12] that the crossed offspring of a trumpeter were crossed

---

[6] Quoted by Bronn, ' Geschichte der Natur,' b. ii. s. 170. *See* Sturm, ' Ueber Racen,' 1825, s. 104-107. For the niata cattle, *see* my ' Journal of Researches,' 1845, p. 146.

[7] Lucas, 'l'Hérédité Nat.,' tom. ii. p. 112.

[8] Mr. Orton, ' Physiology of Breed-ing,' 1855, p. 9.

[9] Boitard and Corbié, ' Les Pigeons,' 1824, p. 224.

[10] ' Les Pigeons,' pp. 168, 198.

[11] ' Das Ganze,' &c., 1837, s. 39.

[12] ' The Pigeon Book,' p. 46.

with trumpeters for three generations, by which time the mongrels had 7-8ths of this blood in their veins, yet the tuft over the beak did not appear. At the fourth generation the tuft appeared, but the birds, though now having 15-16ths trumpeter's blood, still did not trumpet. This case well shows the wide difference between inheritance and prepotency; for here we have a well-established old race which transmits its characters faithfully, but which, when crossed with any other race, has the feeblest power of transmitting its two chief characteristic qualities.

I will give one other instance with fowls and pigeons of weakness and strength in the transmission of the same character to their crossed offspring. The Silk-fowl breeds true, and there is reason to believe is a very ancient race ; but when I reared a large number of mongrels from a Silk-hen by a Spanish cock, not one exhibited even a trace of the so-called silkiness. Mr. Hewitt also asserts that in no instance are the silky feathers transmitted by this breed when crossed with any other variety. But three birds out of many raised by Mr. Orton from a cross between a silk-cock and a bantam-hen, had silky feathers.[13] So that it is certain that this breed very seldom has the power of transmitting its peculiar plumage to its crossed progeny. On the other hand, there is a silk sub-variety of the fantail pigeon, which has its feathers in nearly the same state as in the Silk-fowl : now we have already seen that fantails, when crossed, possess singularly weak power in transmitting their general qualities ; but the silk sub-variety when crossed with any other small-sized race invariably transmits its silky feathers![14]

The law of prepotency comes into action when species are crossed, as with races and individuals. Gärtner has unequivocally shown[15] that this is the case with plants. To give one instance: when *Nicotiana paniculata* and *vincæflora* are crossed, the character of *N. paniculata* is almost completely lost in the hybrid; but if *N. quadrivalvis* be crossed with *N. vincæflora*, this latter species, which was before so prepotent, now in its turn almost disappears under the power of *N. quadrivalvis*. It is remarkable that the prepotency of one species over another in transmission is quite independent, as shown by Gärtner, of the greater or less facility with which the one fertilises the other.

With animals, the jackal is prepotent over the dog, as is stated by Flourens who made many crosses between these animals; and this was likewise the case with a hybrid which I once saw between a jackal and terrier. I cannot doubt, from the observations of Colin and others, that the ass is prepotent over the horse; the prepotency in this instance running more strongly through the male than through the female ass; so that the mule resembles the ass more closely than does the hinny.[16] The

---

[13] 'Physiology of Breeding,' p. 22; Mr, Hewitt, in 'The Poultry Book,' by Tegetmeier, 1866, p. 224.

[14] Boitard and Corbié, 'Les Pigeons,' 1824, p. 226.

[15] 'Bastarderzeugung,' s. 256, 290, &c. Naudin ('Nouvelles Archives du Muséum,' tom. i. p. 149) gives a striking instance of prepotency in *Datura stramonium* when crossed with two other species.

[16] Flourens, 'Longévité Humaine,' p. 144, on crossed jackals. With respect to the difference between the

male pheasant, judging from Mr. Hewitt's descriptions,[17] and from the hybrids which I have seen, preponderates over the domestic fowl; but the latter, as far as colour is concerned, has considerable power of transmission, for hybrids raised from five differently coloured hens differed greatly in plumage. I formerly examined some curious hybrids in the Zoological Gardens, between the Penguin variety of the common duck and the Egyptian goose (*Tadorna Ægyptiaca*); and although I will not assert that the domesticated variety preponderated over the natural species, yet it had strongly impressed its unnatural upright figure on these hybrids.

I am aware that such cases as the foregoing have been ascribed by various authors, not to one species, race, or individual being prepotent over the other in impressing its character on its crossed offspring, but to such rules as that the father influences the external characters and the mother the internal or vital organs. But the great diversity of the rules given by various authors almost proves their falseness. Dr. Prosper Lucas has fully discussed this point, and has shown[18] that none of the rules (and I could add others to those quoted by him) apply to all animals. Similar rules have been enounced for plants, and have been proved by Gärtner[19] to be all erroneous. If we confine our view to the domesticated races of a single species, or perhaps even to the species of the same genus, some such rules may hold good; for instance, it seems that in reciprocally crossing various breeds of fowls the male generally gives colour;[20] but conspicuous exceptions have passed under my own eyes. In sheep it seems that the ram usually gives its peculiar horns and fleece to its crossed offspring, and the bull the presence or absence of horns.

In the following chapter on Crossing I shall have occasion to show that certain characters are rarely or never blended by crossing, but are trans-

---

mule and the hinny, I am aware that this has generally been attributed to the sire and dam transmitting their characters differently; but Colin, who has given in his 'Traité Phys. Comp.,' tom. ii. pp. 537-539, the fullest description which I have met with of these reciprocal hybrids, is strongly of opinion that the ass preponderates in both crosses, but in an unequal degree. This is likewise the conclusion of Flourens, and of Bechstein in his 'Naturgeschichte Deutschlands,' b. i. s. 294. The tail of the hinny is much more like that of the horse than is the tail of the mule, and this is generally accounted for by the males of both species transmitting with greater power this part of their structure; but a compound hybrid which I saw in the Zoological Gardens, from a mare by a hybrid ass-zebra, closely resembled its mother in its tail.

[17] Mr. Hewitt, who has had such great experience in raising these hybrids, says ('Poultry Book,' by Mr. Tegetmeier, 1866, pp. 165-167) that in all, the head was destitute of wattles, comb, and ear-lappets; and all closely resembled the pheasant in the shape of the tail and general contour of the body. These hybrids were raised from hens of several breeds by a cock-pheasant; but another hybrid, described by Mr. Hewitt, was raised from a hen-pheasant by a silver-laced Bantam cock, and this possessed a rudimental comb and wattles.

[18] 'L'Héréd. Nat.,' tom. ii. book ii. ch. i.

[19] 'Bastarderzeugung,' s. 264-266. Naudin ('Nouvelles Archives du Muséum,' tom. i. p. 148) has arrived at a similar conclusion.

[20] 'Cottage Gardener,' 1856, pp. 101, 137.

mitted in an unmodified state from either parent-form; I refer to this fact here because it is sometimes accompanied on the one side by prepotency, which thus acquires the false appearance of unusual strength. In the same chapter I shall show that the rate at which a species or breed absorbs and obliterates another by repeated crosses, depends in chief part on prepotency in transmission.

In conclusion, some of the cases above given,—for instance, that of the trumpeter pigeon,—prove that there is a wide difference between mere inheritance and prepotency. This latter power seems to us, in our ignorance, to act in most cases quite capriciously. The very same character, even though it be an abnormal or monstrous one, such as silky feathers, may be transmitted by different species, when crossed, either with prepotent force or singular feebleness. It is obvious, that a purely-bred form of either sex, in all cases in which prepotency does not run more strongly in one sex than the other, will transmit its character with prepotent force over a mongrelized and already variable form.[21] From several of the above-given cases we may conclude that mere antiquity of character does not by any means necessarily make it prepotent. In some cases prepotency apparently depends on the same character being present and visible in one of the two breeds which are crossed, and latent or invisible in the other breed; and in this case it is natural that the character which is potentially present in both should be prepotent. Thus, we have reason to believe that there is a latent tendency in all horses to be dun-coloured and striped; and when a horse of this kind is crossed with one of any other colour, it is said that the offspring are almost sure to be striped. Sheep have a similar latent tendency to become dark-coloured, and we have seen with what prepotent force a ram with a few black spots, when crossed with white sheep of various breeds, coloured its offspring. All pigeons have a latent tendency to become slaty-blue, with certain characteristic marks, and it is known that, when a bird thus coloured is crossed with one of any other colour, it is most difficult afterwards to eradicate the blue tint. A nearly parallel case is offered by those black bantams which, as they grow

---

[21] *See* some remarks on this head with respect to sheep by Mr. Wilson, in 'Gardener's Chronicle,' 1863, p. 15.

old, develop a latent tendency to acquire red feathers. But
there are exceptions to the rule: hornless breeds of cattle
possess a latent capacity to reproduce horns, yet when crossed
with horned breeds they do not invariably produce offspring
bearing horns.

We meet with analogous cases with plants. Striped flowers,
though they can be propagated truly by seed, have a latent tendency to become uniformly coloured, but when once crossed by
a uniformly coloured variety, they ever afterwards fail to produce striped seedlings.[22] Another case is in some respects more
curious: plants bearing peloric or regular flowers have so strong
a latent tendency to reproduce their normally irregular flowers,
that this often occurs by buds when a plant is transplanted into
poorer or richer soil.[23] Now I crossed the peloric snapdragon
(*Antirrhinum majus*), described in the last chapter, with pollen
of the common form; and the latter, reciprocally, with peloric
pollen. I thus raised two great beds of seedlings, and not one
was peloric. Naudin[24] obtained the same result from crossing a
peloric Linaria with the common form. I carefully examined the
flowers of ninety plants of the crossed Antirrhinum in the two
beds, and their structure had not been in the least affected by the
cross, except that in a few instances the minute rudiment of the
fifth stamen, which is always present, was more fully or even
completely developed. It must not be supposed that this entire
obliteration of the peloric structure in the crossed plants can
be accounted for by any incapacity of transmission; for I raised
a large bed of plants from the peloric Antirrhinum, artificially
fertilised by its own pollen, and sixteen plants, which alone
survived the winter, were all as perfectly peloric as the parent-
plant. Here we have a good instance of the wide difference
between the inheritance of a character and the power of transmitting it to crossed offspring. The crossed plants, which perfectly resembled the common snapdragon, were allowed to sow
themselves, and, out of a hundred and twenty-seven seedlings,
eighty-eight proved to be common snapdragons, two were in an
intermediate condition between the peloric and normal state,

[22] Verlot, ' Des Variétés,' 1865, p. 66.
[23] Moquin-Tandon, ' Tératologie,' p. 191.
[24] ' Nouvelles Archives du Muséum,' tom. i, p. 137.

and thirty-seven were perfectly peloric, having reverted to the structure of their one grandparent. This case seems at first sight to offer an exception to the rule formerly given, namely, that a character which is present in one form and latent in the other is generally transmitted with prepotent force when the two forms are crossed. For in all the Scrophulariaceæ, and especially in the genera Antirrhinum and Linaria, there is, as was shown in the last chapter, a strong latent tendency to become peloric; and there is also, as we have just seen, a still stronger tendency in all peloric plants to reacquire their normal irregular structure. So that we have two opposed latent tendencies in the same plants. Now, with the crossed Antirrhinums the tendency to produce normal or irregular flowers, like those of the common Snap-dragon, prevailed in the first generation; whilst the tendency to pelorism, appearing to gain strength by the intermission of a generation, prevailed to a large extent in the second set of seedlings. How it is possible for a character to gain strength by the intermission of a generation, will be considered in the chapter on pangenesis.

On the whole, the subject of prepotency is extremely intri-cate,—from its varying so much in strength, even in regard to the same character, in different animals,—from its running either equally in both sexes, or, as frequently is the case with animals, but not with plants, much stronger in the one sex than the other,—from the existence of secondary sexual characters,—from the transmission of certain characters being limited, as we shall immediately see, by sex,—from certain characters not blending together,—and, perhaps, occasionally from the effects of a previous fertilisation on the mother. It is therefore not surprising that every one hitherto has been baffled in drawing up general rules on the subject of prepotency.

Inheritance as limited by Sex.

New characters often appear in one sex, and are afterwards transmitted to the same sex, either exclusively or in a much greater degree than to the other. This subject is important, because with animals of many kinds in a state of nature, both high and low in the scale, secondary sexual characters, not in any way directly connected with the organs of reproduction, are often conspicuously present. With our domesticated animals, also, these same secondary characters are often found to differ greatly from the state in which they exist in the parent-species. And the principle of inheritance as limited by sex shows how such characters might have been first acquired and subsequently modified.

Dr. P. Lucas, who has collected many facts on this subject, shows[25] that when a peculiarity, in no manner connected with the reproductive organs, appears in either parent, it is often transmitted exclusively to the offspring of the same sex, or to a much greater number of them than of the opposite sex. Thus, in the family of Lambert, the horn-like projections on the skin were transmitted from the father to his sons and grandsons alone; so it has been with other cases of ichthyosis, with supernumerary digits, with a deficiency of digits and phalanges, and in a lesser degree with various diseases, especially with colour-blindness, and a hæmorrhagic diathesis, that is, an extreme liability to profuse and uncontrollable bleeding from trifling wounds. On the other hand, mothers have transmitted, during several generations, to their daughters alone, supernumerary and deficient digits, colour-blindness, and other peculiarities. So that we see that the very same peculiarity may become attached to either sex, and be long inherited by that sex alone; but the attachment in certain cases is much more frequent to one than the other sex. The same peculiarities also may be promiscuously transmitted to either sex. Dr. Lucas gives other cases, showing that the male occasionally transmits his peculiarities to his daughters alone, and the mother to her sons alone; but even in this case we see that inheritance is to a certain extent, though inversely, regulated by sex. Dr. Lucas, after weighing the whole evidence, comes to the conclusion that every peculiarity, according to the sex in which it first appears, tends to be transmitted in a greater or lesser degree to that sex.

A few details from the many cases collected by Mr. Sedgwick,[26] may be here given. Colour-blindness, from some unknown cause, shows itself much oftener in males than in females; in upwards of two hundred cases collected by Mr. Sedgwick, nine-tenths related to men; but it is eminently liable to be transmitted through women. In the case given by Dr. Earle, members of eight related families were affected during five generations: these families consisted of sixty-one individuals, namely, of thirty-two males, of whom nine-sixteenths were incapable of distinguishing colour, and of twenty-nine females, of whom only one-fifteenth were thus affected. Al-

---

[25] 'L'Héréd. Nat.,' tom. ii. pp. 137-165. *See*, also, Mr. Sedgwick's four memoirs, immediately to be referred to.
[26] On Sexual Limitation in Heredi-

tary Diseases, 'Brit. and For. Med.-Chirurg. Review,' April, 1861, p. 477; July, p. 198; April, 1863, p. 445; and July, p. 159.

though colour-blindness thus generally clings to the male sex, nevertheless, in one instance in which it first appeared in a female, it was transmitted during five generations to thirteen individuals, all of whom were females. A hæmorrhagic diathesis, often accompanied by rheumatism, has been known to affect the males alone during five generations, being transmitted, however, through the females. It is said that deficient phalanges in the fingers have been inherited by the females alone during ten generations. In another case, a man thus deficient in both hands and feet, transmitted the peculiarity to his two sons and one daughter; but in the third generation, out of nineteen grandchildren, twelve sons had the family defect, whilst the seven daughters were free. In ordinary cases of sexual limitation, the sons or daughters inherit the peculiarity, whatever it may be, from their father or mother, and transmit it to their children of the same sex; but generally with the hæmorrhagic diathesis, and often with colour-blindness, and in some other cases, the sons never inherit the peculiarity directly from their fathers, but the daughters, and the daughters alone, transmit the latent tendency, so that the sons of the daughters alone exhibit it. Thus, the father, grandson, and great-great-grandson will exhibit a peculiarity,—the grandmother, daughter, and great-granddaughter having transmitted it in a latent state. Hence we have, as Mr. Sedgwick remarks, a double kind of atavism or reversion ; each grandson apparently receiving and developing the peculiarity from his grandfather, and each daughter apparently receiving the latent tendency from her grandmother.

From the various facts recorded by Dr. Prosper Lucas, Mr. Sedgwick, and others, there can be no doubt that peculiarities first appearing in either sex, though not in any way necessarily or invariably connected with that sex, strongly tend to be inherited by the offspring of the same sex, but are often transmitted in a latent state through the opposite sex.

Turning now to domesticated animals, we find that certain characters not proper to the parent-species are often confined to, and inherited by, one sex alone; but we do not know the history of the first appearance of such characters. In the chapter on Sheep, we have seen that the males of certain races differ greatly from the females in the shape of their horns, these being absent in the ewes of some breeds, in the development of fat in the tail in certain fat-tailed breeds, and in the outline of the forehead. These differences, judging from the character of the allied wild species, cannot be accounted for by supposing that they have been derived from distinct parent-forms. There is, also, a great difference between the horns of the two sexes in one Indian breed of goats. The bull zebu is said to have a larger hump than the cow. In the Scotch deer-hound the two sexes differ in size more than in any other variety of the dog,[27] and, judging from analogy, more than in the aboriginal parent-species. The peculiar colour called tortoise-shell is very rarely seen in a male cat ; the males of this variety being of a rusty tint. A tendency to baldness in man before the advent of old age is certainly inherited ; and in the European, or at least in the

---

[27] W. Scrope, ' Art of Deer Stalking,' p. 354.

Englishman, is an attribute of the male sex, and may almost be ranked as an incipient secondary sexual character.

In various breeds of the fowl the males and females often differ greatly; and these differences are far from being the same with those which distinguish the two sexes in the parent-species, the *Gallus bankiva*; and consequently have originated under domestication. In certain sub-varieties of the Game race we have the unusual case of the hens differing from each other more than the cocks. In an Indian breed of a white colour stained with soot, the hens invariably have black skins, and their bones are covered by a black periosteum, whilst the cocks are never or most rarely thus characterised. Pigeons offer a more interesting case; for the two sexes rarely differ throughout the whole great family, and the males and females of the parent-form, the *C. livia*, are undistinguishable; yet we have seen that with Pouters the male has the characteristic quality of pouting more strongly developed than the female; and in certain sub-varieties [28] the males alone are spotted or striated with black. When male and female English carrier-pigeons are exhibited in separate pens, the difference in the development of the wattle over the beak and round the eyes is conspicuous. So that here we have instances of the appearance of secondary sexual characters in the domesticated races of a species in which such differences are naturally quite absent.

On the other hand, secondary sexual characters which properly belong to the species are sometimes quite lost, or greatly diminished, under domestication. We see this in the small size of the tusks in our improved breeds of the pig, in comparison with those of the wild boar. There are sub-breeds of fowls in which the males have lost the fine flowing tail-feathers and hackles; and others in which there is no difference in colour between the two sexes. In some cases the barred plumage, which in gallinaceous birds is commonly the attribute of the hen, has been transferred to the cock, as in the cuckoo sub-breeds. In other cases masculine characters have been partly transferred to the female, as with the splendid plumage of the golden-spangled Hamburgh hen, the enlarged comb of the Spanish hen, the pugnacious disposition of the Game hen, and as in the well-developed spurs which occasionally appear in the hens of various breeds. In Polish fowls both sexes are ornamented with a topknot, that of the male being formed of hackle-like feathers, and this is a new male character in the genus Gallus. On the whole, as far as I can judge, new characters are more apt

[28] Boitard and Corbié, ' Les Pigeons,' p. 173; Dr. F. Chapuis, ' Le Pigeon Voyageur Belge,' 1865, p. 87.

to appear in the males of our domesticated animals than in the females, and afterwards to be either exclusively or more strongly inherited by the males. Finally, in accordance with the principle of inheritance as limited by sex, the appearance of secondary sexual characters in natural species offers no especial difficulty, and their subsequent increase and modification, if of any service to the species, would follow through that form of selection which in my ' Origin of Species' I have called sexual selection.

Inheritance at corresponding periods of Life.

This is an important subject. Since the publication of my ' Origin of Species,' I have seen no reason to doubt the truth of the explanation there given of perhaps the most remarkable of all the facts in biology, namely, the difference between the embryo and the adult animal. The explanation is, that variations do not necessarily or generally occur at a very early period of embryonic growth, and that such variations are inherited at a corresponding age. As a consequence of this the embryo, even when the parent-form undergoes a great amount of modification, is left only slightly modified; and the embryos of widely-different animals which are descended from a common progenitor remain in many important respects like each other and their common progenitor. We can thus understand why embryology should throw a flood of light on the natural system of classification, for this ought to be as far as possible genealogical. When the embryo leads an independent life, that is, becomes a larva, it has to be adapted to the surrounding conditions in its structure and instincts, independently of those of its parents; and the principle of inheritance at corresponding periods of life renders this possible.

This principle is, indeed, in one way so obvious that it escapes attention. We possess a number of races of animals and plants, which, when compared with each other and with their parent-forms, present conspicuous differences, both in the immature and mature states. Look at the seeds of the several kinds of peas, beans, maize, which can be propagated truly, and see how they differ in size, colour, and shape, whilst the full-grown plants differ but little. Cabbages on the other hand differ greatly in foliage and manner of growth, but hardly at all in their seeds; and generally it will be found that the differences between cultivated plants at different periods of growth are not necessarily closely connected together, for plants may differ much in their seeds and little when full-grown, and conversely may yield seeds hardly distinguishable, yet differ much when full-grown. In the several breeds of poultry, descended from a single species, differences in the eggs and chickens, in the plumage at the first and subsequent moults, in the comb and wattles during maturity, are all inherited. With man peculiarities in the milk and second teeth, of which I have received the details, are inheritable, and with man longevity is often transmitted. So again with our improved breeds of cattle and sheep, early maturity, including the early development of the teeth, and with certain breeds of fowl the early appearance of secondary sexual characters, all come under the same head of inheritance at corresponding periods.

Numerous analogous facts could be given. The silk-moth, perhaps, offers the best instance; for in the breeds which transmit their characters truly, the eggs differ in size, colour, and shape;—the caterpillars differ, in moulting three or four times, in colour, even in having a dark-coloured mark like an eyebrow, and in the loss of certain instincts;—the cocoons differ in size, shape, and in the colour and quality of the silk; these several differences being followed by slight or barely distinguishable differences in the mature moth.

But it may be said that, if in the above cases a new peculiarity is inherited, it must be at the corresponding stage of development; for an egg or seed can resemble only an egg or seed, and the horn in a full-grown ox can resemble only a horn. The following cases show inheritance at corresponding periods more plainly, because they refer to peculiarities which might have supervened, as far as we can see, earlier or later in life, yet are inherited at the same period at which they first appeared.

In the Lambert family the porcupine-like excrescences appeared in the father and sons at the same age, namely, about nine weeks after birth.[29] In the extraordinary hairy family described by Mr. Crawfurd,[30] children were produced during three generations with hairy ears; in the father the hair began to grow over his body at six years old; in his daughter somewhat earlier, namely, at one year; and in both generations the milk teeth appeared late in life, the permanent teeth being afterwards singularly deficient. Greyness of hair at an unusually early age has been transmitted in some families. These cases border on diseases inherited at corresponding periods of life, to which I shall immediately refer.

It is a well-known peculiarity with almond-tumbler pigeons, that the full beauty and peculiar character of the plumage does not appear until the bird has moulted two or three times. Neumeister describes and figures a breed of pigeons in which the whole body is white except the breast, neck, and head; but before the first moult all the white feathers acquire coloured edges. Another breed is more remarkable: its first plumage is black, with rusty-red wing-bars and a crescent-shaped mark on the breast; these marks then become white, and remain so during three or four moults; but after this period the white spreads over the body, and the bird loses its beauty.[31] Prize canary-birds have their wings and tail black: "this colour, however, is only retained until " the first moult, so that they must be exhibited ere the change takes place. " Once moulted, the peculiarity has ceased. Of course all the birds emanating " from this stock have black wings and tails the first year."[32] A curious and somewhat analogous account has been given [33] of a family of wild pied rooks which were first observed in 1798, near Chalfont, and which every year from that date up to the period of the published notice, viz. 1837, " have several of their brood particoloured, black and white. This " variegation of the plumage, however, disappears with the first moult; " but among the next young families there are always a few pied ones." These changes of plumage, which appear and are inherited at various corresponding periods of life in the pigeon, canary-bird, and rook, are remarkable, because the parent-species undergo no such change.

Inherited diseases afford evidence in some respects of less value than the foregoing cases, because diseases are not necessarily connected with any change in structure; but in other respects of more value, because the periods have been more carefully observed. Certain diseases are communicated to the child apparently by a process like inoculation, and the child is from the first affected; such cases may be here passed over. Large classes of diseases usually appear at certain ages, such as St. Vitus's dance in youth, consumption in early mid-life, gout later, and apoplexy still later; and these are naturally inherited at the same period. But even in diseases of this class, instances have been recorded, as with St. Vitus's

---

[29] Prichard, 'Phys. Hist. of Mankind,' 1851, vol. i. p. 349.

[30] 'Embassy to the Court of Ava,' vol. i. p. 320. The third generation is described by Capt. Yule in his ' Narrative of the Mission to the Court of Ava,' 1855, p. 94.

[31] 'Das Ganze der Taubenzucht,' 1837, s. 21, tab. i., fig. 4; s. 24, tab. iv., fig. 2.

[32] Kidd's 'Treatise on the Canary,' p. 18.

[33] Charlesworth, 'Mag. of Nat. Hist.,' vol. i., 1837, p. 167.

dance, showing that an unusually early or late tendency to the disease is inheritable.[34]   In most cases the appearance of any inherited disease is largely determined by certain critical periods in each person's life, as well as by unfavourable conditions.   There are many other diseases, which are not attached to any particuliar period, but which certainly tend to appear in the child at about the same age at which the parent was first attacked.   An array of high authorities, ancient and modern, could be given in support of this proposition.   The illustrious Hunter believed in it; and Piorry[35] cautions the physician to look closely to the child at the period when any grave inheritable disease attacked the parent. Dr. Prosper Lucas,[36] after collecting facts from every source, asserts that affections of all kinds, though not related to any particular period of life, tend to reappear in the offspring at whatever period of life they first appeared in the progenitor.

As the subject is important, it may be well to give a few instances, simply as illustrations, not as proof; for proof, recourse must be had to the authorities above quoted.   Some of the following cases have been selected for the sake of showing that, when a slight departure from the rule occurs, the child is affected somewhat earlier in life than the parent.   In the family of Le Compte blindness was inherited during three generations, and no less than thirty-seven children and grandchildren were all affected at about the same age, namely seventeen or eighteen.[37]   In another case a father and his four children all became blind at twenty-one years old; in another, a grandmother grew blind at thirty-five, her daughter at nineteen, and three grandchildren at the ages of thirteen and eleven.[38]   So with deafness, two brothers, their father and paternal grandfather, all became deaf at the age of forty.[39]

Esquirol gives several striking instances of insanity coming on at the same age, as that of a grandfather, father, and son, who all committed suicide near their fiftieth year.   Many other cases could be given, as of a whole family who became insane at the age of forty.[40]   Other cerebral affections sometimes follow the same rule,—for instance, epilepsy and apoplexy.   A woman died of the latter disease when sixty-three years old; one of her daughters at forty-three, and the other at sixty-seven : the latter had twelve children, who all died from tubercular meningitis.[41]   I mention this latter case because it illustrates a frequent occurrence, namely, a change in the precise nature of an inherited disease, though still affecting the same organ.

---

[34] Dr. Prosper Lucas, ' Héréd. Nat.,' tom. ii. p. 713.

[35] ' L'Héréd. dans les Maladies,' 1840, p. 135.   For Hunter, *see* Harlan's ' Med. Researches,' p. 530.

[36] ' L'Héréd. Nat.,' tom. ii. p. 850.

[37] Sedgwick, ' Brit. and For. Med.-Chirurg. Review,' April, 1861, p. 485. I have seen three accounts, all taken from the same original authority (which I have not been able to consult), and all differ in the details ! but as they agree in the main facts, I have ventured to quote this case.

[38] Prosper Lucas, ' Héréd. Nat.,' tom. i. p. 400.

[39] Sedgwick, idem, July, 1861, p. 202.

[40] Piorry, p. 109 ; Prosper Lucas, tom. ii. p. 759.

[41] Prosper Lucas, tom. ii. p. 748.

Asthma has attacked several members of the same family when forty years old, and other families during infancy. The most different diseases, as angina pectoris, stone in the bladder, and various affections of the skin, have appeared in successive generations at nearly the same age. The little finger of a man began from some unknown cause to grow inwards, and the same finger in his two sons began at the same age to bend inwards in a similar manner. Strange and inexplicable neuralgic affections have caused parents and children to suffer agonies at about the same period of life.[42]

I will give only two other cases, which are interesting as illustrating the disappearance as well as the appearance of disease at the same age. Two brothers, their father, their paternal uncles, seven cousins, and their paternal grandfather, were all similarly affected by a skin-disease, called pityriasis versicolor; " the disease, strictly limited to the males of the family (though transmitted through the females), usually appeared at puberty, and disappeared at about the age of forty or forty-five years." The second case is that of four brothers, who when about twelve years old suffered almost every week from severe headaches, which were relieved only by a recumbent position in a dark room. Their father, paternal uncles, paternal grandfather, and paternal granduncles all suffered in the same way from headaches, which ceased at the age of fifty-four or fifty-five in all those who lived so long. None of the females of the family were affected.[43]

It is impossible to read the foregoing accounts, and the many others which have been recorded, of diseases coming on during three or even more generations, at the same age in several members of the same family, especially in the case of rare affections in which the coincidence cannot be attributed to chance, and doubt that there is a strong tendency to inheritance in disease at corresponding periods of life. When the rule fails, the disease is apt to come on earlier in the child than in the parent; the exceptions in the other direction being very much rarer. Dr. Lucas [44] alludes to several cases of inherited diseases coming on at an earlier period. I have already given one striking instance with blindness during three generations; and Mr. Bowman remarks that this frequently occurs with cataract. With cancer there seems to be a peculiar liability to earlier inheritance: Mr. Paget, who has particularly

[42] Prosper Lucas, tom. ii. pp. 678, 700, 702 ; Sedgwick, idem, April, 1863, p. 449, and July, 1863, p. 162; Dr. J. Steinan, 'Essay on Hereditary Disease,' 1843, pp. 27, 34.

[43] These cases are given by Mr. Sedgwick, on the authority of Dr. H. Stewart, in 'Med.-Chirurg. Review,' April, 1863, pp. 449, 477.

[44] 'Héréd. Nat.,' tom. ii. p. 852.

attended to this subject, and tabulated a large number of cases, informs me that he believes that in nine cases out of ten the later generation suffers from the disease at an earlier period than the previous generation. He adds, " In the instances in which the opposite relation holds, and the members of later genera- tions have cancer at a later age than their predecessors, I think it will be found that the non-cancerous parents have lived to extreme old ages." So that the longevity of a non-affected parent seems to have the power of determining in the offspring the fatal period; and we thus apparently get another element of complexity in inheritance.

The facts, showing that with certain diseases the period of inheritance occasionally or even frequently advances, are important with respect to the general descent-theory, for they render it in some degree probable that the same thing would occur with ordinary modifications of structure. The final result of a long series of such advances would be the gradual oblite- ration of characters proper to the embryo and larva, which would thus come to resemble more and more closely the mature parent-form. But any structure which was of service to the embryo or larva would be preserved by the destruction at this stage of growth of each individual which manifested any ten- dency to lose at too early an age its own proper character.

Finally, from the numerous races of cultivated plants and domestic animals, in which the seed or eggs, the young or old, differ from each other and from their parent-species;— from the cases in which new characters have appeared at a particular period, and afterwards have been inherited at the same period;—and from what we know with respect to disease, we must believe in the truth of the great principle of inherit- ance at corresponding periods of life.

Summary of the three preceding Chapters.—Strong as is the force of inheritance, it allows the incessant appearance of new characters. These, whether beneficial or injurious, of the most trifling importance, such as a shade of colour in a flower, a coloured lock of hair, or a mere gesture; or of the highest importance, as when affecting the brain or an organ so perfect

and complex as the eye; or of so grave a nature as to deserve
to be called a monstrosity, or so peculiar as not to occur
normally in any member of the same natural class, are all
sometimes strongly inherited by man, the lower animals, and
plants. In numberless cases it suffices for the inheritance of a
peculiarity that one parent alone should be thus characterised.
Inequalities in the two sides of the body, though opposed to
the law of symmetry, may be transmitted. There is a considerable body of evidence showing that even mutilations, and
the effects of accidents, especially or perhaps exclusively when
followed by disease, are occasionally inherited. There can be
no doubt that the evil effects of long-continued exposure in
the parent to injurious conditions are sometimes transmitted
to the offspring. So it is, as we shall see in a future chapter,
with the effects of the use and disuse of parts, and of mental
habits. Periodical habits are likewise transmitted, but generally,
as it would appear, with little force.

Hence we are led to look at inheritance as the rule, and
non-inheritance as the anomaly. But this power often appears
to us in our ignorance to act capriciously, transmitting a character with inexplicable strength or feebleness. The very
same peculiarity, as the weeping habit of trees, silky-feathers,
&c., may be inherited either firmly or not at all by different
members of the same group, and even by different individuals
of the same species, though treated in the same manner. In
this latter case we see that the power of transmission is a
quality which is merely individual in its attachment. As with
single characters, so it is with the several concurrent slight
differences which distinguish sub-varieties or races; for of these,
some can be propagated almost as truly as species, whilst others
cannot be relied on. The same rule holds good with plants,
when propagated by bulbs, offsets, &c., which in one sense still
form parts of the same individual, for some varieties retain or
inherit through successive bud-generations their character far
more truly than others.

Some characters not proper to the parent-species have certainly been inherited from an extremely remote epoch, and may
therefore be considered as firmly fixed. But it is doubtful whether
length of inheritance in itself gives fixedness of character;

though the chances are obviously in favour of any character
which has long been transmitted true or unaltered, still being
transmitted true as long as the conditions of life remain the
same. We know that many species, after having retained the
same character for countless ages, whilst living under their
natural conditions, when domesticated have varied in the most
diversified manner, that is, have failed to transmit their original
form ; so that no character appears to be absolutely fixed. We
can sometimes account for the failure of inheritance by the
conditions of life being opposed to the development of certain
characters; and still oftener, as with plants cultivated by grafts
and buds, by the conditions causing new and slight modifications incessantly to appear. In this latter case it is not that
inheritance wholly fails, but that new characters are continually
superadded. In some few cases, in which both parents are similarly characterised, inheritance seems to gain so much force by
the combined action of the two parents, that it counteracts its
own power, and a new modification is the result.

In many cases the failure of the parents to transmit their
likeness is due to the breed having been at some former period
crossed ; and the child takes after his grandparent or more
remote ancestor of foreign blood. In other cases, in which the
breed has not been crossed, but some ancient character has
been lost through variation, it occasionally reappears through
reversion, so that the parents apparently fail to transmit their
own likeness. In all cases, however, we may safely conclude
that the child inherits all its characters from its parents, in
whom certain characters are latent, like the secondary sexual
characters of one sex in the other. When, after a long succession of bud-generations, a flower or fruit becomes separated
into distinct segments, having the colours or other attributes
of both parent-forms, we cannot doubt that these characters
were latent in the earlier buds, though they could not then
be detected, or could be detected only in an intimately commingled state. So it is with animals of crossed parentage, which
with advancing years occasionally exhibit characters derived
from one of their two parents, of which not a trace could at
first be perceived. Certain monstrosities, which resemble what
naturalists call the typical form of the group in question,

apparently come under the same law of reversion. It is assuredly an astonishing fact that the male and female sexual elements, that buds, and even full-grown animals, should retain characters, during several generations in the case of crossed breeds, and during thousands of generations in the case of pure breeds, written as it were in invisible ink, yet ready at any time to be evolved under the requisite conditions.

What these conditions are, we do not in many cases at all know. But the act of crossing in itself, apparently from causing some disturbance in the organisation, certainly gives a strong tendency to the reappearance of long-lost characters, both corporeal and mental, independently of those derived from the cross. A return of any species to its natural conditions of life, as with feral animals and plants, favours reversion; though it is certain that this tendency exists, we do not know how far it prevails, and it has been much exaggerated. On the other hand, the crossed offspring of plants which have had their organisation disturbed by cultivation, are more liable to reversion than the crossed offspring of species which have always lived under their natural conditions.

When distinguishable individuals of the same family, or races, or species are crossed, we see that the one is often prepotent over the other in transmitting its own character. A race may possess a strong power of inheritance, and yet when crossed, as we have seen with trumpeter-pigeons, yield to the prepotency of every other race. Prepotentcy of transmission may be equal in the two sexes of the same species, but often runs more strongly in one sex. It plays an important part in determining the rate at which one race can be modified or wholly absorbed by repeated crosses with another. We can seldom tell what makes one race or species prepotent over another; but it sometimes depends on the same character being present and visible in one parent, and latent or potentially present in the other.

Characters may first appear in either sex, but oftener in the male than in the female, and afterwards be transmitted to the offspring of the same sex. In this case we may feel confident that the peculiarity in question is really present though latent in the opposite sex; hence the father may transmit through his daughter any character to his grandson; and the mother

conversely to her granddaughter. We thus learn, and the fact
is an important one, that transmission and development are distinct powers. Occasionally these two powers seem to be antagonistic, or incapable of combination in the same individual;
for several cases have been recorded in which the son has not
directly inherited a character from his father, or directly transmitted it to his son, but has received it by transmission through
his non-affected mother, and transmitted it through his non-
affected daughter. Owing to inheritance being limited by sex,
we can see how secondary sexual characters may first have
arisen under nature; their preservation and accumulation being
dependent on their service to either sex.

At whatever period of life a new character first appears,
it generally remains latent in the offspring until a corresponding
age is attained, and then it is developed. When this rule fails,
the child generally exhibits the character at an earlier period
than the parent. On this principle of inheritance at corresponding periods, we can understand how it is that most animals
display from the germ to maturity such a marvellous succession
of characters.

Finally, though much remains obscure with respect to Inheritance, we may look at the following laws as fairly well
established. Firstly, a tendency in every character, new and
old, to be transmitted by seminal and bud generation, though
often counteracted by various known and unknown causes.
Secondly, reversion or atavism, which depends on transmission
and development being distinct powers: it acts in various degrees
and manners through both seminal and bud generation. Thirdly,
prepotency of transmission, which may be confined to one sex,
or be common to both sexes of the prepotent form. Fourthly,
transmission, limited by sex, generally to the same sex in
which the inherited character first appeared. Fifthly, inheritance at corresponding periods of life, with some tendency to the
earlier development of the inherited character. In these laws
of Inheritance, as displayed under domestication, we see an
ample provision for the production, through variability and
natural selection, of new specific forms.

CHAPTER XV.

ON CROSSING.

FREE INTERCROSSING OBLITERATES THE DIFFERENCES BETWEEN ALLIED BREEDS — WHEN THE NUMBERS OF TWO COMMINGLING BREEDS ARE UNEQUAL, ONE ABSORBS THE OTHER — THE RATE OF ABSORPTION DETERMINED BY PREPOTENCY OF TRANS-MISSION, BY THE CONDITIONS OF LIFE, AND BY NATURAL SELECTION — ALL ORGANIC BEINGS OCCASIONALLY INTERCROSS ; APPARENT EXCEPTIONS — ON CERTAIN CHARACTERS INCAPABLE OF FUSION ; CHIEFLY OR EXCLUSIVELY THOSE WHICH HAVE SUDDENLY APPEARED IN THE INDIVIDUAL — ON THE MODIFICATION OF OLD RACES, AND THE FORMATION OF NEW RACES, BY CROSSING — SOME CROSSED RACES HAVE BRED TRUE FROM THEIR FIRST PRODUCTION — ON THE CROSSING OF DISTINCT SPECIES IN RELATION TO THE FORMATION OF DOMESTIC RACES.

In the two previous chapters, when discussing reversion and prepotency, I was necessarily led to give many facts on crossing. In the present chapter I shall consider the part which crossing plays in two opposed directions,—firstly, in obliterating cha-racters, and consequently in preventing the formation of new races ; and secondly, in the modification of old races, or in the formation of new and intermediate races, by a combination of characters. I shall also show that certain characters are incap-able of fusion.

The effects of free or uncontrolled breeding between the members of the same variety or of closely allied varieties are important ; but are so obvious that they need not be dis-cussed at much length. It is free intercrossing which chiefly gives uniformity, both under nature and under domestication, to the individuals of the same species or variety, when they live mingled together and are not exposed to any cause inducing excessive variability. The prevention of free crossing, and the intentional matching of individual animals, are the corner-stones of the breeder's art. No man in his senses would expect to improve or modify a breed in any particular manner, or keep an old breed true and distinct, unless he separated his animals. The killing of inferior animals in each generation comes to the

same thing as their separation. In savage and semi-civilised
countries, where the inhabitants have not the means of separating their animals, more than a single breed of the same
species rarely or never exists. In former times, even in a
country so civilised as North America, there were no distinct
races of sheep, for all had been mingled together.[1] The celebrated agriculturist Marshall[2] remarks that "sheep that are
" kept within fences, as well as shepherded flocks in open
" countries, have generally a similarity, if not a uniformity,
" of character in the individuals of each flock;" for they breed
freely together, and are prevented from crossing with other
kinds; whereas in the unenclosed parts of England the un-
shepherded sheep, even of the same flock, are far from true or
uniform, owing to various breeds having mingled and crossed.
We have seen that the half-wild cattle in the several British
parks are uniform in character in each; but in the different
parks, from not having mingled and crossed during many generations, they differ in a slight degree.

We cannot doubt that the extraordinary number of varieties
and sub-varieties of the pigeon, amounting to at least one hundred and fifty, is partly due to their remaining, differently from
other domesticated birds, paired for life when once matched.
On the other hand, breeds of cats imported into this country
soon disappear, for their nocturnal and rambling habits render
it hardly possible to prevent free crossing. Rengger[3] gives an
interesting case with respect to the cat in Paraguay: in all the
distant parts of the kingdom it has assumed, apparently from
the effects of the climate, a peculiar character, but near the
capital this change has been prevented, owing, as he asserts,
to the native animal frequently crossing with cats imported
from Europe. In all cases like the foregoing, the effects of an
occasional cross will be augmented by the increased vigour and
fertility of the crossed offspring, of which fact evidence will
hereafter be given; for this will lead to the mongrels increasing
more rapidly than the pure parent-breeds.

---

[1] Communications to the Board of
Agriculture, vol. i. p. 367.
[2] 'Review of Reports, North of England,' 1808, p. 200.
[3] 'Säugethiere von Paraguay,' 1830,
s. 212.

When distinct breeds are allowed to cross freely, the result will be a heterogeneous body; for instance, the dogs in Paraguay are far from uniform, and can no longer be affiliated to their parent-races.[4] The character which a crossed body of animals will ultimately assume must depend on several contingencies,—namely, on the relative numbers of the individuals belonging to the two or more races which are allowed to mingle; on the prepotency of one race over the other in the transmission of character; and on the conditions of life to which they are exposed. When two commingled breeds exist at first in nearly equal numbers, the whole will sooner or later become intimately blended, but not so soon, both breeds being equally favoured in all respects, as might have been expected. The following calculation[5] shows that this is the case: if a colony with an equal number of black and white men were founded, and we assume that they marry indiscriminately, are equally prolific, and that one in thirty annually dies and is born; then "in 65 "years the number of blacks, whites, and mulattoes would be "equal. In 91 years the whites would be 1-10th, the blacks "1-10th, and the mulattoes, or people of intermediate degrees of "colour, 8-10ths of the whole number. In three centuries not "1-100th part of the whites would exist."

When one of two mingled races exceeds the other greatly in number, the latter will soon be wholly, or almost wholly, absorbed and lost.[6] Thus European pigs and dogs have been largely introduced into the islands of the Pacific Ocean, and the native races have been absorbed and lost in the course of about fifty or sixty years;[7] but the imported races no doubt were favoured. Rats may be considered as semi-domesticated animals. Some snake-rats (*Mus alexandrinus*) escaped in the Zoological Gardens of London, "and for a long time afterwards "the keepers frequently caught cross-bred rats, at first half-breds, "afterwards with less and less of the character of the snake-rat, "till at length all traces of it disappeared."[8] On the other hand,

[4] Rengger, 'Säugethiere,' &c., s. 154.

[5] White, 'Regular Gradation in Man,' p. 146.

[6] Dr. W. F. Edwards, in his 'Charactères Physiolog. des Races Humaines,' p. 23, first called attention to this subject, and ably discussed it.

[7] Rev. D. Tyerman, and Bennett, 'Journal of Voyages,' 1821-1829, vol. i. p. 300.

[8] Mr. S. J. Salter, 'Journal Linn. Soc.,' vol. vi., 1862, p. 71.

in some parts of London, especially near the docks, where fresh
rats are frequently imported, an endless variety of intermediate
forms may be found between the brown, black, and snake rat,
which are all three usually ranked as distinct species.

How many generations are necessary for one species or race to
absorb another by repeated crosses has often been discussed;[9]
and the requisite number has probably been much exaggerated. Some writers have maintained that a dozen, or score,
or even more generations, are necessary; but this in itself is
improbable, for in the tenth generation there will be only
1-1024th part of foreign blood in the offspring. Gärtner found,[10]
that with plants one species could be made to absorb another in
from three to five generations, and he believes that this could
always be effected in from six to seven generations. In one
instance, however, Kölreuter[11] speaks of the offspring of *Mirabilis vulgaris,* crossed during eight successive generations by *M.
longiflora,* as resembling this latter species so closely, that the
most scrupulous observer could detect " vix aliquam notabilem
differentiam ;"—he succeeded, as he says, " ad plenariam fere
transmutationem." But this expression shows that the act of
absorption was not even then absolutely complete, though these
crossed plants contained only the 1-256th part of *M. vulgaris.*
The conclusions of such accurate observers as Gärtner and
Kölreuter are of far higher worth than those made without
scientific aim by breeders. The most remarkable statement
which I have met with of the persistent endurance of the effects
of a single cross is given by Fleischmann,[12] who, in reference
to German sheep, says " that the original coarse sheep have
" 5500 fibres of wool on a square inch ; grades of the third or
" fourth Merino cross produced about 8000, the twentieth cross
" 27,000, the perfect pure Merino blood 40,000 to 48,000." So that
in this case common German sheep crossed twenty times successively with Merinos have not by any means acquired wool as fine
as that of the pure breed. In all cases, the rate of absorption will

[9] Sturm, ' Ueber Racen, &c.,' 1825, s.
107. Bronn, ' Geschichte der Natur.,'
b. ii. s. 170, gives a table of the proportions of blood after successive crosses.
Dr. P. Lucas, ' l'Hérédité Nat.,' tom. ii.
p. 308.

[10] ' Bastarderzeugung,' s. 463, 470.
[11] ' Nova Acta Petrop.,' 1794, p. 393:
see also previous volume.
[12] As quoted in the ' True Principles
of Breeding,' by C. H. Macknight and
Dr. H. Madden, 1865, p. 11.

depend largely on the conditions of life being favourable to any particular character; and we may suspect that there would be under the climate of Germany a constant tendency to degeneration in the wool of Merinos, unless prevented by careful selection; and thus perhaps the foregoing remarkable case may be explained. The rate of absorption must also depend on the amount of distinguishable difference between the two forms which are crossed, and especially, as Gärtner insists, on prepotency of transmission in the one form over the other. We have seen in the last chapter that one of two French breeds of sheep yielded up its character, when crossed with Merinos, very much slower than the other; and the common German sheep referred to by Fleischmann may present an analogous case. But in all cases there will be during many subsequent generations more or less liability to reversion, and it is this fact which has probably led authors to maintain that a score or more of generations are requisite for one race to absorb another. In considering the final result of the commingling of two or more breeds, we must not forget that the act of crossing in itself tends to bring back long-lost characters not proper to the immediate parent-forms.

With respect to the influence of the conditions of life on any two breeds which are allowed to cross freely, unless both are indigenous and have long been accustomed to the country where they live, they will, in all probability, be unequally affected by the conditions, and this will modify the result. Even with indigenous breeds, it will rarely or never occur that both are equally well adapted to the surrounding circumstances; more especially when permitted to roam freely, and not carefully tended, as will generally be the case with breeds allowed to cross. As a consequence of this, natural selection will to a certain extent come into action, and the best fitted will survive, and this will aid in determining the ultimate character of the commingled body.

How long a time it would require before such a crossed body of animals would assume within a limited area a uniform character no one can say; that they would ultimately become uniform from free intercrossing, and from the survival of the fittest, we may feel assured; but the character thus acquired would rarely or never, as we may infer from the several previous considerations, be exactly intermediate between that of the two parent-breeds. With respect to the very slight differences by which the individuals of the same sub-variety, or even of allied varieties, are characterised, it is obvious that free crossing would soon obliterate such small distinctions. The formation of new varieties, independently of selection, would also thus be prevented; except when the same variation continually recurred from the action of some strongly predisposing cause. Hence we may conclude that free crossing has in all cases played an important part in giving to all the members of the same domestic race, and of the same natural species, uniformity of character, though largely modified by natural selection and by the direct action of the surrounding conditions.

On the possibility of all organic beings occasionally intercrossing. —But it may be asked, can free crossing occur with hermaphrodite animals and plants ? All the higher animals, and the few insects which have been domesticated, have separated sexes, and must inevitably unite for each birth. With respect to the crossing of hermaphrodites, the subject is too large for the present volume, and will be more properly treated in a succeeding work. In my ' Origin of Species,' however, I have given a short abstract of the reasons which induce me to believe that all organic beings occasionally cross, though perhaps in some cases only at long intervals of time.[13] I will here just recall the fact that many plants, though hermaphrodite in structure, are unisexual in function;—such as those called by C. K. Sprengel *dichogamous*, in which the pollen and stigma of the same flower are matured at different periods; or those called by me *reciprocally dimorphic*, in which the flower's own pollen is not fitted to fertilise its own stigma; or again, the many kinds in which curious mechanical contrivances exist, effectually preventing self-fertilisation. There are, however, many hermaphrodite plants which are not in any way specially constructed to favour intercrossing, but which nevertheless commingle almost as freely as animals with separated sexes. This is the case with cabbages, radishes, and onions, as I know from

[13] With respect to plants, an admirable essay on this subject (Die Geschlechter-Vertheilung bei den Pflanzen : 1867) has lately been published by Dr. Hildebrand, who arrives at the same general conclusions as I have done.

having experimented on them: even the peasants of Liguria say that cabbages must be prevented "from falling in love" with each other. In the orange tribe, Gallesio [14] remarks that the amelioration of the various kinds is checked by their continual and almost regular crossing. So it is with numerous other plants.

Nevertheless some cultivated plants can be named which rarely intercross, as the common pea, or which never intercross, as I have reason to believe is the case with the sweet-pea (*Lathyrus odoratus*) ; yet the structure of these flowers certainly favours an occasional cross. The varieties of the tomato and aubergine (*Solanum*) and pimenta (*Pimenta vulgaris?*) are said [15] never to cross, even when growing alongside each other. But it should be observed that these are all exotic plants, and we do not know how they would behave in their native country when visited by the proper insects.

It must also be admitted that some few natural species appear under our present state of knowledge to be perpetually self-fertilised, as in the case of the Bee Ophrys (*O. apifera*), though adapted in its structure to be occasionally crossed. The *Leersia oryzoides* produces minute enclosed flowers which cannot possibly be crossed, and these alone, to the exclusion of the ordinary flowers, have as yet been known to yield seed.[16] A few additional and analogous cases could be advanced. But these facts do not make me doubt that it is a general law of nature that the individuals of the same species occasionally intercross, and that some great advantage is derived from this act. It is well known (and I shall hereafter have to give instances) that some plants, both indigenous and naturalised, rarely or never produce flowers; or, if they flower, never produce seeds. But no one is thus led to doubt that it is a general law of nature that phanerogamic plants should produce flowers, and that these flowers should produce seed. When they fail, we believe that such plants would perform their proper functions under different conditions, or that they formerly did so and will do so again. On analogous grounds, I believe that the few flowers

[14] ' Teoria della Riproduzione Vegetal,' 1816, p. 12.
[15] Verlot, 'Des Variétés,' 1865, p. 72.
[16] Duval-Jouve, 'Bull. Soc. Bot. de France,' tom. x., 1863, p. 194.

which do not now intercross, either would do so under different conditions, or that they formerly fertilised each other at intervals —the means for effecting this being generally still retained— and they will do so again at some future period, unless indeed they become extinct.   On this view alone, many points in the structure and action of the reproductive organs in hermaphrodite plants and animals are intelligible,—for instance, the male and female organs never being so completely enclosed as to render access from without impossible.   Hence we may conclude that the most important of all the means for giving uniformity to the individuals of the same species, namely, the capacity of occa- sionally intercrossing, is present, or has been formerly present, with all organic beings.

On certain Characters not blending.—When two breeds are crossed their characters usually become intimately fused together; but some characters refuse to blend, and are transmitted in an unmodified state either from both parents or from one.   When grey and white mice are paired, the young are not piebald nor of an intermediate tint, but are pure white or of the ordinary grey colour : so it is when white and common collared turtle- doves are paired.   In breeding Game fowls, a great authority, Mr. J. Douglas, remarks, "I may here state a strange fact: if you cross a black with a white game, you get birds of both breeds of the clearest colour." Sir R. Heron crossed during many years white, black, brown, and fawn- coloured Angora rabbits, and never once got these colours mingled in the same animal, but often all four colours in the same litter.[17]   Additional cases could be given, but this form of inheritance is very far from universal even with respect to the most distinct colours.   When turnspit dogs and ancon sheep, both of which have dwarfed limbs, are crossed with common breeds, the offspring are not intermediate in structure, but take after either parent. When tailless or hornless animals are crossed with perfect animals, it frequently, but by no means invariably, happens that the offspring are

---

[17] Extract of a letter from Sir R. Heron, 1838, given me by Mr. Yarrell. With respect to mice, *see* 'Annal. des Sc. Nat.,' tom. i. p. 180; and I have heard of other similar cases.   For turtle-doves, Boitard and Corbié, 'Les Pigeons,' &c., p. 238. For the Game fowl, 'The Poultry Book,' 1866, p. 128.   For crosses of tailless fowls, *see* Bechstein, 'Naturges. Deutsch.' b. iii. s. 403. Bronn, 'Geschichte der Natur,' b. ii. s. 170, gives analogous facts with horses. On the hairless condition of crossed South American dogs, *see* Rengger, 'Säugethiere von Paraguay,' s. 152 : but I saw in the Zoological Gardens mongrels, from a similar cross, which were hairless, quite hairy, or hairy in patches, that is, piebald with hair. For crosses of Dorking and other fowls, *see* 'Poultry Chronicle,' vol. ii. p. 355. About the crossed pigs, extract of letter from Sir R. Heron to Mr. Yarrell. For other cases, *see* P. Lucas, 'Héréd. Nat.,' tom. i. p. 212.

either perfectly furnished with these organs or are quite destitute of them. According to Rengger, the hairless condition of the Paraguay dog is either perfectly or not at all transmitted to its mongrel offspring; but I have seen one partial exception in a dog of this parentage which had part of its skin hairy, and part naked; the parts being distinctly separated as in a piebald animal. When Dorking fowls with five toes are crossed with other breeds, the chickens often have five toes on one foot and four on the other. Some crossed pigs raised by Sir R. Heron between the solid-hoofed and common pig had not all four feet in an intermediate condition, but two feet were furnished with properly divided, and two with united hoofs.

Analogous facts have been observed with plants: Major Trevor Clarke crossed the little, glabrous-leaved, annual stock (*Matthiola*), with pollen of a large, red-flowered, rough-leaved, biennial stock, called *cocardeau* by the French, and the result was that half the seedlings had glabrous and the other half rough leaves, but none had leaves in an intermediate state. That the glabrous seedlings were the product of the rough-leaved variety, and not accidentally of the mother-plant's own pollen, was shown by their tall and strong habit of growth.[18] In the succeeding generations raised from the rough-leaved crossed seedlings, some glabrous plants appeared, showing that the glabrous character, though incapable of blending with and modifying the rough leaves, was all the time latent in this family of plants. The numerous plants formerly referred to, which I raised from reciprocal crosses between the peloric and common Antirrhinum, offer a nearly parallel case; for in the first generation all the plants resembled the common form, and in the next generation, out of one hundred and thirty-seven plants, two alone were in an intermediate condition, the others perfectly resembling either the peloric or common form. Major Trevor Clarke also fertilised the above-mentioned red-flowered stock with pollen from the purple Queen stock, and about half the seedlings scarcely differed in habit, and not at all in the red colour of the flower, from the mother-plant, the other half bearing blossoms of a rich purple, closely like those of the paternal plant. Gärtner crossed many white and yellow-flowered species and varieties of Verbascum; and these colours were never blended, but the offspring bore either pure white or pure yellow blossoms; the former in the larger proportion.[19] Dr. Herbert raised many seedlings, as he informed me, from Swedish turnips crossed by two other varieties, and these never produced flowers of an intermediate tint, but always like one of their parents. I fertilised the purple sweet-pea (*Lathyrus odoratus*), which has a dark reddish-purple standard-petal and violet-coloured wings and keel, with pollen of the painted-lady sweet-pea, which has a pale cherry-coloured standard, and almost white wings and keel; and from the same pod I twice raised plants perfectly resembling both sorts; the greater number resembling the father. So perfect was the resemblance, that I should have thought there had

[18] 'Internat. Hort. and Bot. Congress of London,' 1866.
[19] 'Bastarderzeugung,' s. 307. Kölreuter ('Dritte Fortsetszung,' s. 34, 39), however, obtained intermediate tints from similar crosses in the genus Verbascum. With respect to the turnips, see Herbert's 'Amaryllidaceæ,' 1837, p. 370.

been some mistake, if the plants which were at first identical with the paternal variety, namely, the painted-lady, had not later in the season produced, as mentioned in a former chapter, flowers blotched and streaked with dark purple. I raised grandchildren and great-grandchildren from these crossed plants, and they continued to resemble the painted-lady, but during the later generations became rather more blotched with purple, yet none reverted completely to the original mother-plant, the purple sweet-pea. The following case is slightly different, but still shows the same principle: Naudin [20] raised numerous hybrids between the yellow *Linaria vulgaris* and the purple *L. purpurea*, and during three successive generations the colours kept distinct in different parts of the same flower.

From such cases as the foregoing, in which the offspring of the first generation perfectly resemble either parent, we come by a small step to those cases in which differently coloured flowers borne on the same root resemble both parents, and by another step to those in which the same flower or fruit is striped or blotched with the two parental colours, or bears a single stripe of the colour or other characteristic quality of one of the parent-forms. With hybrids and mongrels it frequently or even generally happens that one part of the body resembles more or less closely one parent and another part the other parent; and here again some resistance to fusion, or, what comes to the same thing, some mutual affinity between the organic atoms of the same nature, apparently comes into play, for otherwise all parts of the body would be equally intermediate in character. So again, when the offspring of hybrids or mongrels, which are themselves nearly intermediate in character, revert either wholly or by segments to their ancestors, the principle of the affinity of similar, or the repulsion of dissimilar atoms, must come into action. To this principle, which seems to be extremely general, we shall recur in the chapter on pangenesis.

It is remarkable, as has been strongly insisted upon by Isidore Geoffroy St. Hilaire in regard to animals, that the transmission of characters without fusion occurs most rarely when species are crossed; I know of one exception alone, namely, with the hybrids naturally produced between the common and hooded crow (*Corvus corone* and *cornix*), which, however, are closely allied species, differing in nothing except colour. Nor have I met with any well-ascertained cases of transmission of this kind, even when one form is strongly prepotent over another, when two races are crossed which have been slowly formed by man's selection, and therefore resemble to a certain extent natural species. Such cases as puppies in the same litter closely resembling two distinct breeds, are probably due to super-fœtation,—that is, to the influence of two fathers. All the characters above enumerated, which are transmitted in a perfect state to some of the offspring and not to others,—such as distinct colours, nakedness of skin, smoothness of leaves, absence of horns or tail, additional toes, pelorism, dwarfed structure, &c.,—have all been known to appear suddenly in individual animals and plants. From this fact, and from the several slight, aggregated differences which distinguish domestic races and species from

---

[20] ' Nouvelles Archives du Muséum,' tom. i. p. 100.

each other, not being liable to this peculiar form of transmission, we may conclude that it is in some way connected with the sudden appearance of the characters in question.

On the Modification of old Races and the Formation of new Races by Crossing.—We have hitherto chiefly considered the effects of crossing in giving uniformity of character; we must now look to an opposite result. There can be no doubt that crossing, with the aid of rigorous selection during several generations, has been a potent means in modifying old races, and in forming new ones. Lord Orford crossed his famous stud of greyhounds once with the bulldog, which breed was chosen from being deficient in scenting powers, and from having what was wanted, courage and perseverance. In the course of six or seven generations all traces of the external form of the bulldog were eliminated, but courage and perseverance remained. Certain pointers have been crossed, as I hear from the Rev. W. D. Fox, with the foxhound, to give them dash and speed. Certain strains of Dorking fowls have had a slight infusion of Game blood; and I have known a great fancier who on a single occasion crossed his turbit-pigeons with barbs, for the sake of gaining greater breadth of beak.

In the foregoing cases breeds have been crossed once, for the sake of modifying some particular character; but with most of the improved races of the pig, which now breed true, there have been repeated crosses,—for instance, the improved Essex owes its excellence to repeated crosses with the Neapolitan, together probably with some infusion of Chinese blood.[21] So with our British sheep: almost all the races, except the Southdown, have been largely crossed; "this, in fact, has been the history of our principal breeds."[22] To give an example, the "Oxfordshire Downs" now rank as an established breed.[23] They were produced about the year 1830 by crossing "Hampshire and in some instances Southdown ewes with Cotswold rams:" now the Hampshire ram was itself produced by repeated crosses between the native

[21] Richardson, 'Pigs,' 1847, pp. 37, 42; S. Sidney's edition of 'Youatt on the Pig,' 1860, p. 3.

[22] *See* Mr. W. C. Spooner's excellent paper on Cross-Breeding, 'Journal Royal Agricult. Soc.,' vol. xx., part ii.:

see also an equally good article by Mr. Ch. Howard, in 'Gardener's Chronicle, 1860, p. 320.

[23] 'Gardener's Chronicle,' 1857, pp. 649, 652.

Hampshire sheep and Southdowns; and the long-woolled
Cotswold were improved by crosses with the Leicester, which
latter again is believed to have been a cross between several
long-woolled sheep. Mr. Spooner, after considering the various
cases which have been carefully recorded, concludes " that from
a judicious pairing of cross-bred animals it is practicable to
establish a new breed." On the Continent the history of several
crossed races of cattle and of other animals has been well ascertained. To give one instance: the King of Wurtemberg, after
twenty-five years' careful breeding, that is after six or seven
generations, made a new breed of cattle from a cross between
a Dutch and Swiss breed, combined with other breeds.[24] The
Sebright bantam, which breeds as true as any other kind of
fowl, was formed about sixty years ago by a complicated cross.[25]
Dark Brahmas, which are believed by some fanciers to constitute a distinct species, were undoubtedly formed [26] in the United
States, within a recent period, by a cross between Chittagongs
and Cochins. With plants I believe there is little doubt that
some kinds of turnips, now extensively cultivated, are crossed
races; and the history of a variety of wheat which was raised
from two very distinct varieties, and which after six years'
culture presented an even sample, has been recorded on good
authority.[27]

Until quite lately, cautious and experienced breeders, though
not averse to a single infusion of foreign blood, were almost
universally convinced that the attempt to establish a new race,
intermediate between two widely distinct races, was hopeless:
" they clung with superstitious tenacity to the doctrine of purity
" of blood, believing it to be the ark in which alone true safety
" could be found." [28] Nor was this conviction unreasonable: when
two distinct races are crossed, the offspring of the first generation are generally nearly uniform in character; but even this
sometimes fails to be the case, especially with crossed dogs and
fowls, the young of which from the first are sometimes much

[24] 'Bulletin de la Soc. d'Acclimat.,'
1862, tom. ix. p. 463. See also, for
other cases, MM. Moll and Gayot, 'Du
Bœuf,' 1860, p. xxxii.
[25] 'Poultry Chronicle,' vol. ii., 1854,
p. 36.

[26] 'The Poultry Book,' by W. B.
Tegetmeier, 1866, p. 58.
[27] 'Gardener's Chronicle, 1852, p.
765.
[28] Spooner, in 'Journal Royal Agricult. Soc.,' vol. xx., part ii.

diversified. As cross-bred animals are generally of large size and vigorous, they have been raised in great numbers for immediate consumption. But for breeding they are found to be utterly useless; for though they may be themselves uniform in character, when paired together they yield during many generations offspring astonishingly diversified. The breeder is driven to despair, and concludes that he will never form an intermediate race. But from the cases already given, and from others which have been recorded, it appears that patience alone is necessary; as Mr. Spooner remarks, "nature opposes no barrier to successful admixture; in the course of time, by the aid of selection and careful weeding, it is practicable to establish a new breed." After six or seven generations the hoped-for result will in most cases be obtained; but even then an occasional reversion, or failure to keep true, may be expected. The attempt, however, will assuredly fail if the conditions of life be decidedly unfavourable to the characters of either parent-breed.[29]

Although the grandchildren and succeeding generations of cross-bred animals are generally variable in an extreme degree, some curious exceptions to the rule have been observed, both with crossed races and species. Thus Boitard and Corbié[30] assert that from a Pouter and a Runt "a Cavalier will appear, which we have classed amongst pigeons of pure race, because it transmits all its qualities to its posterity." The editor of the 'Poultry Chronicle'[31] bred some bluish fowls from a black Spanish cock and a Malay hen; and these remained true to colour "generation after generation." The Himalayan breed of rabbits was certainly formed by crossing two sub-varieties of the silver-grey rabbit; although it suddenly assumed its present character, which differs much from that of either parent-breed, yet it has ever since been easily and truly propagated. I crossed some Labrador and Penguin ducks, and recrossed the mongrels with Penguins; afterwards, most of the ducks reared during three generations were nearly uniform in character, being brown with a white crescentic mark on the lower part of the breast,

[29] See Colin's 'Traité de Phys. Comp. des Animaux Domestiques,' tom. ii. p. 536, where this subject is well treated.

[30] 'Les Pigeons,' p. 37.
[31] Vol. i., 1854, p. 101.

and with some white spots at the base of the beak ; so that by the aid of a little selection a new breed might easily have been formed. In regard to crossed varieties of plants, Mr. Beaton remarks[32] that "Melville's extraordinary cross between the Scotch kale and an early cabbage is as true and genuine as any on record;" but in this case no doubt selection was practised. Gärtner[33] has given five cases of hybrids, in which the progeny kept constant; and hybrids between *Dianthus armoria* and *deltoides* remained true and uniform to the tenth generation. Dr. Herbert likewise showed me a hybrid from two species of Loasa which from its first production had kept constant during several generations.

We have seen in the earlier chapters, that some of our domesticated animals, such as dogs, cattle, pigs, &c., are almost certainly descended from more than one species, or wild race, if any one prefers to apply this latter term to forms which were enabled to keep distinct in a state of nature. Hence the crossing of aboriginally distinct species probably came into play at an early period in the formation of our present races. From Rütimeyer's observations there can be little doubt that this occurred with cattle; but in most cases some one of the forms which were allowed to cross freely, will, it is probable, have absorbed and obliterated the others. For it is not likely that semi-civilized men would have taken the necessary pains to modify by selection their commingled, crossed, and fluctuating stock. Nevertheless, those animals which were best adapted to their conditions of life would have survived through natural selection; and by this means crossing will often have indirectly aided in the formation of primeval domesticated breeds.

Within recent times, as far as animals are concerned, the crossing of distinct species has done little or nothing in the formation or modification of our races. It is not yet known whether the species of silk-moth which have been recently crossed in France will yield permanent races. In the fourth chapter I alluded with some hesitation to the statement that a new breed, between the hare and rabbit, called leporides, had been formed in France, and was found capable of propagating

[32] Cottage Gardener,' 1856, p. 110.   [33] 'Bastarderzeugung,' s. 553.

itself; but it is now positively affirmed[34] that this is an error. With plants which can be multiplied by buds and cuttings, hybridisation has done wonders, as with many kinds of Roses, Rhododendrons, Pelargoniums, Calceolarias, and Petunias. Nearly all these plants can be propagated by seed; most of them freely; but extremely few or none come true by seed.

Some authors believe that crossing is the chief cause of variability,—that is, of the appearance of absolutely new characters. Some have gone so far as to look at it as the sole cause; but this conclusion is disproved by some of the facts given in the chapter on Bud-variation. The belief that characters not present in either parent or in their ancestors frequently originate from crossing is doubtful; that they occasionally thus arise is probable; but this subject will be more conveniently discussed in a future chapter on the causes of Variability.

A condensed summary of this and of the three following chapters, together with some remarks on Hybridism, will be given in the nineteenth chapter.

[34] Dr. Pigeaux, in 'Bull. Soc. d'Acclimat.,' tom. iii., July 1866, as quoted in 'Annals and Mag. of Nat. Hist.,' 1867, vol. xx. p. 75.

CHAPTER XVI.

CAUSES WHICH INTERFERE WITH THE FREE CROSSING OF
VARIETIES — INFLUENCE OF DOMESTICATION ON FERTILITY.

DIFFICULTIES IN JUDGING OF THE FERTILITY OF VARIETIES WHEN CROSSED —
VARIOUS CAUSES WHICH KEEP VARIETIES DISTINCT, AS THE PERIOD OF BREEDING
AND SEXUAL PREFERENCE — VARIETIES OF WHEAT SAID TO BE STERILE WHEN
CROSSED — VARIETIES OF MAIZE, VERBASCUM, HOLLYHOCK, GOURDS, MELONS, AND
TOBACCO, RENDERED IN SOME DEGREE MUTUALLY STERILE — DOMESTICATION
ELIMINATES THE TENDENCY TO STERILITY NATURAL TO SPECIES WHEN CROSSED
— ON THE INCREASED FERTILITY OF UNCROSSED ANIMALS AND PLANTS FROM
DOMESTICATION AND CULTIVATION.

THE domesticated races of both animals and plants, when
crossed, are with extremely few exceptions quite prolific,—in
some cases even more so than the purely bred parent-races.
The offspring, also, raised from such crosses are likewise, as we
shall see in the following chapter; generally more vigorous and
fertile than their parents.   On the other hand, species when
crossed, and their hybrid offspring, are almost invariably in some
degree sterile; and here there seems to exist a broad and in-
superable distinction between races and species.   The importance of this subject as bearing on the origin of species is
obvious; and we shall hereafter recur to it.

It is unfortunate how few precise observations have been
made on the fertility of mongrel animals and plants during
several successive generations.   Dr. Broca[1] has remarked that
no one has observed whether, for instance, mongrel dogs, bred
inter se, are indefinitely fertile; yet, if a shade of infertility
be detected by careful observation in the offspring of natural
forms when crossed, it is thought that their specific distinction
is proved.   But so many breeds of sheep, cattle, pigs, dogs, and
poultry, have been crossed and recrossed in various ways, that
any sterility, if it had existed, would from being injurious

[1] 'Journal de Physiolog.,' tom. ii., 1859, p. 385.

almost certainly have been observed. In investigating the
fertility of crossed varieties many sources of doubt occur.
Whenever the least trace of sterility between two plants,
however closely allied, was observed by Kölreuter, and more
especially by Gärtner, who counted the exact number of seed
in each capsule, the two forms were at once ranked as distinct species; and if this rule be followed, assuredly it will
never be proved that varieties when crossed are in any degree
sterile. We have formerly seen that certain breeds of dogs do
not readily pair together; but no observations have been made
whether, when paired, they produce the full number of young,
and whether the latter are perfectly fertile *inter se;* but, supposing that some degree of sterility were found to exist,
naturalists would simply infer that these breeds were descended
from aboriginally distinct species; and it would be scarcely
possible to ascertain whether or not this explanation was the
true one.

The Sebright Bantam is much less prolific than any other
breed of fowls, and is descended from a cross between two very
distinct breeds, recrossed by a third sub-variety. But it would
be extremely rash to infer that the loss of fertility was in any
manner connected with its crossed origin, for it may with more
probability be attributed either to long-continued close interbreeding, or to an innate tendency to sterility correlated with
the absence of hackles and sickle tail-feathers.

Before giving the few recorded cases of forms, which must be
ranked as varieties, being in some degree sterile when crossed,
I may remark that other causes sometimes interfere with
varieties freely intercrossing. Thus they may differ too greatly
in size, as with some kinds of dogs and fowls: for instance,
the editor of the ' Journal of Horticulture, &c.,' [2] says that
he can keep Bantams with the larger breeds without much
danger of their crossing, but not with the smaller breeds, such
as Games, Hamburgs, &c. With plants a difference in the
period of flowering serves to keep varieties distinct, as with the
various kinds of maize and wheat: thus Colonel Le Couteur [3]
remarks, "the Talavera wheat, from flowering much earlier than
any other kind, is sure to continue pure." In different parts of

[2] Dec. 1863, p. 484.   [3] On the Varieties of Wheat, p. 66.

the Falkland Islands the cattle are breaking up into herds of different colours; and those on the higher ground, which are generally white, usually breed, as I am informed by Admiral Sulivan, three months earlier than those on the lowlands; and this would manifestly tend to keep the herds from blending.

Certain domestic races seem to prefer breeding with their own kind; and this is a fact of some importance, for it is a step towards that instinctive feeling which helps to keep closely allied species in a state of nature distinct. We have now abundant evidence that, if it were not for this feeling, many more hybrids would be naturally produced than is the case. We have seen in the first chapter that the alco dog of Mexico dislikes dogs of other breeds; and the hairless dog of Paraguay mixes less readily with the European races, than the latter do with each other. In Germany the female Spitz-dog is said to receive the fox more readily than will other dogs; a female Australian Dingo in England attracted the wild male foxes. But these differences in the sexual instinct and attractive power of the various breeds may be wholly due to their descent from distinct species. In Paraguay the horses have much freedom, and an excellent observer [4] believes that the native horses of the same colour and size prefer associating with each other, and that the horses which have been imported from Entre Rios and Banda Oriental into Paraguay likewise prefer associating together. In Circassia six sub-races of the horse are known and have received distinct names; and a native proprietor of rank [5] asserts that horses of three of these races, whilst living a free life, almost always refuse to mingle and cross, and will even attack each other.

It has been observed, in a district stocked with heavy Lincolnshire and light Norfolk sheep, that both kinds, though bred together, when turned out, "in a short time separate to a sheep;" the Lincolnshires drawing off to the rich soil, and the Norfolks to their own dry light soil; and as long as there is plenty of grass, "the two breeds keep themselves as distinct as rooks and pigeons." In this case different habits of

[4] Rengger, 'Säugethiere von Paraguay,' s. 336.
[5] See a memoir by MM. Lherbette and De Quatrefages, in 'Bull. Soc. d'Acclimat.,' tom. viii., July, 1861, p. 312.

life tend to keep the races distinct. On one of the Faroe islands, not more than half a mile in diameter, the half-wild native black sheep are said not to have readily mixed with the imported white sheep. It is a more curious fact that the semi-monstrous ancon sheep of modern origin " have been observed to keep together, separating themselves from the rest of the flock, when put into enclosures with other sheep." [6]   With respect to fallow deer, which live in a semi-domesticated condition, Mr. Bennett [7] states that the dark and pale coloured herds, which have long been kept together in the Forest of Dean, in High Meadow Woods, and in the New Forest, have never been known to mingle: the dark-coloured deer, it may be added, are believed to have been first brought by James I. from Norway, on account of their greater hardiness. I imported from the island of Porto Santo two of the feral rabbits, which differ, as described in the fourth chapter, from common rabbits; both proved to be males, and, though they lived during some years in the Zoological Gardens, the superintendent, Mr. Bartlett, in vain endeavoured to make them breed with various tame kinds; but whether this refusal to breed was due to any change in instinct, or simply to their extreme wildness; or whether confinement had rendered them sterile, as often occurs, cannot be told.

Whilst matching for the sake of experiment many of the most distinct breeds of pigeons, it frequently appeared to me that the birds, though faithful to their marriage vow, retained some desire after their own kind. Accordingly I asked Mr. Wicking, who has kept a larger stock of various breeds together than any man in England, whether he thought that they would prefer pairing with their own kind, supposing that there were males and females enough of each; and he without hesitation answered that he was convinced that this was the case. It has often been noticed that the dovecot pigeon seems to have an actual aversion towards the several fancy breeds; [8] yet all have

[6] For the Norfolk sheep, see Marshall's 'Rural Economy of Norfolk,' vol. ii. p. 133. See Rev. L. Landt's ' Description of Faroe,' p. 66. For the ancon sheep, see ' Phil. Transact.,' 1813, p. 90.

[7] White's 'Nat. Hist. of Selbourne,' edited by Bennett, p. 39. With respect to the origin of the dark-coloured deer, see ' Some Account of English Deer Parks,' by E. P. Shirley, Esq.

[8] ' The Dovecote,' by the Rev. E. S. Dixon. p. 155; Bechstein, ' Naturgesch. Deutschlands,' Band iv., 1795. s. 17.

certainly sprung from a common progenitor. The Rev. W. D. Fox informs me that his flocks of white and common Chinese geese kept distinct.

These facts and statements, though some of them are incapable of proof, resting only on the opinion of experienced observers, show that some domestic races are led by different habits of life to keep to a certain extent separate, and that others prefer coupling with their own kind, in the same manner as species in a state of nature, though in a much less degree.

With respect to sterility from the crossing of domestic races, I know of no well-ascertained case with animals. This fact, seeing the great difference in structure between some breeds of pigeons, fowls, pigs, dogs, &c., is extraordinary, in contrast with the sterility of many closely allied natural species when crossed; but we shall hereafter attempt to show that it is not so extraordinary as it at first appears. And it may be well here to recall to mind that the amount of external difference between two species will not safely guide us in foretelling whether or not they will breed together,—some closely allied species when crossed being utterly sterile, and others which are extremely unlike being moderately fertile. I have said that no case of sterility in crossed races rests on satisfactory evidence; but here is one which at first seems trustworthy. Mr. Youatt,[9] and a better authority cannot be quoted, states, that formerly in Lancashire crosses were frequently made between longhorn and shorthorn cattle; the first cross was excellent, but the produce was uncertain; in the third or fourth generation the cows were bad milkers; " in addition to which, there was much uncertainty whether the cows would conceive; and full one-third of the cows among some of these half-breds failed to be in calf." This at first seems a good case; but Mr. Wilkinson states,[10] that a breed derived from this same cross was actually established in another part of England; and if it had failed in fertility, the fact would surely have been noticed. Moreover, supposing that Mr. Youatt had proved his case, it might be argued that the sterility was wholly due to the two parent-breeds being descended from primordially distinct species.

I will give a case with plants, to show how difficult it is to get sufficient evidence. Mr. Sheriff, who has been so successful in the formation of new races of wheat, fertilised the Hopetoun with the Talavera; in the first and second generations the produce was intermediate in character, but in the fourth generation " it was found to consist of many varieties; nine-tenths of the florets proved barren, and many of the seeds seemed shrivelled abortions, void of vitality, and the whole race was evidently verging to extinction."[11] Now, considering how little these

---

[9] 'Cattle,' p. 202.

[10] Mr. J. Wilkinson, in 'Remarks addressed to Sir J. Sebright,' 1820, p. 38.

[11] 'Gardener's Chronicle,' 1858, p. 771.

varieties of wheat differ in any important character, it seems to me very improbable that the sterility resulted, as Mr. Sheriff thought, from the cross, but from some quite distinct cause. Until such experiments are many times repeated, it would be rash to trust them; but unfortunately they have been rarely tried even once with sufficient care.

Gärtner has recorded a more remarkable and trustworthy case : he fertilised thirteen panicles (and subsequently nine others) on a dwarf maize bearing yellow seed[12] with pollen of a tall maize having red seed; and one head alone produced good seed, only five in number. Though these plants are monœcious, and therefore do not require castration, yet I should have suspected some accident in the manipulation had not Gärtner expressly stated that he had during many years grown these two varieties together, and they did not spontaneously cross; and this, considering that the plants are monœcious and abound with pollen, and are well known generally to cross freely, seems explicable only on the belief that these two varieties are in some degree mutually infertile. The hybrid plants raised from the above five seed were intermediate in structure, extremely variable, and perfectly fertile.[13] No one, I believe, has hitherto suspected that these varieties of maize are distinct species; but had the hybrids been in the least sterile, no doubt Gärtner would at once have so classed them. I may here remark, that with undoubted species there is not necessarily any close relation between the sterility of a first cross and that of the hybrid offspring. Some species can be crossed with facility, but produce utterly sterile hybrids; others can be crossed with extreme difficulty, but the hybrids when produced are moderately fertile. I am not aware, however, of any instance quite like this of the maize with natural species, namely, of a first cross made with difficulty, but yielding perfectly fertile hybrids.

The following case is much more remarkable, and evidently perplexed Gärtner, whose strong wish it was to draw a broad line of distinction between species and varieties. In the genus Verbascum, he made, during eighteen years, a vast number of experiments, and crossed no less than 1085 flowers and counted their seeds. Many of these experiments consisted in crossing white and yellow varieties of both *V. lychnitis* and *V. blattaria* with nine other species and their hybrids. That the white and yellow flowered plants of these two species are really varieties, no one has doubted; and Gärtner actually raised in the case of both species one variety from the seed of the other. Now in two of his works[14] he distinctly asserts that crosses between similarly-coloured flowers yield more seed than between dissimilarly-coloured; so that the yellow-flowered variety of either species (and conversely with the white-flowered variety), when crossed with pollen of its own kind, yields more seed than when crossed with that of the white variety; and so it is when differently coloured species are crossed. The general results may be seen in the Table at the

---

[12] 'Bastarderzeugung,' s. 87, 169. *See* also the Table at the end of volume.

[13] 'Bastarderzeugung,' s. 87, 577.

[14] 'Kenntniss der Befruchtung,' s. 137; 'Bastarderzeugung,' s. 92, 181. On raising the two varieties from seed *see* s. 307.

end of his volume. In one instance he gives[15] the following details; but I must premise that Gärtner, to avoid exaggerating the degree of sterility in his crosses, always compares the *maximum* number obtained from a cross with the *average* number naturally given by the pure mother-plant. The white variety of *V. lychnitis*, naturally fertilised by its own pollen, gave from an *average* of twelve capsules ninety-six good seeds in each; whilst twenty flowers fertilised with pollen from the yellow variety of this same species, gave as the *maximum* only eighty-nine good seed; so that we have the proportion of 1000 to 908, according to Gärtner's usual scale. I should have thought it possible that so small a difference in fertility might have been accounted for by the evil effects of the necessary castration; but Gärtner shows that the white variety of *V. lychnitis*, when fertilised first by the white variety of *V. blattaria*, and then by the yellow variety of this species, yielded seed in the proportion of 622 to 438; and in both these cases castration was performed. Now the sterility which results from the crossing of the differently coloured varieties of the same species, is fully as great as that which occurs in many cases when distinct species are crossed. Unfortunately Gärtner compared the results of the first unions alone, and not the sterility of the two sets of hybrids produced from the white variety of *V. lychnitis* when fertilised by the white and yellow varieties of *V. blattaria*, for it is probable that they would have differed in this respect.

Mr. J. Scott has given me the results of a series of experiments on Verbascum, made by him in the Botanic Gardens of Edinburgh. He repeated some of Gärtner's experiments on distinct species, but obtained only fluctuating results; some confirmatory, but the greater number contradictory; nevertheless these seem hardly sufficient to overthrow the conclusions arrived at by Gärtner from experiments tried on a much larger scale. In the second place Mr. Scott experimented on the relative fertility of unions between similarly and dissimilarly-coloured varieties of the same species. Thus he fertilised six flowers of the yellow variety of *V. lychnitis* by its own pollen, and obtained six capsules, and calling, for the sake of having a standard of comparison, the average number of good seed in each one hundred, he found that this same yellow variety, when fertilised by the white variety, yielded from seven capsules an average of ninety-four seed. On the same principle, the white variety of *V. lychnitis* by its own pollen (from six capsules), and by the pollen of the yellow variety (eight capsules), yielded seed in the proportion of 100 to 82. The yellow variety of *V. thapsus* by its own pollen (eight capsules), and by that of the white variety (only two capsules), yielded seed in the proportion of 100 to 94. Lastly, the white variety of *V. blattaria* by its own pollen (eight capsules), and by that of the yellow variety (five capsules), yielded seed in the proportion of 100 to 79. So that in every case the unions of dissimilarly-coloured varieties of the same species were less fertile than the unions of similarly-coloured varieties; when all the cases are grouped together, the difference of fertility is as 86 to 100. Some additional trials were made, and altogether thirty-six similarly-coloured unions yielded thirty-five good

---

[15] ' Bastarderzeugung,' s. 216.

capsules; whilst thirty-five dissimilarly-coloured unions yielded only twenty-six good capsules. Besides the foregoing experiments, the purple *V. phœ-niceum* was crossed by a rose-coloured and a white variety of the same species; these two varieties were also crossed together, and these several unions yielded less seed than *V. phœniceum* by its own pollen. Hence it follows from Mr. Scott's experiments, that in the genus Verbascum the similarly and dissimilarly-coloured varieties of the same species behave, when crossed, like closely allied but distinct species.[16]

This remarkable fact of the sexual affinity of similarly-coloured varieties, as observed by Gärtner and Mr. Scott, may not be of very rare occurrence; for the subject has not been attended to by others. The following case is worth giving, partly to show how difficult it is to avoid error. Dr. Herbert[17] has remarked that variously-coloured double varieties of the hollyhock (*Althæa rosea*) may be raised with certainty by seed from plants growing close together. I have been informed that nurserymen who raise seed for sale do not separate their plants; accordingly I procured seed of eighteen named varieties; of these, eleven varieties produced sixty-two plants all perfectly true to their kind; and seven produced forty-nine plants, half of which were true and half false. Mr. Masters of Canterbury has given me a more striking case; he saved seed from a great bed of twenty-four named varieties planted in closely adjoining rows, and each variety reproduced itself truly with only sometimes a shade of difference in tint. Now in the hollyhock the pollen, which is abundant, is matured and nearly all shed before the stigma of the same flower is ready to receive it;[18] and as bees covered with pollen incessantly fly from plant to plant, it would appear that adjoining varieties could not escape being crossed. As, however, this does not occur, it appeared to me probable that the pollen

---

[16] The following facts, given by Kölreuter in his 'Dritte Fortsetzung,' s. 34, 39, appear at first sight strongly to confirm Mr. Scott's and Gärtner's statements; and to a certain limited extent they do so. Kölreuter asserts, from innumerable observations, that insects incessantly carry pollen from one species and variety of Verbascum to another; and I can confirm this assertion; yet he found that the white and yellow varieties of *Verbascum lychnitis* often grew wild mingled together: moreover, he cultivated these two varieties in considerable numbers during four years in his garden, and they kept true by seed; but when he crossed them, they produced flowers of an intermediate tint. Hence it might have been thought that both varieties must have a stronger elective affinity for the pollen of their own variety than for that of the other; this elective affinity,

I may add, of each species for its own pollen (Kölreuter, 'Dritte Forts.,' s. 39, and Gärtner, 'Bastarderz.,' *passim*) being a perfectly well-ascertained power. But the force of the foregoing facts is much lessened by Gärtner's numerous experiments, for, differently from Kölreuter, he never once got ('Bastarderz.,' s. 307) an intermediate tint when he crossed the yellow and white flowered varieties of Verbascum. So that the fact of the white and yellow varieties keeping true to their colour by seed does not prove that they were not mutually fertilised by the pollen carried by insects from one to the other.

[17] 'Amaryllidaceæ,' 1837, p. 366. Gärtner has made a similar observation.

[18] Kölreuter first observed this fact. 'Mém. de l'Acad. St. Petersburg,' vol. iii. p. 197. *See* also C. K. Sprengel, 'Das Entdeckte Geheimniss,' s. 345.

of each variety was prepotent on its own stigma over that of all other varieties. But Mr. C. Turner of Slough, well known for his success in the cultivation of this plant, informs me that it is the doubleness of the flowers which prevents the bees gaining access to the pollen and stigma; and he finds that it is difficult even to cross them artificially. Whether this explanation will fully account for varieties in close proximity propagating themselves so truly by seed, I do not know.

The following cases are worth giving, as they relate to monœcious forms, which do not require, and consequently have not been injured by, castration. Girou de Buzareingues crossed what he designates three varieties of gourd,[19] and asserts that their mutual fertilisation is less easy in proportion to the difference which they present. I am aware how imperfectly the forms in this group were until recently known; but Sageret,[20] who ranked them according to their mutual fertility, considers the three forms above alluded to as varieties, as does a far higher authority, namely, M. Naudin.[21] Sageret[22] has observed that certain melons have a greater tendency, whatever the cause may be, to keep true than others; and M. Naudin, who has had such immense experience in this group, informs me that he believes that certain varieties intercross more readily than others of the same species; but he has not proved the truth of this conclusion; the frequent abortion of the pollen near Paris being one great difficulty. Nevertheless, he has grown close together, during seven years, certain forms of Citrullus, which, as they could be artificially crossed with perfect facility and produced fertile offspring, are ranked as varieties; but these forms when not artificially crossed kept true. Many other varieties, on the other hand, in the same group cross with such facility, as M. Naudin repeatedly insists, that without being grown far apart they cannot be kept in the least true.

Another case, though somewhat different, may be here given, as it is highly remarkable, and is established on excellent evidence. Kölreuter minutely describes five varieties of the common tobacco,[23] which were reciprocally crossed, and the offspring were intermediate in character and as fertile as their parents: from this fact Kölreuter inferred that they are really varieties; and no one, as far as I can discover, seems to have doubted that such is the case. He also crossed reciprocally these five varieties with N. glutinosa, and they yielded very sterile hybrids; but those raised from the var. perennis, whether used as the father or mother plant, were not so sterile as the hybrids from the four other varieties.[24] So that the sexual

---

[19] Namely, Barbarines, Pastissons, Giraumous : 'Annal. des Sc. Nat.,' tom. xxx., 1833, pp. 398 and 405.

[20] 'Mémoire sur les Cucurbitaceæ,' 1826, pp. 46, 55.

[21] 'Annales des Sc. Nat.,' 4th series, tom. vi. M. Naudin considers these forms as undoubtedly varieties of Cucurbita pepo.

[22] 'Mém. Cucurb.,' p. 8.

[23] 'Zweite Forts.,' s. 53, namely, Nicotiana major vulgaris; (2) perennis;

(3) Transylvanica; (4) a sub-var. of the last; (5) major latifol. fl. alb.

[24] Kölreuter was so much struck with this fact that he suspected that a little pollen of N. glutinosa in one of his experiments might have accidentally got mingled with that of var. perennis, and thus aided its fertilising power. But we now know conclusively from Gärtner ('Bastarderz.,' s. 34, 43) that two kinds of pollen never act conjointly on a third species; still less will the pollen of a

capacity of this one variety has certainly been in some degree modified, so as to approach in nature that of *N. glutinosa*.[25]

These facts with respect to plants show that in some few cases certain varieties have had their sexual powers so far modified, that they cross together less readily and yield less seed than other varieties of the same species. We shall presently see that the sexual functions of most animals and plants are eminently liable to be affected by the conditions of life to which they are exposed; and hereafter we shall briefly discuss the conjoint bearing of this and other facts on the difference in fertility between crossed varieties and crossed species.

Domestication eliminates the tendency to Sterility which is general with Species when crossed.

This hypothesis was first propounded by Pallas,[26] and has been adopted by several authors. I can find hardly any direct facts in its support; but unfortunately no one has compared, in the case of either animals or plants, the fertility of anciently domesticated varieties, when crossed with a distinct species, with that of the wild parent-species when similarly crossed. No one has compared, for instance, the fertility of *Gallus bankiva* and of the domesticated fowl, when crossed with a distinct species of Gallus or Phasianus; and the

distinct species, mingled with a plant's own pollen, if the latter be present in sufficient quantity, have any effect. The sole effect of mingling two kinds of pollen is to produce in the same capsule seeds which yield plants, some taking after the one and some after the other parent.

[25] Mr. Scott has made some observations on the absolute sterility of a purple and white primrose (*Primula vulgaris*) when fertilised by pollen from the common primrose (' Journal of Proc. of Linn. Soc.,' vol. viii., 1864, p. 98); but these observations require confirmation. I raised a number of purple-flowered long-styled seedlings from seed kindly sent me by Mr. Scott, and, though they were all in some degree sterile, they were much more fertile with pollen taken from the common primrose than with their own

pollen. Mr. Scott has likewise described a red equal-styled cowslip (*P. veris*, idem, p. 106), which was found by him to be highly sterile when crossed with the common cowslip; but this was not the case with several equal-styled red seedlings raised by me from his plant. This variety of the cowslip presents the remarkable peculiarity of combining male organs in every respect like those of the short-styled form, with female organs resembling in function and partly in structure those of the long-styled form; so that we have the singular anomaly of the two forms combined in the same flower. Hence it is not surprising that these flowers should be spontaneously self-fertile in a high degree.

[26] ' Act. Acad. St. Petersburg,' 1780, part ii., pp. 84, 100.

experiment would in all cases be surrounded by many difficulties. Dureau de la Malle, who has so closely studied classical
literature, states [27] that in the time of the Romans the common
mule was produced with more difficulty than at the present
day; but whether this statement may be trusted I know not. A
much more important, though somewhat different, case is given
by M. Groenland,[28] namely, that plants, known from their intermediate character and sterility to be hybrids between Ægilops
and wheat, have perpetuated themselves under culture since
1857, *with a rapid but varying increase of fertility in each generation.* In the fourth generation the plants, still retaining their
intermediate character, had become as fertile as common
cultivated wheat.

The indirect evidence in favour of the Pallasian doctrine
appears to me to be extremely strong. In the earlier chapters
I have attempted to show that our various breeds of dogs are
descended from several wild species; and this probably is the
case with sheep. There can no longer be any doubt that
the Zebu or humped Indian ox belongs to a distinct species from
European cattle: the latter, moreover, are descended from two
or three forms, which may be called either species or wild races,
but which co-existed in a state of nature and kept distinct. We
have good evidence that our domesticated pigs belong to at
least two specific types, *S. scrofa* and *Indica,* which probably
lived together in a wild state in South-eastern Europe. Now, a
widely-extended analogy leads to the belief that if these several
allied species, in the wild state or when first reclaimed, had
been crossed, they would have exhibited, both in their first
unions and in their hybrid offspring, some degree of sterility.
Nevertheless the several domesticated races descended from
them are now all, as far as can be ascertained, perfectly fertile
together. If this reasoning be trustworthy, and it is apparently
sound, we must admit the Pallasian doctrine that long-continued
domestication tends to eliminate that sterility which is natural
to species when crossed in their aboriginal state.

[27] 'Annales des Sc. Nat.,' tom. xxi. (1st series), p. 61.
[28] 'Bull. Bot. Soc. de France,' Dec. 27th, 1861, tom. viii. p. 612.

On increased Fertility from Domestication and Cultivation.

Increased fertility from domestication, without any reference to crossing, may be here briefly considered.   This subject
bears indirectly on two or three points connected with the modification of organic beings.   As Buffon long ago remarked,[29]
domestic animals breed oftener in the year and produce more
young at a birth than wild animals of the same species; they,
also, sometimes breed at an earlier age.   The case would hardly
have deserved further notice, had not some authors lately
attempted to show that fertility increases and decreases in an
inverse ratio with the amount of food.   This strange doctrine
has apparently arisen from individual animals when supplied
with an inordinate quantity of food, and from plants of many
kinds when grown on excessively rich soil, as on a dunghill,
becoming sterile; but to this latter point I shall have occasion
presently to return.   With hardly an exception, our domesticated
animals, which have long been habituated to a regular and
copious supply of food, without the labour of searching for it,
are more fertile than the corresponding wild animals.   It is
notorious how frequently cats and dogs breed, and how many
young they produce at a birth.   The wild rabbit is said
generally to breed four times yearly, and to produce from
four to eight young; the tame rabbit breeds six or seven times
yearly, and produces from four to eleven young.   The ferret,
though generally so closely confined, is more prolific than its
supposed wild prototype.   The wild sow is remarkably prolific,
for she often breeds twice in the year, and produces from four to
eight and sometimes even twelve young at a birth; but the
domestic sow regularly breeds twice a year, and would breed
oftener if permitted; and a sow that produces less than eight at
a birth " is worth little, and the sooner she is fattened for the
butcher the better."   The amount of food affects the fertility
even of the same individual: thus sheep, which on mountains
never produce more than one lamb at a birth, when brought

[29] Quoted by Isid. Geoffroy St.
Hilaire, 'Hist. Naturelle Générale,' tom.
iii. p. 476.   Since this MS. has been sent
to press a full discussion on the present
subject has appeared in Mr. Herbert
Spencer's 'Principles of Biology,' vol.
ii. 1867, p. 457 *et seq.*

down to lowland pastures frequently bear twins. This difference apparently is not due to the cold of the higher land, for
sheep and other domestic animals are said to be extremely
prolific in Lapland. Hard living, also, retards the period at
which animals conceive; for it has been found disadvantageous
in the northern islands of Scotland to allow cows to bear calves
before they are four years old.[30]

Birds offer still better evidence of increased fertility from domestication:
the hen of the wild *Gallus bankiva* lays from six to ten eggs, a number
which would be thought nothing of with the domestic hen. The wild
duck lays from five to ten eggs; the tame one in the course of the year
from eighty to one hundred. The wild grey-lag goose lays from five to
eight eggs; the tame from thirteen to eighteen, and she lays a second
time; as Mr. Dixon has remarked, "high-feeding, care, and moderate
warmth induce a habit of prolificacy which becomes in some measure
hereditary." Whether the semi-domesticated dovecot pigeon is more
fertile than the wild rock-pigeon, *C. livia*, I know not; but the more
thoroughly domesticated breeds are nearly twice as fertile as dovecots:
the latter, however, when caged and highly fed, become equally fertile
with house pigeons. The peahen alone of domesticated birds is rather
more fertile, according to some accounts, when wild in its native Indian
home, than when domesticated in Europe and exposed to our much colder
climate.[31]

With respect to plants, no one would expect wheat to tiller more, and
each ear to produce more grain, in poor than in rich soil; or to get in
poor soil a heavy crop of peas or beans. Seeds vary so much in number

---

[30] For cats and dogs, &c., *see* Bellingeri, in 'Annal. des Sc. Nat.,' 2nd
series, Zoolog., tom. xii. p. 155. For
ferrets, Bechstein, 'Naturgeschichte
Deutschlands,' Band i., 1801, s. 786,
795. For rabbits, ditto, s. 1123, 1131;
and Bronn's 'Geschichte der Natur,'
B. ii. s. 99. For mountain sheep, ditto,
s. 102. For the fertility of the wild
sow, *see* Bechstein's 'Naturgesch.
Deutschlands,' B. i., 1801, s. 534; for
the domestic pig, Sidney's edit. of
Youatt on the Pig, 1860, p. 62. With
respect to Lapland, *see* Acerbi's 'Travels
to the North Cape,' Eng. translat., vol. ii.
p. 222. About the Highland cows, *see*
Hogg on Sheep, p. 263.

[31] For the eggs of *Gallus bankiva*, see
Blyth, in 'Annals and Mag. of Nat.
Hist.,' 2nd series, vol. i., 1848, p. 456.
For wild and tame ducks, Macgillivray,
'British Birds,' vol. v. p. 37; and 'Die
Enten,' s. 87. For wild geese, L. Lloyd,
'Scandinavian Adventures,' vol. ii.
1854, p. 413; and for tame geese,
'Ornamental Poultry,' by Rev. E. S.
Dixon, p. 139. On the breeding of
pigeons, Pistor, 'Das Ganze der Taubenzucht,' 1831, s. 46; and Boitard and
Corbié, 'Les Pigeons,' p. 158. With
respect to peacocks, according to Temminck ('Hist. Nat. Gén. des Pigeons,'
&c., 1813, tom. ii. p. 41, the hen lays
in India even as many as twenty eggs;
but according to Jerdon and another
writer (quoted in Tegetmeier's 'Poultry
Book,' 1866, pp. 280, 282), she there lays
only from four to nine or ten eggs: in
England she is said, in the 'Poultry
Book,' to lay five or six, but another
writer says from eight to twelve eggs.

that it is difficult to estimate them; but on comparing beds of carrots saved for seed in a nursery garden with wild plants, the former seemed to produce about twice as much seed. Cultivated cabbages yielded thrice as many pods by measure as wild cabbages from the rocks of South Wales. The excess of berries produced by the cultivated Asparagus in comparison with the wild plant is enormous. No doubt many highly cultivated plants, such as pears, pineapples, bananas, sugar-cane, &c., are nearly or quite sterile; and I am inclined to attribute this sterility to excess of food and to other unnatural conditions; but to this subject I shall presently recur.

In some cases, as with the pig, rabbit, &c., and with those plants which are valued for their seed, the direct selection of the more fertile individuals has probably much increased their fertility; and in all cases this may have occurred indirectly, from the better chance of the more numerous offspring produced by the more fertile individuals having survived. But with cats, ferrets, and dogs, and with plants like carrots, cabbages, and asparagus, which are not valued for their prolificacy, selection can have played only a subordinate part; and their increased fertility must be attributed to the more favourable conditions of life under which they have long existed.

CHAPTER XVII.

ON THE GOOD EFFECTS OF CROSSING, AND ON THE EVIL EFFECTS OF CLOSE INTERBREEDING.

DEFINITION OF CLOSE INTERBREEDING — AUGMENTATION OF MORBID TENDENCIES — GENERAL EVIDENCE ON THE GOOD EFFECTS DERIVED FROM CROSSING, AND ON THE EVIL EFFECTS FROM CLOSE INTERBREEDING — CATTLE, CLOSELY INTERBRED ; HALF-WILD CATTLE LONG KEPT IN THE SAME PARKS — SHEEP — FALLOW-DEER — DOGS — RABBITS — PIGS — MAN, ORIGIN OF HIS ABHORRENCE OF INCESTUOUS MARRIAGES — FOWLS — PIGEONS — HIVE-BEES — PLANTS, GENERAL CONSIDERATIONS ON THE BENEFITS DERIVED FROM CROSSING — MELONS, FRUIT-TREES, PEAS, CABBAGES, WHEAT, AND FOREST-TREES — ON THE INCREASED SIZE OF HYBRID PLANTS, NOT EXCLUSIVELY DUE TO THEIR STERILITY — ON CERTAIN PLANTS WHICH EITHER NORMALLY OR ABNORMALLY ARE SELF-IMPOTENT, BUT ARE FERTILE, BOTH ON THE MALE AND FEMALE SIDE, WHEN CROSSED WITH DISTINCT INDIVIDUALS EITHER OF THE SAME OR ANOTHER SPECIES — CONCLUSION.

THE gain in constitutional vigour, derived from an occasional cross between individuals of the same variety, but belonging to distinct families, or between distinct varieties, has not been so largely or so frequently discussed, as have the evil effects of too close interbreeding. But the former point is the more important of the two, inasmuch as the evidence is more decisive. The evil results from close interbreeding are difficult to detect, for they accumulate slowly, and differ much in degree with different species; whilst the good effects which almost invariably follow a cross are from the first manifest. It should, however, be clearly understood that the advantage of close interbreeding, as far as the retention of character is concerned, is indisputable, and often outweighs the evil of a slight loss of constitutional vigour. In relation to the subject of domestication, the whole question is of some importance, as too close interbreeding interferes with the improvement of old races, and especially with the formation of new ones. It is important as indirectly bearing on Hybridism ; and perhaps on the extinction of species, when any form has become so rare that only a few individuals

remain within a confined area. It bears in an important
manner on the influence of free intercrossing, in obliterating
individual differences, and thus giving uniformity of character
to the individuals of the same race or species; for if additional
vigour and fertility be thus gained, the crossed offspring will
multiply and prevail, and the ultimate result will be far greater
than otherwise would have occurred. Lastly, the question is
of high interest, as bearing on mankind. Hence I shall discuss
this subject at full length. As the facts which prove the evil
effects of close interbreeding are more copious, though less
decisive, than those on the good effects of crossing, I shall, under
each group of beings, begin with the former.

There is no difficulty in defining what is meant by a cross;
but this is by no means easy in regard to " breeding in and in "
or " too close interbreeding," because, as we shall see, different
species of animals are differently affected by the same degree of
interbreeding. The pairing of a father and daughter, or mother
and son, or brothers and sisters, if carried on during several
generations, is the closest possible form of interbreeding. But
some good judges, for instance Sir J. Sebright, believe that the
pairing of a brother and sister is closer than that of parents
and children; for when the father is matched with his daughter
he crosses, as is said, with only half his own blood. The consequences of close interbreeding carried on for too long a time,
are, as is generally believed, loss of size, constitutional vigour,
and fertility, sometimes accompanied by a tendency to malformation. Manifest evil does not usually follow from pairing
the nearest relations for two, three, or even four generations; but several causes interfere with our detecting the evil
—such as the deterioration being very gradual, and the difficulty of distinguishing between such direct evil and the inevitable augmentation of any morbid tendencies which may be
latent or apparent in the related parents. On the other hand,
the benefit from a cross, even when there has not been any very
close interbreeding, is almost invariably at once conspicuous.
There is reason to believe, and this was the opinion of that most
experienced observer Sir J. Sebright,[1] that the evil effects of
close interbreeding may be checked by the related individuals

---

[1] 'The Art of Improving the Breed, &c.,' 1809, p. 16.

being separated during a few generations and exposed to different
conditions of life.

That evil directly follows from any degree of close interbreeding has been denied by many persons; but rarely by any
practical breeder; and never, as far as I know, by one who has
largely bred animals which propagate their kind quickly. Many
physiologists attribute the evil exclusively to the combination
and consequent increase of morbid tendencies common to both
parents: that this is an active source of mischief there can be
no doubt. It is unfortunately too notorious that men and
various domestic animals endowed with a wretched constitution, and with a strong hereditary disposition to disease, if not
actually ill, are fully capable of procreating their kind. Close
interbreeding, on the other hand, induces sterility; and this
indicates something quite distinct from the augmentation of
morbid tendencies common to both parents. The evidence
immediately to be given convinces me that it is a great law of
nature, that all organic beings profit from an occasional cross
with individuals not closely related to them in blood; and
that, on the other hand, long-continued close interbreeding is
injurious.

Various general considerations have had much influence in
leading me to this conclusion; but the reader will probably rely
more on special facts and opinions. The authority of experienced observers, even when they do not advance the grounds of
their belief, is of some little value. Now almost all men who
have bred many kinds of animals and have written on the
subject, such as Sir J. Sebright, Andrew Knight, &c.,[2] have
expressed the strongest conviction on the impossibility of long-
continued close interbreeding. Those who have compiled works
on agriculture, and have associated much with breeders, such as
the sagacious Youatt, Low, &c., have strongly declared their
opinion to the same effect. Prosper Lucas, trusting largely
to French authorities, has come to a similar conclusion. The
distinguished German agriculturist Hermann von Nathusius,
who has written the most able treatise on this subject which
I have met with, concurs; and as I shall have to quote from

---

[2] For Andrew Knight, *see* A. Walker, on ' Intermarriage,' 1838, p. 227. Sir J.
Sebright's Treatise has just been quoted.

this treatise, I may state that Nathusius is not only intimately acquainted with works on agriculture in all languages, and knows the pedigrees of our British breeds better than most Englishmen, but has imported many of our improved animals, and is himself an experienced breeder.

Evidence of the evil effects of close interbreeding can most readily be acquired in the case of animals, such as fowls, pigeons, &c., which propagate quickly, and, from being kept in the same place, are exposed to the same conditions. Now I have inquired of very many breeders of these birds, and I have hitherto not met with a single man who was not thoroughly convinced that an occasional cross with another strain of the same sub-variety was absolutely necessary. Most breeders of highly-improved or fancy birds value their own strain, and are most unwilling, at the risk, in their opinion, of deterioration, to make a cross. The purchase of a first-rate bird of another strain is expensive, and exchanges are troublesome; yet all breeders, as far as I can hear, excepting those who keep large stocks at different places for the sake of crossing, are driven after a time to take this step.

Another general consideration which has had great influence on my mind is, that with all hermaphrodite animals and plants, which it might have been thought would have perpetually ferti- lised themselves, and thus have been subjected for long ages to the closest interbreeding, there is no single species, as far as I can discover, in which the structure ensures self-fertilisation. On the contrary, there are in a multitude of cases, as briefly stated in the fifteenth chapter, manifest adaptations which favour or inevit- ably lead to an occasional cross between one hermaphrodite and another of the same species; and these adaptive structures are utterly purposeless, as far as we can see, for any other end.

With *Cattle* there can be no doubt that extremely close interbreeding may be long carried on, advantageously with respect to external characters and with no manifestly apparent evil as far as constitution is concerned. The same remark is applicable to sheep. Whether these animals have gradually been rendered less susceptible than others to this evil, in order to permit them to live in herds,—a habit which leads the old and vigorous males to expel all intruders, and in consequence often to pair with their own daughters, I will not pretend to decide. The case of Bakewell's Long- horns, which were closely interbred for a long period, has often been quoted; yet Youatt says[3] the breed "had acquired a delicacy of consti-tution inconsistent with common management," and "the propagation of the species was not always certain." But the Shorthorns offer the most striking case of close interbreeding; for instance, the famous bull Favourite (who was himself the offspring of a half-brother and sister from Foljambe) was matched with his own daughter, granddaughter, and great-granddaughter; so that the produce of this last union, or the great-great-granddaughter, had 15-16ths, or 93·75 per cent. of the blood of Favourite in her veins. This cow was matched with the bull Well-ington, having 62·5 per cent. of Favourite blood in his veins, and pro-duced Clarissa; Clarissa was matched with the bull Lancaster, having 68·75 of the same blood, and she yielded valuable offspring.[4] Nevertheless Collings, who reared these animals, and was a strong advocate for close breeding, once crossed his stock with a Galloway, and the cows from this cross realised the highest prices. Bates's herd was esteemed the most cele-brated in the world. For thirteen years he bred most closely in and in; but during the next seventeen years, though he had the most exalted notion of the value of his own stock, he thrice infused fresh blood into his herd: it is said that he did this, not to improve the form of his animals, but on account of their lessened fertility. Mr. Bates's own view, as given by a celebrated breeder,[5] was, that "to breed in and in from a bad stock was ruin and devastation; yet that the practice may be safely followed within certain limits when the parents so related are descended from first-rate animals." We thus see that there has been extremely close inter-breeding with Shorthorns; but Nathusius, after the most careful study of their pedigrees, says that he can find no instance of a breeder who has strictly followed this practice during his whole life. From this study and his own experience, he concludes that close interbreeding is necessary to ennoble the stock; but that in effecting this the greatest care is necessary, on account of the tendency to infertility and weakness. It may be added, that another high authority[6] asserts that many more calves are born cripples from Shorthorns than from other and less closely inter-bred races of cattle.

Although by carefully selecting the best animals (as Nature effectually does by the law of battle) close interbreeding may be long carried on with cattle, yet the good effects of a cross between almost any two breeds is at once shown by the greater size and vigour of the offspring; as Mr. Spooner writes to me, "crossing distinct breeds certainly improves cattle for the butcher." Such crossed animals are of course of no value to the breeder; but they have been raised during many years in several

[3] 'Cattle,' p. 199.
[4] Nathusius, 'Ueber Shorthorn Rind-vieh,' 1857, s. 71: see also 'Gardener's Chronicle,' 1860, p. 270. Many analo-gous cases are given in a pamphlet re-cently published by Mr. C. Macknight and Dr. H. Madden, 'On the True Principles of Breeding;' Melbourne, Australia, 1865.
[5] Mr. Willoughby Wood, in 'Gar-dener's Chronicle,' 1855, p. 411; and 1860, p. 270. See the very clear tables and pedigrees given in Nathusius' 'Rindvieh,' s. 72-77.
[6] Mr. Wright, 'Journal of Royal Agricult. Soc.,' vol. vii., 1846, p. 204.

parts of England to be slaughtered;[7] and their merit is now so fully recognised, that at fat-cattle shows a separate class has been formed for their reception. The best fat ox at the great show at Islington in 1862 was a crossed animal.

The half-wild cattle, which have been kept in British parks probably for 400 or 500 years, or even for a longer period, have been advanced by Culley and others as a case of long-continued interbreeding within the limits of the same herd without any consequent injury. With respect to the cattle at Chillingham, the late Lord Tankerville owned that they were bad breeders.[8] The agent, Mr. Hardy, estimates (in a letter to me, dated May, 1861) that in the herd of about fifty the average number annually slaughtered, killed by fighting, and dying, is about ten, or one in five. As the herd is kept up to nearly the same average number, the annual rate of increase must be likewise about one in five. The bulls, I may add, engage in furious battles, of which battles the present Lord Tankerville has given me a graphic description, so that there will always be rigorous selection of the most vigorous males. I procured in 1855 from Mr. D. Gardner, agent to the Duke of Hamilton, the following account of the wild cattle kept in the Duke's park in Lanarkshire, which is about 200 acres in extent. The number of cattle varies from sixty-five to eighty; and the number annually killed (I presume by all causes) is from eight to ten; so that the annual rate of increase can hardly be more than one in six. Now in South America, where the herds are half-wild, and therefore offer a nearly fair standard of comparison, according to Azara the natural increase of the cattle on an estancia is from one-third to one-fourth of the total number, or one in between three and four; and this, no doubt, applies exclusively to adult animals fit for consumption. Hence the half-wild British cattle which have long interbred within the limits of the same herd are relatively far less fertile. Although in an unenclosed country like Paraguay there must be some crossing between the different herds, yet even there the inhabitants believe that the occasional introduction of animals from distant localities is necessary to prevent " degeneration in size and diminution of fertility."[9] The decrease in size from ancient times in the Chillingham and Hamilton cattle must have been prodigious, for Professor Rütimeyer has shown that they are almost certainly the descendants of the gigantic *Bos primigenius*. No doubt this decrease in size may be largely attributed to less favourable conditions of life; yet animals roaming over large parks, and fed during severe winters, can hardly be considered as placed under very unfavourable conditions.

With *Sheep* there has often been long-continued interbreeding within the limits of the same flock; but whether the nearest relations have been matched so frequently as in the case of Shorthorn cattle, I do not know. The Messrs. Brown during fifty years have never infused fresh blood into their excellent flock of Leicesters. Since 1810 Mr. Barford has acted on the same principle with the Foscote flock. He asserts that half a century

---

[7] Youatt on Cattle, p. 202.

[8] Report British Assoc., Zoolog. Sect., 1838.

[9] Azara, 'Quadrupèdes du Paraguay, tom. ii. pp. 354, 368.

of experience has convinced him that when two nearly related animals are quite sound in constitution, in-and-in breeding does not induce degeneracy; but he adds that he "does not pride himself on breeding from the nearest affinities." In France the Naz flock has been bred for sixty years without the introduction of a single strange ram.[10] Nevertheless, most great breeders of sheep have protested against close interbreeding prolonged for too great a length of time.[11] The most celebrated of recent breeders, Jonas Webb, kept five separate families to work on, thus "retaining the requisite distance of relationship between the sexes."[12]

Although by the aid of careful selection the near interbreeding of sheep may be long continued without any manifest evil, yet it has often been the practice with farmers to cross distinct breeds to obtain animals for the butcher, which plainly shows that good is derived from this practice. Mr. Spooner sums up his excellent Essay on Crossing by asserting that there is a direct pecuniary advantage in judicious cross-breeding, especially when the male is larger than the female. A former celebrated breeder, Lord Somerville, distinctly states that his half-breeds from Ryelands and Spanish sheep were larger animals than either the pure Ryelands or pure Spanish sheep.[13]

As some of our British parks are ancient, it occurred to me that there must have been long-continued close interbreeding with the fallow deer (*Cervus dama*) kept in them; but on inquiry I find that it is a common practice to infuse new blood by procuring bucks from other parks. Mr. Shirley,[14] who has carefully studied the management of deer, admits that in some parks there has been no admixture of foreign blood from a time beyond the memory of man. But he concludes "that in the end "the constant breeding in-and-in is sure to tell to the disadvantage of "the whole herd, though it may take a very long time to prove it; "moreover, when we find, as is very constantly the case, that the intro- "duction of fresh blood has been of the very greatest use to deer, both "by improving their size and appearance, and particularly by being of "service in removing the taint of 'rickback,' if not of other diseases, to "which deer are sometimes subject when the blood has not been "changed, there can, I think, be no doubt but that a judicious cross "with a good stock is of the greatest consequence, and is indeed essential, "sooner or later, to the prosperity of every well-ordered park."

Mr. Meynell's famous foxhounds have been adduced, as showing that no ill effects follow from close interbreeding; and Sir J. Sebright ascertained from him that he frequently bred from father and daughter, mother and

[10] For the case of the Messrs. Brown, see 'Gard. Chronicle,' 1855, p. 26. For the Foscote flock, 'Gard. Chron.,' 1860, p. 416. For the Naz flock, 'Bull. de la Soc. d'Acclimat.,' 1860, p. 477.

[11] Nathusius, 'Rindvieh,' s. 65; Youatt on Sheep, p. 495.

[12] 'Gard. Chronicle,' 1861, p. 631.

[13] Lord Somerville, 'Facts on Sheep

and Husbandry,' p. 6. Mr. Spooner, in 'Journal of Royal Agricult. Soc. of England,' vol. xx., part ii. See also an excellent paper on the same subject in 'Gard. Chronicle,' 1860, p. 321, by Mr. Charles Howard.

[14] 'Some Account of English Deer Parks,' by Evelyn P. Shirley, 1867.

son, and sometimes even from brothers and sisters. Sir J. Sebright, however, declares,[15] that by breeding *in-and-in*, by which he means matching brothers and sisters, he has actually seen strong spaniels become weak and diminutive lapdogs. The Rev. W. D. Fox has communicated to me the case of a small lot of bloodhounds, long kept in the same family, which had become very bad breeders, and nearly all had a bony enlargement in the tail. A single cross with a distinct strain of bloodhounds restored their fertility, and drove away the tendency to malformation in the tail. I have heard the particulars of another case with bloodhounds, in which the female had to be held to the male. Considering how rapid is the natural increase of the dog, it is difficult to understand the high price of most highly improved breeds, which almost implies long-continued close interbreeding, except on the belief that this process lessens fertility and increases liability to distemper and other diseases. A high authority, Mr. Scrope, attributes the rarity and deterioration in size of the Scotch deerhound (the few individuals now existing throughout the country being all related) in large part to close interbreeding.

With all highly-bred animals there is more or less difficulty in getting them to procreate quickly, and all suffer much from delicacy of constitution; but I do not pretend that these effects ought to be wholly attributed to close interbreeding. A great judge of rabbits[16] says, " the long-eared does are often too highly bred or forced in their youth to be of much value as breeders, often turning out barren or bad mothers." Again : " Very long-eared bucks will also sometimes prove barren." These highly-bred rabbits often desert their young, so that it is necessary to have nurse-rabbits.

With *Pigs* there is more unanimity amongst breeders on the evil effects of close interbreeding than, perhaps, with any other large animal. Mr. Druce, a great and successful breeder of the Improved Oxfordshires (a crossed race), writes, " without a change of boars of a different tribe, but of the same breed, constitution cannot be preserved." Mr. Fisher Hobbs, the raiser of the celebrated Improved Essex breed, divided his stock into three separate families, by which means he maintained the breed for more than twenty years, " by judicious selection from the *three distinct families*."[17] Lord Western was the first importer of a Neapolitan boar and sow. " From this pair he bred in-and-in, until the breed was in danger of becoming extinct, a sure result (as Mr. Sidney remarks) of in-and-in breeding." Lord Western then crossed his Neapolitan pigs with the old Essex, and made the first great step towards the Improved Essex breed. Here is a more interesting case. Mr. J. Wright, well known as a breeder, crossed[18] the same boar with the daughter, granddaughter, and great-granddaughter, and so on for seven generations. The result was, that in many instances the offspring failed to breed; in others they produced few that lived; and of the latter many were idiotic, without sense

---

[15] 'The Art of Improving the Breed,' &c., p. 13. With respect to Scotch deer-hounds, *see* Scrope's 'Art of Deer Stalking,' pp. 350-353.

[16] 'Cottage Gardener,' 1861, p. 327.

[17] Sidney's edit. of Youatt on the Pig, 1860, p. 30; p. 33, quotation from Mr. Druce; p. 29, on Lord Western's case.

[18] 'Journal, Royal Agricult. Soc. of England,' 1846, vol. vii. p. 205.

even to suck, and when attempting to move could not walk straight. Now it deserves especial notice, that the two last sows produced by this long course of interbreeding were sent to other boars, and they bore several litters of healthy pigs. The best sow in external appearance produced during the whole seven generations was one in the last stage of descent; but the litter consisted of this one sow. She would not breed to her sire, yet bred at the first trial to a stranger in blood. So that, in Mr. Wright's case, long-continued and extremely close interbreeding did not affect the external form or merit of the young; but with many of them the general constitution and mental powers, and especially the reproductive functions, were seriously affected.

Nathusius gives[19] an analogous and even more striking case: he imported from England a pregnant sow of the large Yorkshire breed, and bred the product closely in-and-in for three generations: the result was unfavourable, as the young were weak in constitution, with impaired fertility. One of the latest sows, which he esteemed a good animal, produced, when paired with her own uncle (who was known to be productive with sows of other breeds), a litter of six, and a second time a litter of only five weak young pigs. He then paired this sow with a boar of a small black breed, which he had likewise imported from England, and which boar, when matched with sows of his own breed, produced from seven to nine young: now, the sow of the large breed, which was so unproductive when paired with her own uncle, yielded to the small black boar, in the first litter twenty-one, and in the second litter eighteen young pigs; so that in one year she produced thirty-nine fine young animals!

As in the case of several other animals already mentioned, even when no injury is perceptible from moderately close interbreeding, yet, to quote the words of Mr. Coate, a most successful breeder (who five times won the annual gold medal of the Smithfield Club Show for the best pen of pigs), "Crosses answer well for profit to the farmer, as you get more constitution and quicker growth; but for me, who sell a great number of pigs for breeding purposes, I find it will not do, as it requires many years to get anything like purity of blood again."[20]

Before passing on to Birds, I ought to refer to man, though I am unwilling to enter on this subject, as it is surrounded by natural prejudices. It has moreover been discussed by various authors under many points of view.[21] Mr. Tylor[22] has shown

[19] 'Ueber Rindvieh,' &c., s. 78.

[20] Sidney on the Pig, p. 36. *See* also note, p. 34. Also Richardson on the Pig, 1847, p. 26.

[21] Dr. Dally has published an excellent article (translated in the 'Anthropolog. Review,' May, 1864, p. 65), criticising all writers who have maintained that evil follows from consanguineous marriages. No doubt on this side of the question many advocates

have injured their cause by inaccuracies: thus it has been stated (Devay, 'Du Danger des Mariages,' &c., 1862, p. 141) that the marriages of cousins have been prohibited by the legislature of Ohio; but I have been assured, in answer to inquiries made in the United States, that this statement is a mere fable.

[22] *See* his most interesting work on the 'Early History of Man,' 1865, chap. x.

that with widely different races, in the most distant quarters of the world, marriages between relations—even between distant relations—have been strictly prohibited. A few exceptional cases can be specified, especially with royal families; and these have been enlarged on in a learned article[23] by Mr. W. Adam, and formerly in 1828 by Hofacker. Mr. Tylor is inclined to believe that the almost universal prohibition of closely-related marriages has arisen from their evil effects having been observed, and he ingeniously explains some apparent anomalies in the prohibition not extending equally to the relations on both the male and female side. He admits, however, that other causes, such as the extension of friendly alliances, may have come into play. Mr. W. Adam, on the other hand, concludes that related marriages are prohibited and viewed with repugnance from the confusion which would thus arise in the descent of property, and from other still more recondite reasons; but I cannot accept this view, seeing that the savages of Australia and South America,[24] who have no property to bequeath or fine moral feelings to confuse, hold the crime of incest in abhorrence.

It would be interesting to know, if it could be ascertained, as throwing light on this question with respect to man, what occurs with the higher anthropomorphous apes—whether the young males and females soon wander away from their parents, or whether the old males become jealous of their sons and expel them, or whether any inherited instinctive feeling, from being beneficial, has been generated, leading the young males and females of the same family to prefer pairing with distinct families, and to dislike pairing with each other. A considerable body of evidence has already been advanced, showing that the offspring from parents which are not related are more vigorous and fertile than those from parents which are closely related; hence any slight feeling, arising from the sexual excitement of novelty or other cause, which led to the former rather than to the latter unions, would be augmented through natural selection, and thus might become instinctive; for those individuals which had an innate preference of this kind would increase in number. It seems more probable, that degraded savages should

[23] On Consanguinity in Marriage, in the 'Fortnightly Review,' 1865, p. 710; Hofacker, 'Ueber die Eigenschaften,' &c.

[24] Sir G. Grey's 'Journal of Expeditions into Australia,' vol. ii. p. 243; and Dobrizhoffer, 'On the Abipones of South America.'

thus unconsciously have acquired their dislike and even abhorrence of incestuous marriages, rather than that they should have discovered by reasoning and observation the evil results. The abhorrence occasionally failing is no valid argument against the feeling being instinctive, for any instinct may occasionally fail or become vitiated, as sometimes occurs with parental love and the social sympathies. In the case of man, the question whether evil follows from close interbreeding will probably never be answered by direct evidence, as he propagates his kind so slowly and cannot be subjected to experiment; but the almost universal practice of all races at all times of avoiding closely-related marriages is an argument of considerable weight; and whatever conclusion we arrive at in regard to the higher animals may be safely extended to man.

Turning now to Birds: in the case of the *Fowl* a whole array of authorities could be given against too close interbreeding. Sir J. Sebright positively asserts that he made many trials, and that his fowls, when thus treated, became long in the legs, small in the body, and bad breeders.[25] He produced the famous Sebright Bantams by complicated crosses, and by breeding in-and-in; and since his time there has been much close interbreeding with these Bantams; and they are now notoriously bad breeders. I have seen Silver Bantams, directly descended from his stock, which had become almost as barren as hybrids; for not a single chicken had been that year hatched from two full nests of eggs. Mr. Hewitt says that with these Bantams the sterility of the male stands, with rare exceptions, in the closest relation with their loss of certain secondary male characters: he adds, " I have noticed, as a general rule, that even the slightest deviation " from feminine character in the tail of the male Sebright — say the " elongation by only half an inch of the two principal tail-feathers — " brings with it improved probability of increased fertility."[26]
Mr. Wright states[27] that Mr. Clark, " whose fighting-cocks were so " notorious, continued to breed from his own kind till they lost their dis- " position to fight, but stood to be cut up without making any resistance, " and were so reduced in size as to be under those weights required for " the best prizes; but on obtaining a cross from Mr. Leighton, they " again resumed their former courage and weight." It should be borne in mind that game-cocks before they fought were always weighed, so that nothing was left to the imagination about any reduction or increase of

[25] 'The Art of Improving the Breed,' p. 13.

[26] 'The Poultry Book,' by W. B. Tegetmeier, 1866, p. 245.

[27] 'Journal Royal Agricult. Soc.' 1846, vol. vii. p. 205 ; see also Ferguson on the Fowl, pp. 83, 317 ; see also 'The Poultry Book,' by Tegetmeier, 1866, p. 135, with respect to the extent to which cock-fighters found that they could venture to breed in-and-in, viz., occasionally a hen with her own son; "but they were cautious not to repeat the in-and-in breeding."

weight. Mr. Clark does not seem to have bred from brothers and sisters, which is the most injurious kind of union; and he found, after repeated trials, that there was a greater reduction in weight in the young from a father paired with his daughter, than from a mother with her son. I may add that Mr. Eyton, of Eyton, the well-known ornithologist, who is a large breeder of Grey Dorkings, informs me that they certainly diminish in size, and become less prolific, unless a cross with another strain is occasionally obtained. So it is with Malays, according to Mr. Hewitt, as far as size is concerned.[28]

An experienced writer[29] remarks that the same amateur, as is well known, seldom long maintains the superiority of his birds; and this, he adds, undoubtedly is due to all his stock "being of the same blood;" hence it is indispensable that he should occasionally procure a bird of another strain. But this is not necessary with those who keep a stock of fowls at different stations. Thus, Mr. Ballance, who has bred Malays for thirty years, and has won more prizes with these birds than any other fancier in England, says that breeding in-and-in does not necessarily cause deterioration; "but all depends upon how this is managed." "My plan " has been to keep about five or six distinct runs, and to rear about " two hundred or three hundred chickens each year, and select the best " birds from each run for crossing. I thus secure sufficient crossing to " prevent deterioration."[30]

We thus see that there is almost complete unanimity with poultry-breeders that, when fowls are kept at the same place, evil quickly follows from interbreeding carried on to an extent which would be disregarded in the case of most quadrupeds. On the other hand, it is a generally received opinion that cross-bred chickens are the hardiest and most easily reared.[31] Mr. Tegetmeier, who has carefully attended to poultry of all breeds, says[32] that Dorking hens, allowed to run with Houdan or Crevecœur cocks, "produce in the early spring chickens that for size, hardihood, " early maturity, and fitness for the market, surpass those of any pure " breed that we have ever raised." Mr. Hewitt gives it as a general rule with fowls, that crossing the breed increases their size. He makes this remark after stating that hybrids from the pheasant and fowl are considerably larger than either progenitor: so again, hybrids from the male golden pheasant and hen common pheasant "are of far larger size than either parent-bird."[33] To this subject of the increased size of hybrids I shall presently return.

With *Pigeons*, breeders are unanimous, as previously stated, that it is absolutely indispensable, notwithstanding the trouble and expense thus caused, occasionally to cross their much-prized birds with individuals of another strain, but belonging, of course, to the same variety. It deserves

[28] 'The Poultry Book,' by W. B. Tegetmeier, 1866, p. 79.
[29] 'The Poultry Chronicle,' 1854, vol. i. p. 43.
[30] 'The Poultry Book,' by W. B. Tegetmeier, 1866, p. 79.

[31] 'The Poultry Chronicle,' vol. i. p. 89.
[32] 'The Poultry Book,' 1866, p. 210.
[33] Ibid, 1866, p. 167; and 'Poultry Chronicle,' vol. iii.. 1855, p. 15.

notice that, when large size is one of the desired characters, as with pouters,[34] the evil effects of close interbreeding are much sooner perceived than when small birds, such as short-faced tumblers, are valued. The extreme delicacy of the high fancy breeds, such as these tumblers and improved English carriers, is remarkable; they are liable to many diseases, and often die in the egg or during the first moult; and their eggs have generally to be hatched under foster-mothers. Although these highly-prized birds have invariably been subjected to much close interbreeding, yet their extreme delicacy of constitution cannot perhaps be thus fully explained. Mr. Yarrell informed me that Sir J. Sebright continued closely interbreeding some owl-pigeons, until from their extreme sterility he as nearly as possible lost the whole family. Mr. Brent[35] tried to raise a breed of trumpeters, by crossing a common pigeon, and recrossing the daughter, granddaughter, great-granddaughter, and great-great-granddaughter, with the same male trumpeter, until he obtained a bird with $\frac{15}{16}$ of trumpeter's blood; but then the experiment failed, for " breeding so close stopped re-production." The experienced Neumeister[36] also asserts that the offspring from dovecotes and various other breeds are " generally very fertile and hardy birds:" so again, MM. Boitard and Corbié,[37] after forty-five years' experience, recommend persons to cross their breeds for amusement; for, if they fail to make interesting birds, they will succeed under an economical point of view, " as it is found that mongrels are more fertile than pigeons of pure race."

I will refer only to one other animal, namely, the Hive-bee, because a distinguished entomologist has advanced this as a case of inevitable close interbreeding. As the hive is tenanted by a single female, it might have been thought that her male and female offspring would always have bred together, more especially as bees of different hives are hostile to each other; a strange worker being almost always attacked when trying to enter another hive. But Mr. Tegetmeier has shown[38] that this instinct does not apply to drones, which are permitted to enter any hive; so that there is no *à priori* improbability of a queen receiving a foreign drone. The fact of the union invariably and necessarily taking place on the wing, during the queen's nuptial flight, seems to be a special provision against continued interbreeding. However this may be, experience has shown, since the introduction of the yellow-banded Ligurian race into Germany and England, that bees freely cross: Mr. Woodbury, who introduced Ligurian bees into Devonshire, found during a single season that three stocks, at distances of from one to two miles from his hives, were crossed by his drones. In one case the Ligurian drones must have flown over the city of Exeter, and over several intermediate hives. On another occasion several common black queens were crossed by Ligurian drones at a distance of from one to three and a half miles.[39]

---

[34] 'A Treatise on Fancy Pigeons,' by J. M. Eaton, p. 56.

[35] 'The Pigeon Book,' p. 46.

[36] ' Das Ganze der Taubenzucht,' 1837, s. 18.

[37] ' Les Pigeons,' 1824, p. 35.

[38] 'Proc. Entomolog. Soc.,' Aug. 6th, 1860, p. 126.

[39] ' Journal of Horticulture,' 1861, pp. 39, 77, 158; and 1864, p. 206.

Plants.

When a single plant of a new species is introduced into any country, if propagated by seed, many individuals will soon be raised, so that if the proper insects be present there will be crossing. With newly-introduced trees or other plants not propagated by seed we are not here concerned. With old-established plants it is an almost universal practice occasionally to make exchanges of seed, by which means individuals which have been exposed to different conditions of life,—and this, as we have seen, diminishes the evil from close interbreeding,—will occasionally be introduced into each district.

Experiments have not been tried on the effects of fertilising flowers with their own pollen during *several* generations. But we shall presently see that certain plants, either normally or abnormally, are more or less sterile, even in the first generation, when fertilised by their own pollen. Although nothing is directly known on the evil effects of long-continued close inter-breeding with plants, the converse proposition that great good is derived from crossing is well established.

With respect to the crossing of individuals belonging to the same sub-variety, Gärtner, whose accuracy and experience exceeded that of all other hybridisers, states [40] that he has many times observed good effects from this step, especially with exotic genera, of which the fertility is somewhat impaired, such as Passiflora, Lobelia, and Fuchsia. Herbert also says,[41] " I am inclined to think that I have derived advantage from impreg-
" nating the flower from which I wished to obtain seed with pollen from
" another individual of the same variety, or at least from another flower,
" rather than with its own." Again, Professor Lecoq asserts that he has ascertained that crossed offspring are more vigorous and robust than their parents.[42]

General statements of this kind, however, can seldom be fully trusted; consequently I have begun a series of experiments, which, if they continue to give the same results as hitherto, will for ever settle the question of the good effects of crossing two distinct plants of the same variety, and of the evil effects of self-fertilisation. A clear light will thus also be thrown on the fact that flowers are invariably constructed so as to permit, or favour, or necessitate the union of two individuals. We shall clearly understand why monœcious and diœcious,—why dimorphic and trimorphic plants exist, and many other such cases. The plan which I have followed in my expe-riments is to grow plants in the same pot, or in pots of the same size, or close together in the open ground; to carefully exclude insects; and then to fertilise some of the flowers with pollen from the same flower, and others on the same plant with pollen from a distinct but adjoining plant. In many, but not all, of these experiments, the crossed plants yielded much more seed than the self-fertilised plants; and I have never seen the

---

[40] ' Beiträge zur Kenntniss der Befruchtung,' 1844, s. 366.

[41] ' Amaryllidaceæ,' p. 371.

[42] ' De la Fécondation,' 2nd edit., 1862, p. 79.

reversed case. The self-fertilised and crossed seeds thus obtained were allowed to germinate in the same glass vessel on damp sand; and as the seeds successively germinated, they were planted in pairs on opposite sides of the same pot, with a superficial partition between them, and were placed so as to be equally exposed to the light. In other cases the self-fertilised and crossed seeds were simply sown on opposite sides of the same small pot. I have, in short, followed different plans, but in every case have taken all the precautions which I could think of, so that the two lots should be equally favoured. Now, I have carefully observed the growth of plants raised from crossed and self-fertilised seed, from their germination to maturity, in species of the following genera, namely, Brassica, Lathyrus, Lupinus, Lobelia, Lactuca, Dianthus, Myosotis, Petunia, Linaria, Calceolaria, Mimulus, and Ipomœa, and the difference in their powers of growth, and of withstanding in certain cases unfavourable conditions, was most manifest and strongly marked. It is of importance that the two lots of seed should be sown or planted on opposite sides of the same pot, so that the seedlings may struggle against each other; for if sown separately in ample and good soil, there is often but little difference in their growth.

I will briefly describe the two most striking cases as yet observed by me. Six crossed and six self-fertilised seeds of *Ipomœa purpurea*, from plants treated in the manner above described, were planted as soon as they had germinated, in pairs on opposite sides of two pots, and rods of equal thickness were given them to twine up. Five of the crossed plants grew from the first more quickly than the opposed self-fertilised plants; the sixth, however, was weakly and was for a time beaten, but at last its sounder constitution prevailed and it shot ahead of its antagonist. As soon as each crossed plant reached the top of its seven-foot rod its fellow was measured, and the result was that, when the crossed plants were seven feet high, the self-fertilised had attained the average height of only five feet four and a half inches. The crossed plants flowered a little before, and more profusely than the self-fertilised plants. On opposite sides of another *small* pot a large number of crossed and self-fertilised seeds were sown, so that they had to struggle for bare existence; a single rod was given to each lot: here again the crossed plants showed from the first their advantage; they never quite reached the summit of the seven-foot rod, but relatively to the self-fertilised plants their average height was as seven feet to five feet two inches. The experiment was repeated in the two following generations with plants raised from the self-fertilised and crossed plants, treated in exactly the same manner, and with nearly the same result. In the second generation, the crossed plants, which were again crossed, produced 121 seed-capsules, whilst the self-fertilised plants, again self-fertilised, produced only 84 capsules.

Some flowers of the *Mimulus luteus* were fertilised with their own pollen, and others were crossed with pollen from distinct plants growing in the same pot. The seeds after germinating were thickly planted on opposite sides of a pot. The seedlings were at first equal in height; but when the young crossed- plants were exactly half an inch, the self-fertilised plants were only a quarter of an inch high. But this inequality did not continue, for, when the crossed plants were four and a half inches high, the self-fertilised were three inches; and they retained the same relative difference till their growth was complete. The crossed plants looked far more vigorous than the uncrossed, and flowered before them; they produced also a far greater number of flowers, which yielded capsules (judging, however, from only a few) containing more seeds. As in the former case, the experiment was repeated in the same manner during the next two generations, and with exactly the same result. Had I not watched these plants of the Mimulus and Ipomœa during their whole growth, I could not have believed it possible, that a difference apparently so slight, as that of the pollen being taken from the same flower, and from a distinct plant growing in the same small pot, could have made so wonderful a difference in the growth and vigour of the plants thus produced. This, under a physiological point of view, is a most remarkable phenomenon.

With respect to the benefit derived from crossing distinct varieties, plenty of evidence has been published. Sageret[43] repeatedly speaks in strong terms of the vigour of melons raised by crossing different varieties, and adds that they are more easily fertilised than common melons, and produce numerous good seed. Here follows the evidence of an English gardener:[44] " I have this summer met with better success in my culti-" vation of melons, in an unprotected state, from the seeds of hybrids " (*i.e.* mongrels) obtained by cross impregnation, than with old varieties. " The offspring of three different hybridisations (one more especially, of " which the parents were the two most dissimilar varieties I could select) " each yielded more ample and finer produce than any one of between " twenty and thirty established varieties."

Andrew Knight[45] believed that his seedlings from crossed varieties of the apple exhibited increased vigour and luxuriance; and M. Chevreul[46] alludes to the extreme vigour of some of the crossed fruit-trees raised by Sageret.

By crossing reciprocally the tallest and shortest peas, Knight[47] says, " I had in this experiment a striking instance of the stimulative effects " of crossing the breeds; for the smallest variety, whose height rarely " exceeded two feet, was increased to six feet; whilst the height of the " large and luxuriant kind was very little diminished." Mr. Laxton gave me seed-peas produced from crosses between four distinct kinds; and the plants thus raised were extraordinarily vigorous, being in each case from one to two or three feet taller than the parent-forms growing close alongside them.

---

[43] 'Mémoire sur les Cucurbitacées,' pp. 36, 28, 30.

[44] Loudon's 'Gard. Mag.,' vol. viii., 1832, p. 52.

[45] 'Transact. Hort. Soc.,' vol. i. p. 25.

[46] 'Annal. des Sc. Nat.,' 3rd series, Bot., tom. vi. p. 189.

[47] 'Philosophical Transactions,' 1799, p. 200.

Wiegmann[48] made many crosses between several varieties of cabbage; and he speaks with astonishment of the vigour and height of the mongrels, which excited the amazement of all the gardeners who beheld them. Mr. Chaundy raised a great number of mongrels by planting together six distinct varieties of cabbage. These mongrels displayed an infinite diversity of character; " But the most remarkable circumstance was, " that, while all the other cabbages and borecoles in the nursery were " destroyed by a severe winter, these hybrids were little injured, and " supplied the kitchen when there was no other cabbage to be had."

Mr. Maund exhibited before the Royal Agricultural Society[49] specimens of crossed wheat, together with their parent varieties; and the editor states that they were intermediate in character, " united with that greater vigour of growth, which it appears, in the vegetable as in the animal world, is the result of a first cross." Knight also crossed several varieties of wheat,[50] and he says " that in the years 1795 and 1796, when almost " the whole crop of corn in the island was blighted, the varieties thus " obtained, and these only, escaped in this neighbourhood, though sown " in several different soils and situations."

Here is a remarkable case: M. Clotzsch[51] crossed *Pinus sylvestris* and *nigricans, Quercus robur* and *pedunculata, Alnus glutinosa* and *incana, Ulmus campestris* and *effusa*; and the cross-fertilised seeds, as well as seeds of the pure parent-trees, were all sown at the same time and in the same place. The result was, that after an interval of eight years, the hybrids were one-third taller than the pure trees!

The facts above given refer to undoubted varieties, excepting the trees crossed by Clotzsch, which are ranked by various botanists as strongly-marked races, sub-species, or species. That true hybrids raised from entirely distinct species, though they lose in fertility, often gain in size and constitutional vigour, is certain. It would be superfluous to quote any facts; for all experimenters, Kölreuter, Gärtner, Herbert, Sageret, Lecoq, and Naudin, have been struck with the wonderful vigour, height, size, tenacity of life, precocity, and hardiness of their hybrid productions. Gärtner[52] sums up his conviction on this head in the strongest terms. Kölreuter[53] gives numerous precise measurements of the weight and height of his hybrids in comparison with measurements of both parent-forms; and speaks with astonishment of their " *statura portentosa*," their " *ambitus vastissimus ac altitudo valde conspicua*." Some exceptions to the rule in the case of very sterile hybrids have, however, been noticed by Gärtner and

---

[48] 'Ueber die Bastarderzeugung,' 1828, s. 32, 33. For Mr. Chaundy's case, *see* Loudon's ' Gard. Mag.,' vol. vii., 1831, p. 696.

[49] ' Gardener's Chron.,' 1846, p. 601.

[50] 'Philosoph. Transact.,' 1799, p. 201.

[51] Quoted in ' Bull. Bot. Soc. France,' vol. ii., 1855, p. 327.

[52] Gärtner, ' Bastarderzeugung,' s. 259, 518, 526 *et seq.*

[53] ' Fortsetzung,' 1763, s. 29 ; 'Dritte Fortsetzung,' s. 44, 96 ; ' Act. Acad. St. Petersburg,' 1782, part ii., p. 251; ' Nova Acta,' 1793, pp. 391, 394 ; ' Nova Acta,' 1795, pp. 316, 323.

Herbert; but the most striking exceptions are given by Max Wichura,[54] who found that hybrid willows were generally tender in constitution, dwarf, and short-lived.

Kölreuter explains the vast increase in the size of the roots, stems, &c., of his hybrids, as the result of a sort of compensation due to their sterility, in the same way as many emasculated animals are larger than the perfect males. This view seems at first sight extremely probable, and has been accepted by various authors;[55] but Gärtner[56] has well remarked that there is much difficulty in fully admitting it; for with many hybrids there is no parallelism between the degree of their sterility and their increased size and vigour. The most striking instances of luxuriant growth have been observed with hybrids which were not sterile in any extreme degree. In the genus Mirabilis, certain hybrids are unusually fertile, and their extraordinary luxuriance of growth, together with their enormous roots,[57] have been transmitted to their progeny. The increased size of the hybrids produced between the fowl and pheasant, and between distinct species of pheasants, has been already noticed. The result in all cases is probably in part due to the saving of nutriment and vital force through the sexual organs not acting, or acting imperfectly, but more especially to the general law of good being derived from a cross. For it deserves especial attention that mongrel animals and plants, which are so far from being sterile that their fertility is often actually augmented, have, as previously shown, their size, hardiness, and constitutional vigour generally increased. It is not a little remarkable that an accession of vigour and size should thus arise under the opposite contingencies of increased and diminished fertility.

It is a perfectly well ascertained fact[58] that hybrids will invariably breed more readily with either pure parent, and not rarely with a distinct species, than with each other. Herbert is inclined to explain even this fact by the advantage derived from a cross; but Gärtner more justly accounts for it by the pollen of the hybrid, and probably its ovules, being in some degree vitiated, whereas the pollen and ovules of both pure parents and of any third species are sound. Nevertheless there are some well-ascertained and remarkable facts, which, as we shall immediately see, show that the act of crossing in itself undoubtedly tends to increase or re-establish the fertility of hybrids.

On certain Hermaphrodite Plants which, either normally or abnormally, require to be fertilised by pollen from a distinct individual or species.

The facts now to be given differ from those hitherto detailed, as the self-sterility does not here result from long-continued,

---

[54] 'Die Bastardbefruchtung,' &c., 1865, s. 31, 41, 42.

[55] Max Wichura fully accepts this view ('Bastardbefruchtung,' s. 43), as does the Rev. M. J. Berkeley, in 'Journal of Hort. Soc.,' Jan. 1866, p. 70.

[56] 'Bastarderzeugung,' s. 394, 526, 528.

[57] Kölreuter, 'Nova Acta,' 1795, p. 316.

[58] Gärtner, 'Bastarderzeugung,' s. 430.

close interbreeding. These facts are, however, connected with our present subject, because a cross with a distinct individual is shown to be either necessary or advantageous. Dimorphic and trimorphic plants, though they are hermaphrodites, must be reciprocally crossed, one set of forms by the other, in order to be fully fertile, and in some cases to be fertile in any degree. But I should not have noticed these plants, had it not been for the following cases given by Dr. Hildebrand:[59]—

Primula sinensis is a reciprocally dimorphic species: Dr. Hildebrand fertilised twenty-eight flowers of both forms, each by pollen of the other form, and obtained the full number of capsules containing on an average 42·7 seed per capsule; here we have complete and normal fertility. He then fertilised forty-two flowers of both forms with pollen of the same form, but taken from a distinct plant, and all produced capsules containing on an average only 19·6 seed. Lastly, and here we come to our more immediate point, he fertilised forty-eight flowers of both forms with pollen of the same form, taken from the same flower, and now he obtained only thirty-two capsules, and these contained on an average 18·6 seed, or one less per capsule than in the former case. So that, with these illegitimate unions, the act of impregnation is less assured, and the fertility slightly less, when the pollen and ovules belong to the same flower, than when belonging to two distinct individuals of the same form. Dr. Hildebrand has recently made analogous experiments on the long-styled form of *Oxalis rosea*, with the same result.[60]

It has recently been discovered that certain plants, whilst growing in their native country under natural conditions, cannot be fertilised with pollen from the same plant. They are sometimes so utterly self-impotent, that, though they can readily be fertilised by the pollen of a distinct species or even distinct genus, yet, wonderful as the fact is, they never produce a single seed by their own pollen. In some cases, moreover, the plant's own pollen and stigma mutually act on each other in a deleterious manner. Most of the facts to be given relate to Orchids, but I will commence with a plant belonging to a widely different family.

Sixty-three flowers of *Corydalis cava*, borne on distinct plants, were fertilised by Dr. Hildebrand[61] with pollen from other plants of the same species; and fifty-eight capsules were obtained, including on an average

[59] 'Botanische Zeitung,' Jan. 1864, s. 3.
[60] 'Monatsbericht Akad. Wissen,' Berlin, 1866, s. 372.
[61] International Hort. Congress, London, 1866.

4·5 seed in each. He then fertilised sixteen flowers produced by the same raceme, one with another, but obtained only three capsules, one of which alone contained any good seeds, namely, two in number. Lastly, he fertilised twenty-seven flowers, each with its own pollen; he left also fifty-seven flowers to be spontaneously fertilised, and this would certainly have ensued if it had been possible, for the anthers not only touch the stigma, but the pollen-tubes were seen by Dr. Hildebrand to penetrate it; nevertheless these eighty-four flowers did not produce a single seed-capsule! This whole case is highly instructive, as it shows how widely different the action of the same pollen is, according as it is placed on the stigma of the same flower, or on that of another flower on the same raceme, or on that of a distinct plant.

With exotic Orchids several analogous cases have been observed, chiefly by Mr. John Scott.[62] *Oncidium sphacelatum* has effective pollen, for with it Mr. Scott fertilised two distinct species; its ovules are likewise capable of impregnation, for they were readily fertilised by the pollen of *O. divaricatum*; nevertheless, between one and two hundred flowers fertilised by their own pollen did not produce a single capsule, though the stigmas were penetrated by the pollen-tubes. Mr. Robinson Munro, of the Royal Botanic Gardens of Edinburgh, also informs me (1864) that a hundred and twenty flowers of this same species were fertilised by him with their own pollen, and did not produce a capsule, but eight flowers fertilised by the pollen of *O. divaricatum* produced four fine capsules: again, between two and three hundred flowers of *O. divaricatum*, fertilised by their own pollen, did not set a capsule, but twelve flowers fertilised by *O. flexuosum* produced eight fine capsules: so that here we have three utterly self-impotent species, with their male and female organs perfect, as shown by their mutual fertilisation. In these cases fertilisation was effected only by the aid of a distinct species. But, as we shall presently see, distinct plants, raised from seed, of *Oncidium flexuosum*, and probably of the other species, would have been perfectly capable of fertilising each other, for this is the natural process. Again, Mr. Scott found that the pollen of a plant of *O. microchilum* was good, for with it he fertilised two distinct species; he found its ovules good, for they could be fertilised by the pollen of one of these species, and by the pollen of a distinct plant of *O. microchilum*; but they could not be fertilised by pollen of the same plant, though the pollen-tubes penetrated the stigma. An analogous case has been recorded by M. Rivière,[63] with two plants of *O. Cavendishianum*, which were both self-sterile, but reciprocally fertilised each other. All these cases refer to the genus Oncidium, but Mr. Scott found that *Maxillaria atro-rubens* was "totally insusceptible of fertilisation with its own pollen," but fertilised, and was fertilised by, a widely distinct species, viz. *M. squalens*.

As these orchids had grown under unnatural conditions, in hot-

[62] 'Proc. Bot. Soc. of Edinburgh,' May, 1863 : these observations are given in abstract, and others are added, in the 'Journal of Proc. of Linn. Soc.,' vol. viii. Bot., 1864, p. 162.

[63] Prof. Lecoq, 'De la Fécondation,' 2nd edit., 1862, p. 76.

houses, I concluded without hesitation that their self-sterility was due to this cause. But Fritz Müller informs me that at Desterro, in Brazil, he fertilised above one hundred flowers of the above-mentioned *Oncidium flexuosum*, which is there endemic, with its own pollen, and with that taken from distinct plants; all the former were sterile, whilst those fertilised by pollen from any *other plant* of the same species were fertile. During the first three days there was no difference in the action of the two kinds of pollen: that placed on the stigma of the same plant separated in the usual manner into grains, and emitted tubes which penetrated the column, and the stigmatic chamber shut itself; but the flowers alone which had been fertilised by pollen taken from a distinct plant produced seed-capsules. On a subsequent occasion these experiments were repeated on a large scale with the same result. Fritz Müller found that four other endemic species of Oncidium were in like manner utterly sterile with their own pollen, but fertile with that from any other plant: some of them likewise produced seed-capsules when impregnated with pollen of widely distinct genera, such as Leptotes, Cyrtopodium, and Rodriguezia! *Oncidium crispum*, however, differs from the foregoing species in varying much in its self-sterility; some plants producing fine pods with their own pollen, others failing to do so; in two or three instances, Fritz Müller observed that the pods produced by pollen taken from a distinct flower on the same plant, were larger than those produced by the flower's own pollen. In *Epidendrum cinnabarinum*, an orchid belonging to another division of the family, fine pods were produced by the plant's own pollen, but they contained by weight only about half as much seed as the capsules which had been fertilized by pollen from a distinct plant, and in one instance from a distinct species; moreover, a very large proportion, and in some cases nearly all the seed produced by the plant's own pollen, was embryonless and worthless. Some self-fertilized capsules of a Maxillaria were in a similar state.

Another observation made by Fritz Müller is highly remarkable, namely, that with various orchids the plant's own pollen not only fails to impregnate the flower, but acts on the stigma, and is acted on, in an injurious or poisonous manner. This is shown by the surface of the stigma in contact with the pollen, and by the pollen itself, becoming in from three to five days dark brown, and then decaying. The discolouration and decay are not caused by parasitic cryptogams, which were observed by Fritz Müller in only a single instance. These changes are well shown by placing on the same stigma, at the same time, the plant's own pollen and that from a distinct plant of the same species, or of another species, or even of another and widely remote genus. Thus, on the stigma of *Oncidium flexuosum*, the plant's own pollen and that from a distinct plant were placed side by side, and in five days' time the latter was perfectly fresh, whilst the plant's own pollen was brown. On the other hand, when the pollen of a distinct plant of the *Oncidium flexuosum*, and of the *Epidendrum zebra* (*nov. spec.?*), were placed together on the same stigma, they behaved in exactly the same manner, the grains separating, emitting tubes, and penetrating the stigma, so that the two pollen-masses, after an interval of eleven days, could not be distinguished except by the difference of their caudicles, which, of course, undergo no change. Fritz Müller has, moreover, made a large number of crosses between orchids belonging to distinct species and genera, and he finds that in all cases when the flowers are not fertilised their footstalks first begin to wither; and the withering slowly spreads upwards until the germens fall off, after an interval of one or two weeks, and in one instance of between six and seven weeks; but even in this latter case, and in most other cases, the pollen and stigma remained in appearance fresh. Occasionally, however, the pollen becomes brownish, generally on the external surface, and not in contact with the stigma, as is invariably the case when the plant's own pollen is applied.

Fritz Müller observed the poisonous action of the plant's own pollen in the above-mentioned *Oncidium flexuosum*, *O. unicorne*, *pubes* (?), and in two other unnamed species. Also in two species of Rodriguezia, in two of Notylia, in one of Burlingtonia, and of a fourth genus in the same group. In all these cases, except the last, it was proved that the flowers were, as might have been expected, fertile with pollen from a distinct plant of the same species. Numerous flowers of one species of Notylia were fertilized with pollen from the same raceme; in two days' time they all withered, the germens began to shrink, the pollen-masses became dark brown, and not one pollen-grain emitted a tube. So that in this orchid the injurious action of the plant's own pollen is more rapid than with *Oncidium flexuosum*. Eight other flowers on the same raceme were fertilized with pollen from a distinct plant of the same species : two of these were dissected, and their stigmas were found to be penetrated by numberless pollen-tubes; and the germens of the other six flowers became well developed. On a subsequent occasion many other flowers were fertilized with their own pollen, and all fell off dead in a few days; whilst some flowers on the same raceme which had been left simply unfertilised adhered and long remained fresh. We have seen that in cross-unions between extremely distinct orchids the pollen long remains undecayed; but Notylia behaved in this respect differently; for when its pollen was placed on the stigma of *Oncidium flexuosum*, both the stigma and pollen quickly became dark brown, in the same manner as if the plant's own pollen had been applied.

Fritz Müller suggests that, as in all these cases the plant's own pollen is not only impotent (thus effectually preventing self-fertilization), but likewise prevents, as was ascertained in the case of the Notylia and *Oncidium flexuosum*, the action of subsequently applied pollen from a distinct individual, it would be an advantage to the plant to have its own pollen rendered more and more deleterious; for the germens would thus quickly be killed, and, dropping off, there would be no further waste in nourishing a part which ultimately could be of no avail. Fritz Müller's discovery that a plant's own pollen and stigma in some cases act on each other as if mutually poisonous, is certainly most remarkable.

We now come to cases closely analogous with those just

given, but different, inasmuch as individual plants alone of the species are self-impotent. This self-impotence does not depend on the pollen or ovules being in a state unfit for fertilisation, for both have been found effective in union with other plants of the same or of a distinct species. The fact of these plants having spontaneously acquired so peculiar a constitution, that they can be fertilised more readily by the pollen of a distinct species than by their own, is remarkable. These abnormal cases, as well as the foregoing normal cases, in which certain orchids, for instance, can be much more easily fertilised by the pollen of a distinct species than by their own, are exactly the reverse of what occurs with all ordinary species. For in these latter the two sexual elements of the same individual plant are capable of freely acting on each other; but are so constituted that they are more or less impotent when brought into union with the sexual elements of a distinct species, and produce more or less sterile hybrids. It would appear that the pollen or ovules, or both, of the individual plants which are in this abnormal state, have been affected in some strange manner by the conditions to which they themselves or their parents have been exposed; but whilst thus rendered self-sterile, they have retained the capacity common to most species of partially fertilizing and being partially fertilized by allied forms. However this may be, the subject, to a certain extent, is related to our general conclusion that good is derived from the act of crossing.

Gärtner experimented on two plants of *Lobelia fulgens*, brought from separate places, and found[64] that their pollen was good, for he fertilised with it *L. cardinalis* and *syphilitica;* their ovules were likewise good, for they were fertilised by the pollen of these same two species; but these two plants of *L. fulgens* could not be fertilised by their own pollen, as can generally be effected with perfect ease with this species. Again, the pollen of a plant of *Verbascum nigrum* grown in a pot was found by Gärtner[65] capable of fertilising *V. lychnitis* and *V. Austriacum;* the ovules could be fertilised by the pollen of *V. thapsus;* but the flowers could not be fertilised by their own pollen. Kölreuter, also,[66] gives the case of three

---

[64] 'Bastarderzeugung,' s. 64, 357.   [65] Idem, s. 357.
[66] 'Zweite Fortsetzung,' s. 10; 'Dritte Fort.,' s. 40.

garden plants of *Verbascum phœniceum*, which bore during two years many flowers ; these he successfully fertilised by the pollen of no less than four distinct species, but they produced not a seed with their own apparently good pollen; subsequently these same plants, and others raised from seed, assumed a strangely fluctuating condition, being temporarily sterile on the male or female side, or on both sides, and sometimes fertile on both sides; but two of the plants were perfectly fertile throughout the summer.

It appears[67] that certain flowers on certain plants of *Lilium candidum* can be fertilised more easily by pollen from a distinct individual than by their own. So, again, with the varieties of the potato. Tinzmann,[68] who made many trials with this plant, says that pollen from another variety sometimes " exerts a powerful influence, and I have found sorts of potatoes " which would not bear seed from impregnation with the pollen of their " own flowers, would bear it when impregnated with other pollen." It does not, however, appear to have been proved that the pollen which failed to act on the flower's own stigma was in itself good.

In the genus Passiflora it has long been known that several species do not produce fruit, unless fertilised by pollen taken from distinct species : thus, Mr. Mowbray[69] found that he could not get fruit from *P. alata* and *racemosa* except by reciprocally fertilising them with each other's pollen. Similar facts have been observed in Germany and France;[70] and I have received two authentic accounts of *P. quadrangularis*, which never produced fruit with its own pollen, but would do so freely when fertilised in one case with the pollen of *P. cœrulea*, and in another case with that of *P. edulis*. So again, with respect to *P. laurifolia*, a cultivator of much experience has recently remarked[71] that the flowers "must be fertilised with the pollen of *P. cœrulea*, or of some other common kind, as their own pollen will not fertilise them." But the fullest details on this subject have been given by Mr. Scott :[72] plants of *Passiflora racemosa, cœrulea*, and *alata* flowered profusely during many years in the Botanic Gardens of Edinburgh, and, though repeatedly fertilised by Mr. Scott and by others with their own pollen, never produced any seed ; yet this occurred at once with all three species when they were crossed together in various ways. But in the case of *P. cœrulea*, three plants, two of which grew in the Botanic Gardens, were all rendered fertile, merely by impregnating the one with pollen of the other. The same result was attained in the same manner with *P. alata*, but only with one plant out of three. As so many self-sterile species have been mentioned, it may be stated that in the case of *P. gracilis*, which is an annual, the flowers are nearly as fertile with their own pollen as with that from a distinct plant ; thus sixteen flowers sponta-

[67] Duvernoy, quoted by Gärtner, ' Bastarderzeugung,' s. 334.

[68] ' Gardener's Chronicle,' 1846, p. 183.

[69] 'Transact. Hort. Soc.,' vol. vii., 1830, p. 95.

[70] Prof, Lecoq, 'De la Fécondation,' 1845, p. 70 ; Gärtner, ' Bastarderzeugung,' s. 64.

[71] 'Gardener's Chron.,' 1866, p. 1068.

[72] 'Journal of Proc. of Linn. Soc.,' vol. viii., 1864, p. 168.

neously self-fertilised produced fruit, each containing on an average 21·3 seed, whilst fruit from fourteen crossed flowers contained 24·1 seed.

Returning to *P. alata*, I have received (1866) some interesting details from Mr. Robinson Munro. Three plants, including one in England, have already been mentioned which were inveterately self-sterile, and Mr. Munro informs me of several others which, after repeated trials during many years, have been found in the same predicament. At some other places, however, this species fruits readily when fertilised with its own pollen. At Taymouth Castle there is a plant which was formerly grafted by Mr. Donaldson on a distinct species, name unknown, and ever since the operation it has produced fruit in abundance by its own pollen; so that this small and unnatural change in the state of this plant has restored its self-fertility! Some of the seedlings from the Taymouth Castle plant were found to be not only sterile with their own pollen, but with each other's pollen, and with the pollen of distinct species. Pollen from the Taymouth plant failed to fertilise certain plants of the same species, but was successful on one plant in the Edinburgh Botanic Gardens. Seedlings were raised from this latter union, and some of their flowers were fertilised by Mr. Munro with their own pollen; but they were found to be as self-impotent as the mother-plant had always proved, except when fertilised by the grafted Taymouth plant, and except, as we shall see, when fertilised by her own seedlings. For Mr. Munro fertilised eighteen flowers on the self-impotent mother-plant with pollen from these her own self-impotent seedlings, and obtained, remarkable as the fact is, eighteen fine capsules full of excellent seed! I have met with no case in regard to plants which shows so well as this of *P. alata*, on what small and mysterious causes complete fertility or complete sterility depends.

The facts hitherto given relate to the much-lessened or completely destroyed fertility of pure species when impregnated with their own pollen, in comparison with their fertility when impregnated by distinct individuals or distinct species; but closely analogous facts have been observed with hybrids.

Herbert states[73] that having in flower at the same time nine hybrid Hippeastrums, of complicated origin, descended from several species, he found that "almost every flower touched with pollen from another cross produced " seed abundantly, and those which were touched with their own pollen " either failed entirely, or formed slowly a pod of inferior size, with fewer " seeds." In the 'Horticultural Journal' he adds that, " the admission of " the pollen of another cross-bred Hippeastrum (however complicated the " cross) to any *one* flower of the number, is almost sure to check the fruc- " tification of the others." In a letter written to me in 1839, Dr. Herbert says that he had already tried these experiments during five consecutive years, and he subsequently repeated them, with the same invariable result.

[73] ' Amaryllidaceæ,' 1837, p. 371 ; 'Journal of Hort. Soc.,' vol. ii., 1847, p. 19.

He was thus led to make an analogous trial on a pure species, namely, on the *Hippeastrum aulicum*, which he had lately imported from Brazil : this bulb produced four flowers, three of which were fertilised by their own pollen, and the fourth by the pollen of a triple cross between *H. bulbosum, reginæ*, and *vittatum ;* the result was, that " the ovaries of the three first flowers " soon ceased to grow, and after a few days perished entirely : whereas the " pod impregnated by the hybrid made vigorous and rapid progress to " maturity, and bore good seed, which vegetated freely." This is, indeed, as Herbert remarks, "a strange truth," but not so strange as it then appeared.

As a confirmation of these statements, I may add that Mr. M. Mayes,[74] after much experience in crossing the species of Amaryllis (Hippeastrum), says, "neither the species nor the hybrids will, we are well aware, produce seed so abundantly from their own pollen as from that of others." So, again, Mr. Bidwell, in New South Wales,[75] asserts that *Amaryllis belladonna* bears many more seeds when fertilised by the pollen of *Brunswigia (Amaryllis* of some authors) *Josephinæ* or of *B. multiflora*, than when fertilised by its own pollen. Mr. Beaton dusted four flowers of a Cyrtanthus with their own pollen, and four with the pollen of *Vallota (Amaryllis) purpurea ;* on the seventh day "those which received their own pollen slackened " their growth, and ultimately perished; those which were crossed with " the Vallota held on."[76] These latter cases, however, relate to uncrossed species, like those before given with respect to Passiflora, Orchids, &c., and are here referred to only because the plants belong to the same group of Amaryllidaceæ.

In the experiments on the hybrid Hippeastrums, if Herbert had found that the pollen of two or three kinds alone had been more efficient on certain kinds than their own pollen, it might have been argued that these, from their mixed parentage, had a closer mutual affinity than the others; but this explanation is inadmissible, for the trials were made reciprocally backwards and forwards on nine different hybrids; and a cross, whichever way taken, always proved highly beneficial. I can add a striking and analogous case from experiments made by the Rev. A. Rawson, of Bromley Common, with some complex hybrids of Gladiolus. This skilful horticulturist possessed a number of French varieties, differing from each other only in the colour and size of the flowers, all descended from Gandavensis, a well-known old hybrid, said to be descended from *G. Natalensis* by the pollen of *G. oppositiflorus.*[77] Mr. Rawson, after repeated trials, found that none of the varieties would set seed with their own pollen, although

---

[74] Loudon's ' Gardener's Magazine,' vol. xi., 1835, p. 260.

[75] ' Gardener's Chronicle,' 1850, p. 470.

[76] ' Journal Hort. Soc.,' vol. v. p. 135. The seedlings thus raised were given to the Hort. Soc. ; but I find, on inquiry, that they unfortunately died the following winter.

[77] Mr. D. Beaton, in ' Journal of Hort.,' 1861, p. 453. Lecoq, however (' De la Fécond.,' 1862, p. 369), states that this hybrid is descended from *G. psittacinus* and *cardinalis ;* but this is opposed to Herbert's experience, who found that the former species could not be crossed.

taken from distinct plants of the same variety, which had, of course, been propagated by bulbs, but that they all seeded freely with pollen from any other variety. To give two examples : Ophir did not produce a capsule with its own pollen, but when fertilised with that of Janire, Brenchleyensis, Vulcain, and Linné, it produced ten fine capsules ; but the pollen of Ophir was good, for when Linné was fertilised by it seven capsules were produced. This latter variety, on the other hand, was utterly barren with its own pollen, which we have seen was perfectly efficient on Ophir. Altogether, Mr. Rawson, in the year 1861, fertilised twenty-six flowers borne by four varieties with pollen taken from other varieties, and every single flower produced a fine seed-capsule ; whereas fifty-two flowers on the same plants, fertilised at the same time with their own pollen, did not yield a single seed-capsule. Mr. Rawson fertilised, in some cases, the alternate flowers, and in other cases all those down one side of the spike, with pollen of other varieties, and the remaining flowers with their own pollen ; I saw these plants when the capsules were nearly mature, and their curious arrangement at once brought full conviction to the mind that an immense advantage had been derived from crossing these hybrids.

Lastly, I have heard from Dr. E. Bornet, of Antibes, who has made numerous experiments in crossing the species of Cistus, but has not yet published the results, that, when any of these hybrids are fertile, they may be said to be, in regard to function, diœcious ; "for the flowers " are always sterile when the pistil is fertilised by pollen taken from the " same flower or from flowers on the same plant. But they are often fertile " if pollen be employed from a distinct individual of the same hybrid " nature, or from a hybrid made by a reciprocal cross."

Conclusion.—The facts just given, which show that certain plants are self-sterile, although both sexual elements are in a fit state for reproduction when united with distinct individuals of the same or other species, appear at first sight opposed to all analogy. The sexual elements of the same flower have become, as already remarked, differentiated in relation to each other, almost like those of two distinct species.

With respect to the species which, whilst living under their natural conditions, have their reproductive organs in this peculiar state, we may conclude that it has been naturally acquired for the sake of effectually preventing self-fertilisation. The case is closely analogous with dimorphic and trimorphic plants, which can be fully fertilised only by plants belonging to the opposite form, and not, as in the foregoing cases, indifferently by any other plant. Some of these dimorphic plants are completely sterile with pollen taken from the same plant or from the same form. It is interesting to observe the graduated series from plants which, when fertilised by their own pollen, yield the full number of seed, but with the seedlings a little dwarfed in stature—to plants which when self-fertilised yield few seeds—to those which yield none—and, lastly, to those in which the plant's own pollen and stigma act on each other like poison. This peculiar state of the reproductive organs, when occurring in certain individuals alone, is evidently abnormal; and as it chiefly affects exotic plants, or indigenous plants cultivated in pots, we may attribute it to some change in the conditions of life, acting on the plants themselves or on their parents. The self-impotent *Passiflora alata*, which recovered its self-fertility after having been grafted on a distinct stock, shows how small a change is sufficient to act powerfully on the reproductive system. The possibility of a plant becoming under culture self-impotent is interesting as throwing light on the occurrence of this same condition in natural species. A cultivated plant in this state generally remains so during its whole life; and from this fact we may infer that the state is probably congenital.

Kölreuter, however, has described some plants of Verbascum which varied in this respect even during the same season. As in all the normal cases, and in many, probably in most, of the abnormal cases, any two self-impotent plants can reciprocally fertilize each other, we may infer that a very slight difference in the nature of their sexual elements suffices to give fertility; but in other instances, as with some Passifloras and the hybrid Gladioli, a greater degree of differentiation appears to be necessary, for with these plants fertility is gained only by the union of distinct species, or of hybrids of distinct parentage. These facts all point to the same general conclusion, namely, that good is derived from a cross between individuals, which either innately, or from exposure to dissimilar conditions, have come to differ in sexual constitution.

Exotic animals confined in menageries are sometimes in nearly the same state as the above-described self-impotent plants; for, as we shall see in the following chapter, certain monkeys, the larger carnivora, several finches, geese, and pheasants, cross together, quite as freely as, or even more freely than, the individuals of the same species breed together. Cases will, also, be given of sexual incompatibility between certain male and female domesticated animals, which, nevertheless, are fertile when matched with any other individual of the same kind.

In the early part of this chapter it was shown that the crossing of distinct forms, whether closely or distantly allied, gives increased size and constitutional vigour, and, except in the case of crossed species, increased fertility, to the offspring. The evidence rests on the universal testimony of breeders (for it should be observed that I am not here speaking of the evil results of close interbreeding), and is practically exemplified in the higher value of cross-bred animals for immediate consumption. The good results of crossing have also been demonstrated, in the case of some animals and of numerous plants, by actual weight and measurement. Although animals of pure blood will obviously be deteriorated by crossing, as far as their characteristic qualities are concerned, there seems to be no exception to the rule that advantages of the kind just mentioned are thus gained, even when there has not been any previous close interbreeding. The rule applies to all animals, even to cattle and sheep, which can long resist breeding in-and-in between the nearest blood-relations. It applies to individuals of the same sub-variety but of distinct families, to varieties or races, to sub-species, as well as to quite distinct species.

In this latter case, however, whilst size, vigour, precocity, and hardiness are, with rare exceptions, gained, fertility, in a greater or less degree, is lost; but the gain cannot be exclusively attributed to the principle of compensation; for there is no close parallelism between the increased size and vigour of the offspring and their sterility. Moreover it has been clearly proved that mongrels which are perfectly fertile gain these same advantages as well as sterile hybrids.

The evil consequences of long-continued close interbreeding are not so easily recognised as the good effects from crossing, for the deterioration is gradual. Nevertheless it is the general opinion of those who have had most experience, especially with animals which propagate quickly, that evil does inevitably follow sooner or later, but at different rates with different animals. No doubt a false belief may widely prevail like a superstition; yet it is difficult to suppose that so many acute and original observers have all been deceived at the expense of much cost and trouble. A male animal may sometimes be paired with his daughter, granddaughter, and so on, even for seven generations, without any manifest bad result; but the experiment has never been tried of matching brothers and sisters, which is considered the closest form of interbreeding, for an equal number of generations. There is good reason to believe that by keeping the members of the same family in distinct bodies, especially if exposed to somewhat different conditions of life, and by occasionally crossing these families, the evil results may be much diminished, or quite eliminated. These results are loss of constitutional vigour, size, and fertility; but there is no necessary deterioration in the general form of the body, or in other good qualities. We have seen that with pigs first-rate animals have been produced after long-continued close interbreeding, though they had become extremely infertile when paired with their near relations. The loss of fertility, when it occurs, seems never to be absolute, but only relative to animals of the same blood; so that this sterility is to a certain extent analogous with that of self-impotent plants which cannot be fertilised by their own pollen, but are perfectly fertile with pollen of any other plant of the same species. The fact of infertility of this peculiar nature being one of the results of long-continued interbreeding, shows that interbreeding does not act merely by combining and augmenting various morbid tendencies common to both parents; for animals with such tendencies, if not at the time actually ill, can generally propagate their kind. Although offspring descended from the nearest blood-relations are not necessarily deteriorated in structure, yet some authors[78] believe that they are eminently liable to malformations; and this is not improbable, as everything which lessens the vital powers acts in this manner. Instances of this kind have been recorded in the case of pigs, bloodhounds, and some other animals.

Finally, when we consider the various facts now given which plainly show that good follows from crossing, and less plainly

[78] This is the conclusion of Prof. Devay, 'Du Danger des Mariages Consang.,' 1862, p. 97. Virchow quotes, in the 'Deutsche Jahrbücher,' 1863, s. 354, some curious evidence on half the cases of a peculiar form of blindness occurring in the offspring from near relations.

that evil follows from close interbreeding, and when we bear in mind that throughout the whole organic world elaborate provision has been made for the occasional union of distinct individuals, the existence of a great law of nature is, if not proved, at least rendered in the highest degree probable; namely, that the crossing of animals and plants which are not closely related to each other is highly beneficial or even necessary, and that interbreeding prolonged during many generations is highly injurious.

CHAPTER XVIII.

ON THE ADVANTAGES AND DISADVANTAGES OF CHANGED
CONDITIONS OF LIFE : STERILITY FROM VARIOUS CAUSES.

ON THE GOOD DERIVED FROM SLIGHT CHANGES IN THE CONDITIONS OF LIFE —
STERILITY FROM CHANGED CONDITIONS, IN ANIMALS, IN THEIR NATIVE COUNTRY
AND IN MENAGERIES — MAMMALS, BIRDS, AND INSECTS — LOSS OF SECONDARY
SEXUAL CHARACTERS AND OF INSTINCTS — CAUSES OF STERILITY — STERILITY
OF DOMESTICATED ANIMALS FROM CHANGED CONDITIONS — SEXUAL INCOMPATI-
BILITY OF INDIVIDUAL ANIMALS — STERILITY OF PLANTS FROM CHANGED CONDI-
TIONS OF LIFE — CONTABESCENCE OF THE ANTHERS — MONSTROSITIES AS A CAUSE
OF STERILITY — DOUBLE FLOWERS — SEEDLESS FRUIT — STERILITY FROM THE
EXCESSIVE DEVELOPMENT OF THE ORGANS OF VEGETATION — FROM LONG-CONTINUED
PROPAGATION BY BUDS — INCIPIENT STERILITY THE PRIMARY CAUSE OF DOUBLE
FLOWERS AND SEEDLESS FRUIT.

*On the Good derived from slight Changes in the Conditions of
Life.*—IN considering whether any facts were known which
might throw light on the conclusion arrived at in the last
chapter, namely, that benefits ensue from crossing, and that it is
a law of nature that all organic beings should occasionally cross,
it appeared to me probable that the good derived from slight
changes in the conditions of life, from being an analogous phenomenon, might serve this purpose. No two individuals, and
still less no two varieties, are absolutely alike in constitution
and structure; and when the germ of one is fertilised by the
male element of another, we may believe that it is acted on in
a somewhat similar manner as an individual when exposed to
slightly changed conditions. Now, every one must have observed the remarkable influence on convalescents of a change of
residence, and no medical man doubts the truth of this fact.
Small farmers who hold but little land are convinced that their
cattle derive great benefit from a change of pasture. In the
case of plants, the evidence is strong that a great advantage is
derived from exchanging seeds, tubers, bulbs, and cuttings from
one soil or place to another as different as possible.

The belief that plants are thus benefited, whether or not well founded, has been firmly maintained from the time of Columella, who wrote shortly after the Christian era, to the present day; and it now prevails in England, France, and Germany.[1] A sagacious observer, Bradley, writing in 1724,[2] says, "When we once become Masters of a good Sort of Seed, we should at " least put it into Two or Three Hands, where the Soils and Situations are " as different as possible; and every Year the Parties should change with " one another; by which Means, I find the Goodness of the Seed will be " maintained for several Years. For Want of this Use many Farmers " have failed in their Crops and been great Losers." He then gives his own practical experience on this head. A modern writer[3] asserts, "Nothing " can be more clearly established in agriculture than that the continual " growth of any one variety in the same district makes it liable to dete- " rioration either in quality or quantity." Another writer states that he sowed close together in the same field two lots of wheat-seed, the product of the same original stock, one of which had been grown on the same land, and the other at a distance, and the difference in favour of the crop from the latter seed was remarkable. A gentleman in Surrey who has long made it his business to raise wheat to sell for seed, and who has constantly realised in the market higher prices than others, assures me that he finds it indispensable continually to change his seed; and that for this purpose he keeps two farms differing much in soil and elevation.

With respect to the tubers of the potato, I find that at the present day the practice of exchanging sets is almost everywhere followed. The great growers of potatoes in Lancashire formerly used to get tubers from Scotland, but they found that " a change from the moss-lands, and vice versâ, was generally sufficient." In former times in France the crop of potatoes in the Vosges had become reduced in the course of fifty or sixty years in the proportion from 120-150 to 30-40 bushels; and the famous Oberlin attributed the surprising good which he effected in large part to changing the sets.[4]

A well-known practical gardener, Mr. Robson[5] positively states that he has himself witnessed decided advantage from obtaining bulbs of the onion, tubers of the potato, and various seeds, all of the same kind, from different soils and distant parts of England. He further states that with

---

[1] For England, see below. For Germany, see Metzger, 'Getreidearten,' 1841, s. 63. For France, Loiseleur-Deslongchamps (' Consid. sur les Céreales,' 1843, p. 200) gives numerous references on this subject. For Southern France, see Godron, 'Florula Juvenalis,' 1854, p. 28.

[2] 'A general Treatise of Husbandry,' vol. iii. p. 58.

[3] 'Gardener's Chronicle and Agricult. Gazette,' 1858, p. 247; and for the second statement, idem, 1850, p. 702. On this same subject, see also Rev. D.

Walker's 'Prize Essay of Highland Agricult. Soc.,' vol. ii. p. 200. Also Marshall's ' Minutes of Agriculture,' November, 1775.

[4] Oberlin's 'Memoirs,' Eng. translat., p. 73. For Lancashire, see Marshall's 'Review of Reports,' 1808, p. 295.

[5] 'Cottage Gardener,' 1856, p. 186. For Mr. Robson's subsequent statements, see ' Journal of Horticulture,' Feb. 18, 1866, p. 121. For Mr. Abbey's remarks on grafting, &c., idem, July 18, 1865, p. 44.

plants propagated by cuttings, as with the Pelargonium, and especially the Dahlia, manifest advantage is derived from getting plants of the same variety, which have been cultivated in another place; or, " where the " extent of the place allows, to take cuttings from one description of soil " to plant on another, so as to afford the change that seems so necessary " to the well-being of the plants." He maintains that after a time an exchange of this nature is " forced on the grower, whether he be pre- " pared for it or not." Similar remarks have been made by another excellent gardener, Mr. Fish, namely, that cuttings of the same variety of Cal- ceolaria, which he obtained from a neighbour, " showed much greater " vigour than some of his own that were treated in exactly the same " manner," and he attributed this solely to his own plants having become " to a certain extent worn out or tired of their quarters." Something of this kind apparently occurs in grafting and budding fruit-trees; for, according to Mr. Abbey, grafts or buds generally take on a distinct variety or even species, or on a stock previously grafted, with greater facility than on stocks raised from seeds of the variety which is to be grafted; and he believes this cannot be altogether explained by the stocks in question being better adapted to the soil and climate of the place. It should, however, be added, that varieties grafted or budded on very distinct kinds, though they may take more readily and grow at first more vigorously than when grafted on closely allied stocks, afterwards often become unhealthy.

I have studied M. Tessier's careful and elaborate experiments,[6] made to disprove the common belief that good is derived from a change of seed ; and he certainly shows that the same seed may with care be cultivated on the same farm (it is not stated whether on exactly the same soil) for ten consecutive years without loss. Another excellent observer, Colonel Le Couteur,[7] has come to the same conclusion ; but then he expressly adds, if the same seed be used, " that which is grown on land manured from the " mixen one year becomes seed for land prepared with lime, and that " again becomes seed for land dressed with ashes, then for land dressed " with mixed manure, and so on." But this in effect is a systematic exchange of seed, within the limits of the same farm.

On the whole the belief, which has long been held by many skilful cultivators, that good follows from exchanging seed, tubers, &c., seems to be fairly well founded. Considering the small size of most seeds, it seems hardly credible that the ad- vantage thus derived can be due to the seeds obtaining in one soil some chemical element deficient in the other soil. As plants after once germinating naturally become fixed to the same spot, it might have been anticipated that they would show the good effects of a change more plainly than animals, which continually wander about; and this apparently is the

6 ' Mém. de l'Acad. des Sciences,' 1790, p. 209.
7 ' On the Varieties of Wheat,' p. 52.

case. Life depending on, or consisting in, an incessant play of
the most complex forces, it would appear that their action is in
some way stimulated by slight changes in the circumstances to
which each organism is exposed.   All forces throughout nature,
as Mr. Herbert Spencer [8] remarks, tend towards an equilibrium, and for the life of each being it is necessary that this
tendency should be checked.   If these views and the foregoing
facts can be trusted, they probably throw light, on the one hand,
on the good effects of crossing the breed, for the germ will be
thus slightly modified or acted on by new forces; and on the
other hand, on the evil effects of close interbreeding prolonged
during many generations, during which the germ will be acted
on by a male having almost identically the same constitution.

Sterility from changed Conditions of Life.

I will now attempt to show that animals and plants, when removed from their natural conditions, are often rendered in some
degree infertile or completely barren ; and this occurs even when
the conditions have not been greatly changed.   This conclusion
is not necessarily opposed to that at which we have just arrived,
namely, that lesser changes of other kinds are advantageous to
organic beings.   Our present subject is of some importance, from
having an intimate connexion with the causes of variability.
Indirectly it perhaps bears on the sterility of species when
crossed : for as, on the one hand, slight changes in the conditions
of life are favourable to plants and animals, and the crossing of
varieties adds to the size, vigour, and fertility of their offspring;
so, on the other hand, certain other changes in the conditions
of life cause sterility ; and as this likewise ensues from
crossing much-modified forms or species, we have a parallel
and double series of facts, which apparently stand in close relation to each other.

It is notorious that many animals, though perfectly tamed,

[8] Mr. Spencer has fully and ably discussed this whole subject in his ' Principles of Biology,' 1864, vol. ii. ch. x.
In the first edition of my ' Origin of
Species,' 1859, p. 267, I spoke of the
good effects from slight changes in
the conditions of life and from cross-
breeding, and of the evil effects from
great changes in the conditions and
from crossing widely distinct forms, as
a series of facts " connected together by
some common but unknown bond, which
is essentially related to the principle of
life."

refuse to breed in captivity.   Isidore Geoffroy St. Hilaire[9]
consequently has drawn a broad distinction between tamed
animals which will not breed under captivity, and truly domesticated animals which breed freely — generally more freely,
as shown in the sixteenth chapter, than in a state of nature.
It is possible and generally easy to tame most animals; but
experience has shown that it is difficult to get them to breed
regularly, or even at all.   I shall discuss this subject in detail;
but will give only those cases which seem most illustrative.
My materials are derived from notices scattered through various
works, and especially from a Report, drawn up for me by the
kindness of the officers of the Zoological Society of London,
which has especial value, as it records all the cases, during nine
years from 1838-46, in which the animals were seen to couple
but produced no offspring, as well as the cases in which they never,
as far as known, coupled.   This MS. Report I have corrected
by the annual Reports subsequently published.   Many facts are
given on the breeding of the animals in that magnificent work,
' Gleanings from the Menageries of Knowsley Hall,' by Dr. Gray.
I made, also, particular inquiries from the experienced keeper
of the birds in the old Surrey Zoological Gardens.   I should
premise that a slight change in the treatment of animals sometimes makes a great difference in their fertility; and it is
probable that the results observed in different menageries would
differ.   Indeed some animals in our Zoological Gardens have
become more productive since the year 1846.   It is, also, manifest from F. Cuvier's account of the Jardin des Plantes,[10] that
the animals formerly bred much less freely there than with us;
for instance, in the Duck tribe, which is highly prolific, only one
species had at that period produced young.

The most remarkable cases, however, are afforded by animals kept in
their native country, which, though perfectly tamed, quite healthy, and
allowed some freedom, are absolutely incapable of breeding. Rengger,[11] who
in Paraguay particularly attended to this subject, specifies six quadrupeds
in this condition; and he mentions two or three others which most rarely

---

[9] ' Essais de Zoologie Générale,' 1841,
p. 256.
[10] Du Rut, ' Annales du Muséum,'
1807, tom. ix. p. 120.

[11] ' Säugethiere von Paraguay.' 1830,
s. 49, 106, 118, 124, 201, 208, 249, 265,
327.

breed. Mr. Bates, in his admirable work on the Amazons, strongly insists on similar cases;[12] and he remarks, that the fact of thoroughly tamed native mammals and birds not breeding when kept by the Indians, cannot be wholly accounted for by their negligence or indifference, for the turkey is valued by them, and the fowl has been adopted by the remotest tribes. In almost every part of the world—for instance, in the interior of Africa, and in several of the Polynesian islands—the natives are extremely fond of taming the indigenous quadrupeds and birds; but they rarely or never succeed in getting them to breed.

The most notorious case of an animal not breeding in captivity is that of the elephant. Elephants are kept in large numbers in their native Indian home, live to old age, and are vigorous enough for the severest labour; yet, with one or two exceptions, they have never been known even to couple, though both males and females have their proper periodical seasons. If, however, we proceed a little eastward to Ava, we hear from Mr. Crawfurd[13] that their " breeding in the domestic state, or at least in the half-domestic state in which the female elephants are generally kept, is of every-day occurrence;" and Mr. Crawfurd informs me that he believes that the difference must be attributed solely to the females being allowed to roam the forests with some degree of freedom. The captive rhinoceros, on the other hand, seems from Bishop Heber's account[14] to breed in India far more readily than the elephant. Four wild species of the horse genus have bred in Europe, though here exposed to a great change in their natural habits of life; but the species have generally been crossed one with another. Most of the members of the pig family breed readily in our menageries: even the Red River hog (*Potamochœrus penicillatus*), from the sweltering plains of West Africa, has bred twice in the Zoological Gardens. Here also the Peccary (*Dicotyles torquatus*) has bred several times; but another species, the *D. labiatus*, though rendered so tame as to be half-domesticated, breeds so rarely in its native country of Paraguay, that according to Rengger[15] the fact requires confirmation. Mr. Bates remarks that the tapir, though often kept tame in Amazonia by the Indians, never breeds.

Ruminants generally breed quite freely in England, though brought from widely different climates, as may be seen in the Annual Reports of the Zoological Gardens, and in the Gleanings from Lord Derby's menagerie.

The Carnivora, with the exception of the Plantigrade division, generally breed (though with capricious exceptions) almost as freely as ruminants. Many species of Felidæ have bred in various menageries, although imported from various climates and closely confined. Mr. Bartlett, the present superintendent of the Zoological Gardens,[16] remarks that the lion appears to breed more frequently and to bring forth more young at a birth than any other species of the family. He adds that the tiger has rarely bred;

---

[12] 'The Naturalist on the Amazons,' 1863, vol. i. pp. 99, 193; vol. ii. p. 113.

[13] 'Embassy to the Court of Ava,' vol. i. p. 534.

[14] 'Journal,' vol. i. p. 213.

[15] 'Säugethiere,' s. 327.

[16] On the Breeding of the larger Felidæ, 'Proc. Zoolog. Soc.,' 1861, p. 140.

" but there are several well-authenticated instances of the female tiger breeding with the lion." Strange as the fact may appear, many animals under confinement unite with distinct species and produce hybrids quite as freely as, or even more freely than, with their own species. On inquiring from Dr. Falconer and others, it appears that the tiger when confined in India does not breed, though it has been known to couple. The cheetah (*Felis jubata*) has never been known by Mr. Bartlett to breed in England, but it has bred at Frankfort; nor does it breed in India, where it is kept in large numbers for hunting; but no pains would be taken to make them breed, as only those animals which have hunted for themselves in a state of nature are serviceable and worth training.[17] According to Rengger, two species of wild cats in Paraguay, though thoroughly tamed, have never bred. Although so many of the Felidæ breed readily in the Zoological Gardens, yet conception by no means always follows union: in the nine-year Report, various species are specified which were observed to couple seventy-three times, and no doubt this must have passed many times unnoticed; yet from the seventy-three unions only fifteen births ensued. The Carnivora in the Zoological Gardens were formerly less freely exposed to the air and cold than at present, and this change of treatment, as I was assured by the former superintendent, Mr. Miller, greatly increased their fertility. Mr. Bartlett, and there cannot be a more capable judge, says, " it is remarkable that lions breed more freely " in travelling collections than in the Zoological Gardens; probably the " constant excitement and irritation produced by moving from place to " place, or change of air, may have considerable influence in the matter."

Many members of the Dog family breed readily when confined. The Dhole is one of the most untameable animals in India, yet a pair kept there by Dr. Falconer produced young. Foxes, on the other hand, rarely breed, and I have never heard of such an occurrence with the European fox: the silver fox of North America (*Canis argentatus*), however, has bred several times in the Zoological Gardens. Even the otter has bred there. Every one knows how readily the semi-domesticated ferret breeds, though shut up in miserably small cages; but other species of Viverra and Paradoxurus absolutely refuse to breed in the Zoological Gardens. The Genetta has bred both here and in the Jardin des Plantes, and produced hybrids. The *Herpestes fasciatus* has likewise bred; but I was formerly assured that the *H. griseus*, though many were kept in the Gardens, never bred.

The Plantigrade Carnivora breed under confinement much less freely, without our being able to assign any reason, than other members of the group. In the nine-year Report it is stated that the bears had been seen in the Zoological Gardens to couple freely, but previously to 1848 had most rarely conceived. In the Reports published since this date three species have produced young (hybrids in one case), and, wonderful to relate, the white Polar bear has produced young. The badger (*Meles taxus*) has bred several times in the Gardens; but I have not heard of this

---

[17] Sleeman's ' Rambles in India,' vol. ii. p. 10.

occurring elsewhere in England, and the event must be very rare, for an instance in Germany has been thought worth recording.[18]   In Paraguay the native Nasua, though kept in pairs during many years and perfectly tamed, has never been known, according to Rengger, to breed or show any sexual passion; nor, as I hear from Mr. Bates, does this animal, or the Cercoleptes, breed in the region of the Amazons.   Two other plantigrade genera, Procyon and Gulo, though often kept tame in Paraguay, never breed there.   In the Zoological Gardens species of Nasua and Procyon have been seen to couple; but they did not produce young.

As domesticated rabbits, guinea-pigs, and white mice breed so abundantly when closely confined under various climates, it might have been thought that most other members of the Rodent order would have bred in captivity, but this is not the case.   It deserves notice, as showing how the capacity to breed sometimes goes by affinity, that the one native rodent of Paraguay, which there breeds *freely* and has yielded successive generations, is the *Cavia aperea*; and this animal is so closely allied to the guinea-pig, that it has been erroneously thought to be the parent-form.[19] In the Zoological Gardens, some rodents have coupled, but have never produced young; some have neither coupled nor bred; but a few have bred, as the porcupine more than once, the Barbary mouse, lemming, chinchilla, and the agouti (*Dasyprocta aguti*), several times.   This latter animal has also produced young in Paraguay, though they were born dead and ill-formed; but in Amazonia, according to Mr. Bates, it never breeds, though often kept tame about the houses.   Nor does the paca (*Cœlogenys paca*) breed there.   The common hare when confined has, I believe, never bred in Europe;[20] though, according to a recent statement, it has crossed with the rabbit.   I have never heard of the dormouse breeding in confinement.   But squirrels offer a more curious case: with one exception, no species has ever bred in the Zoological Gardens, yet as many as fourteen individuals of *S. palmarum* were kept together during several years.   The *S. cinerea* has been seen to couple, but it did not produce young; nor has this species, when rendered extremely tame in its native country, North America, been ever known to breed.[21]   At Lord Derby's menagerie squirrels of many kinds were kept in numbers, but Mr. Thompson, the superintendent, told me that none had ever bred there, or elsewhere as far as he knew.   I have never heard of the English squirrel breeding in confinement.   But the species which has bred more than once in the Zoological Gardens is the one which perhaps might have been least expected, namely, the flying squirrel (*Sciuropterus volucella*): it has, also, bred several times

[18] Wiegmann's 'Archif für Naturgesch.,' 1837, s. 162.

[19] Rengger, 'Säugethiere,' &c., s. 276. On the parentage of the guinea-pig, *see* also Isid. Geoffroy St. Hilaire, ' Hist. Nat. Gén.'

[20] Although the existence of the *Leporides*, as described by Dr. Broca (' Journal de Phys.,' tom. ii. p. 370),

is now positively denied, yet Dr. Pigeaux ('Annals and Mag. of Nat. Hist.,' vol. xx., 1867, p. 75) affirms that the hare and rabbit have produced hybrids.

[21] 'Quadrupeds of North America,' by Audubon and Bachman, 1846, p. 268.

near Birmingham; but the female never produced more than two young at a birth, whereas in its native American home she bears from three to six young.[22]

Monkeys, in the nine-year Report from the Zoological Gardens, are stated to unite most freely, but during this period, though many individuals were kept, there were only seven births. I have heard of one American monkey alone, the Ouistiti, breeding in Europe.[23] A Macacus, according to Flourens, bred in Paris; and more than one species of this genus has produced young in London, especially the *Macacus rhesus*, which everywhere shows a special capacity to breed under confinement. Hybrids have been produced both in Paris and London from this same genus. The Arabian baboon, or *Cynocephalus hamadryas*,[24] and a Cercopithecus have bred in the Zoological Gardens, and the latter species at the Duke of Northumberland's. Several members of the family of Lemurs have produced hybrids in the Zoological Gardens. It is much more remarkable that monkeys very rarely breed when confined in their native country; thus the Cay (*Cebus azaræ*) is frequently and completely tamed in Paraguay, but Rengger[25] says that it breeds so rarely, that he never saw more than two females which had produced young. A similar observation has been made with respect to the monkeys which are frequently tamed by the aborigines in Brazil.[26] In the region of the Amazons, these animals are so often kept in a tame state, that Mr. Bates in walking through the streets of Parà counted thirteen species; but, as he asserts, they have never been known to breed in captivity.[27]

Birds.

Birds offer in some respects better evidence than quadrupeds, from their breeding more rapidly and being kept in greater numbers. We have seen that carnivorous animals are more fertile under confinement than most other mammals. The reverse holds good with carnivorous birds. It is said[28] that as many as eighteen species have been used in Europe for hawking, and several others in Persia and India;[29] they have been kept in their native country in the finest condition, and have been flown during six, eight, or nine years;[30] yet there is no record of their having ever produced young. As these birds were formerly caught whilst young, at great expense, being imported from Iceland, Norway, and Sweden, there can

---

[22] Loudon's 'Mag. of Nat. Hist.,' vol. ix., 1836, p. 571; Audubon and Bachman's 'Quadrupeds of North America,' p. 221.

[23] Flourens, ' De l'Instinct,' &c., 1845, p. 88.

[24] *See* ' Annual Reports Zoolog. Soc.,' 1855, 1858, 1863, 1864; 'Times' newspaper, Aug. 10th, 1847; Flourens, ' De l'Instinct,' p. 85.

[25] 'Säugethiere,' &c., s. 34, 49.

[26] Art. Brazil, 'Penny Cyclop.,' p. 363.

[27] 'The Naturalist on the River Amazon,' vol. i. p. 99.

[28] 'Encyclop. of Rural Sports,' p. 691.

[29] According to Sir A. Burnes ('Cabool,' &c., p. 51), eight species are used for hawking in Scinde.

[30] Loudon's 'Mag. of Nat. Hist.,' vol. vi., 1833, p. 110.

be little doubt that, if possible, they would have been propagated. In the Jardin des Plantes, no bird of prey has been known to couple.[31] No hawk, vulture, or owl has ever produced fertile eggs in the Zoological Gardens, or in the old Surrey Gardens, with the exception, in the former place on one occasion, of a condor and a kite (*Milvus niger*). Yet several species, namely, the *Aquila fusca*, *Haliœtus leucocephalus*, *Falco tinnunculus*, *F. subbuteo*, and *Buteo vulgaris*, have been seen to couple in the Zoological Gardens. Mr. Morris[32] mentions as a unique fact that a kestrel (*Falco tinnunculus*) bred in an aviary. The one kind of owl which has been known to couple in the Zoological Gardens was the Eagle Owl (*Bubo maximus*); and this species shows a special inclination to breed in captivity; for a pair at Arundel Castle, kept more nearly in a state of nature " than ever fell to the lot of an animal deprived of its liberty,"[33] actually reared their young. Mr. Gurney has given another instance of this same owl breeding in confinement; and he records the case of a second species of owl, the *Strix passerina*, breeding in captivity.[34]

Of the smaller graminivorous birds, many kinds have been kept tame in their native countries, and have lived long; yet, as the highest authority on cage-birds[35] remarks, their propagation is " uncommonly difficult." The canary-bird shows that there is no inherent difficulty in these birds breeding freely in confinement; and Audubon says[36] that the *Fringilla* (*Spiza*) *ciris* of North America breeds as perfectly as the canary. The difficulty with the many finches which have been kept in confinement is all the more remarkable as more than a dozen species could be named which have yielded hybrids with the canary; but hardly any of these, with the exception of the siskin (*Fringilla spinus*), have reproduced their own kind. Even the bullfinch (*Loxia pyrrhula*) has bred as frequently with the canary, though belonging to a distinct genus, as with its own species.[37] With respect to the skylark (*Alauda arvensis*), I have heard of birds living for seven years in an aviary, which never produced young; and a great London bird-fancier assured me that he had never known an instance of their breeding; nevertheless one case has been recorded.[38] In the nine-year Report from the Zoological Society, twenty-four incessorial species are enumerated which had not bred, and of these only four were known to have coupled.

Parrots are singularly long-lived birds; and Humboldt mentions the curious fact of a parrot in South America, which spoke the language of

---

[31] F. Cuvier, 'Annal. du Muséum,' tom. ix. p. 128.

[32] 'The Zoologist,' vol. vii.-viii., 1849-50, p. 2648.

[33] Knox, 'Ornithological Rambles in Sussex,' p. 91.

[34] 'The Zoologist,' vol. vii.-viii., 1849-50, p. 2566; vol. ix.-x., 1851-2, p. 3207.

[35] Bechstein, 'Naturgesch. der Stubenvögel,' 1840, s. 20.

[36] 'Ornithological Biography,' vol. v. p. 517.

[37] A case is recorded in 'The Zoologist,' vol. i.-ii., 1843-45, p. 453. For the siskin breeding, vol. iii.-iv., 1845-46, p. 1075. Bechstein, 'Stubenvögel,' s. 139, speaks of bullfinches making nests, but rarely producing young.

[38] Yarrell's 'Hist. British Birds,' 1839, vol. i. p. 412.

an extinct Indian tribe, so that this bird preserved the sole relic of a lost language. Even in this country there is reason to believe [39] that parrots have lived to the age of nearly one hundred years; yet, though many have been kept in Europe, they breed so rarely that the event has been thought worth recording in the gravest publications.[40] According to Bechstein [41] the African *Psittacus erithacus* breeds oftener than any other species: the *P. macoa* occasionally lays fertile eggs, but rarely succeeds in hatching them; this bird, however, has the instinct of incubation sometimes so strongly developed, that it will hatch the eggs of fowls or pigeons. In the Zoological Gardens and in the old Surrey Gardens some few species have coupled, but, with the exception of three species of parrakeets, none have bred. It is a much more remarkable fact that in Guiana parrots of two kinds, as I am informed by Sir R. Schomburgk, are often taken from the nests by the Indians and reared in large numbers; they are so tame that they fly freely about the houses, and come when called to be fed, like pigeons; yet he has never heard of a single instance of their breeding.[42] In Jamaica, a resident naturalist, Mr. R. Hill, [43] says, " no birds more readily " submit to human dependence than the parrot-tribe, but no instance of a " parrot breeding in this tame life has been known yet." Mr. Hill specifies a number of other native birds kept tame in the West Indies, which never breed in this state.

The great pigeon family offers a striking contrast with parrots : in the nine-year Report thirteen species are recorded as having bred, and, what is more noticeable, only two were seen to couple without any result. Since the above date every annual Report gives many cases of various pigeons breeding. The two magnificent crowned pigeons (*Goura coronata* and *Victoriæ*) produced hybrids; nevertheless, of the former species more than a dozen birds were kept, as I am informed by Mr. Crawfurd, in a park at Penang, under a perfectly well-adapted climate, but never once bred. The *Columba migratoria* in its native country, North America, invariably lays two eggs, but in Lord Derby's menagerie never more than one. The same fact has been observed with the *C. leucocephala*.[44]

Gallinaceous birds of many genera likewise show an eminent capacity for breeding under captivity. This is particularly the case with pheasants ; yet our English species seldom lays more than ten eggs in confinement; whilst from eighteen to twenty is the usual number in the wild state.[45] With the Gallinaceæ, as with all other orders, there are marked and inex-

---

[39] Loudon's ' Mag. of Nat. History,' vol. ix., 1836, p. 347.

[40] 'Mémoires du Muséum d'Hist. Nat.,' tom. x. p. 314: five cases of parrots breeding in France are here recorded. *See*, also, ' Report Brit. Assoc. Zoolog.,' 1843.

[41] ' Stubenvögel,' s. 105, 83.

[42] Dr. Hancock remarks ('Charlesworth's Mag. of Nat. Hist.,' vol. ii., 1838, p. 492), "it is singular that, amongst the numerous useful birds that are indigenous to Guiana, none are found to propagate among the Indians; yet the common fowl is reared in abundance throughout the country."

[43] ' A Week at Port Royal,' 1855, p. 7.

[44] Audubon, 'American Ornithology,' vol. v. pp. 552, 557.

[45] Moubray on Poultry, 7th edit., p. 133.

plicable exceptions in regard to the fertility of certain species and genera under confinement. Although many trials have been made with the common partridge, it has rarely bred, even when reared in large aviaries; and the hen will never hatch her own eggs.[46] The American tribe of Guans or Cracidæ are tamed with remarkable ease, but are very shy breeders in this country;[47] but with care various species were formerly made to breed rather freely in Holland.[48] Birds of this tribe are often kept in a perfectly tamed condition in their native country by the Indians, but they never breed.[49] It might have been expected that grouse from their habits of life would not have bred in captivity, more especially as they are said soon to languish and die.[50] But many cases are recorded of their breeding: the capercailzie (*Tetrao urogallus*) has bred in the Zoological Gardens; it breeds without much difficulty when confined in Norway, and in Russia five successive generations have been reared: *Tetrao tetrix* has likewise bred in Norway; *T. Scoticus* in Ireland; *T. umbellus* at Lord Derby's; and *T. cupido* in North America.

It is scarcely possible to imagine a greater change in habits than that which the members of the ostrich family must suffer, when cooped up in small enclosures under a temperate climate, after freely roaming over desert and tropical plains or entangled forests. Yet almost all the kinds, even the mooruk (*Casuarinus Bennettii*) from New Ireland, has frequently produced young in the various European menageries. The African ostrich, though perfectly healthy and living long in the South of France, never lays more than from twelve to fifteen eggs, though in its native country it lays from twenty-five to thirty.[51] Here we have another instance of fertility impaired, but not lost, under confinement, as with the flying squirrel, the hen-pheasant, and two species of American pigeons.

Most Waders can be tamed, as the Rev. E. S. Dixon informs me, with remarkable facility; but several of them are short-lived under confinement, so that their sterility in this state is not surprising. The cranes breed more readily than other genera: *Grus montigresia* has bred several times in Paris and in the Zoological Gardens, as has *G. cineria* at the latter place, and *G. antigone* at Calcutta. Of other members of this great order, *Tetrapteryx paradisea* has bred at Knowsley, a Porphyrio in Sicily, and the *Gallinula chloropus* in the Zoological Gardens. On the other hand, several

---

[46] Temminck, 'Hist. Nat. Gén. des Pigeons,' &c., 1813, tom. iii. pp. 288, 382; 'Annals and Mag. of Nat. Hist.,' vol. xii., 1843, p. 453. Other species of partridge have occasionally bred; as the red-legged (*P. rubra*), when kept in a large court in France (*see* Journal de Physique,' tom. xxv. p. 294), and in the Zoological Gardens in 1856.

[47] Rev. E. S. Dixon, 'The Dovecote,' 1851, pp. 243-252.

[48] Temminck, 'Hist. Nat. Gén. des Pigeons,' &c., tom. ii. pp. 456, 458; tom. iii. pp. 2, 13, 17.

[49] Bates, 'The Naturalist on the Amazons,' vol. i. p. 193; vol. ii. p. 112.

[50] Temminck, 'Hist. Nat. Gén.,' &c., tom. iii. p. 125. For *Tetrao urogallus, see* L. Lloyd, 'Field Sports of North of Europe,' vol. i. pp. 287, 314; and 'Bull. de la Soc. d'Acclimat.,' tom. vii., 1860, p. 600. For *T. Scoticus*, Thompson, 'Nat. Hist. of Ireland,' vol. ii., 1850, p. 49. For *T. cupido*, 'Boston Journal of Nat. Hist.,' vol. iii. p. 199.

[51] Marcel de Serres, 'Annales des Sci. Nat.,' 2nd series, Zoolog., tom. xiii. p. 175.

birds belonging to this order will not breed in their native country, Jamaica; and the Psophia, though often kept by the Indians of Guiana about their houses, " is seldom or never known to breed."[52]

No birds breed with such complete facility under confinement as the members of the great Duck family; yet, considering their aquatic and wandering habits, and the nature of their food, this could not have been anticipated. Even some time ago above two dozen species had bred in the Zoological Gardens; and M. Selys-Longchamps has recorded the production of hybrids from forty-four different members of the family; and to these Professor Newton has added a few more cases.[53] " There is not," says Mr. Dixon,[54] " in the wide world, a goose which is not in the strict sense of the word domesticable;" that is, capable of breeding under confinement; but this statement is probably too bold. The capacity to breed sometimes varies in individuals of the same species; thus Audubon[55] kept for more than eight years some wild geese (*Anser Canadensis*), but they would not mate; whilst other individuals of the same species produced young during the second year. I know of but one instance in the whole family of a species which absolutely refuses to breed in captivity, namely, the *Dendrocygna viduata*, although, according to Sir R. Schomburgk,[56] it is easily tamed, and is frequently kept by the Indians of Guiana. Lastly, with respect to Gulls, though many have been kept in the Zoological Gardens and in the old Surrey Gardens, no instance was known before the year 1848 of their coupling or breeding; but since that period the herring gull (*Larus argentatus*) has bred many times in the Zoological Gardens and at Knowsley.

There is reason to believe that insects are affected by confinement like the higher animals. It is well known that the Sphingidæ rarely breed when thus treated. An entomologist[57] in Paris kept twenty-five specimens of *Saturnia pyri*, but did not succeed in getting a single fertile egg. A number of females of *Orthosia munda* and of *Mamestra suasa* reared in confinement were unattractive to the males.[58] Mr. Newport kept nearly a hundred individuals of two species of Vanessa, but not one paired; this, however, might have been due to their habit of coupling on the wing.[59] Mr. Atkinson could never succeed in India in making the Tarroo silk-moth breed in confinement.[60] It appears that a number of moths, especially the Sphingidæ, when hatched in the autumn out of their proper season,

---

[52] Dr. Hancock, in 'Charlesworth's Mag. of Nat. Hist.' vol. ii., 1838, p. 491; R. Hill, 'A Week at Port Royal,' p. 8; 'Guide to the Zoological Gardens,' by P. L. Sclater, 1859, pp. 11, 12; 'The Knowsley Menagerie,' by Dr. Gray, 1846, pl. xiv.; E. Blyth, 'Report Asiatic Soc. of Bengal,' May, 1855.

[53] Prof. Newton, in 'Proc. Zoolog. Soc.,' 1860, p. 336.

[54] 'The Dovecote and Aviary,' p. 428.

[55] 'Ornithological Biography,' vol. iii. p. 9.

[56] 'Geograph. Journal,' vol. xiii., 1844, p. 32.

[57] Loudon's 'Mag. of Nat. Hist.,' vol. v., 1832, p. 153.

[58] 'Zoologist,' vols. v.-vi., 1847-48, p. 1660.

[59] 'Transact. Entomolog. Soc.,' vol. iv., 1845, p. 60.

[60] 'Transact. Linn. Soc.,' vol. vii. p. 40.

are completely barren; but this latter case is still involved in some obscurity.[61]

Independently of the fact of many animals under confinement not coupling, or, if they couple, not producing young, there is evidence of another kind that their sexual functions are thus disturbed. For many cases have been recorded of the loss by male birds when confined of their characteristic plumage. Thus the common linnet (*Linota cannabina*) when caged does not acquire the fine crimson colour on its breast, and one of the buntings (*Emberiza passerina*) loses the black on its head. A Pyrrhula and an Oriolus have been observed to assume the quiet plumage of the hen-bird; and the *Falco albidus* returned to the dress of an earlier age.[62] Mr. Thompson, the superintendent of the Knowsley menagerie, informed me that he had often observed analogous facts. The horns of a male deer (*Cervus Canadensis*) during the voyage from America were badly developed; but subsequently in Paris perfect horns were produced.

When conception takes place under confinement, the young are often born dead, or die soon, or are ill-formed. This frequently occurs in the Zoological Gardens, and, according to Rengger, with native animals confined in Paraguay. The mother's milk often fails. We may also attribute to the disturbance of the sexual functions the frequent occurrence of that monstrous instinct which leads the mother to devour her own offspring,—a mysterious case of perversion, as it at first appears.

Sufficient evidence has now been advanced to prove that animals when first confined are eminently liable to suffer in their reproductive systems. We feel at first naturally inclined to attribute the result to loss of health, or at least to loss of vigour; but this view can hardly be admitted when we reflect how healthy, long-lived, and vigorous many animals are under cap-

[61] *See* an interesting paper by Mr. Newman, in the 'Zoologist,' 1857, p. 5764; and Dr. Wallace, in 'Proc. Entomolog. Soc.,' June 4th, 1860, p. 119.

[62] Yarrell's 'British Birds,' vol. i. p. 506; Bechstein, 'Stubenvögel,' s. 185; 'Philosoph. Transact.,' 1772, p. 271. Bronn ('Geschichte der Natur,' Band ii. s. 96) has collected a number of cases. For the case of the deer, *see* 'Penny Cyclop.,' vol. viii. p. 350.

tivity, such as parrots, and hawks when used for hawking, chetahs when used for hunting, and elephants. The reproductive organs themselves are not diseased; and the diseases, from which animals in menageries usually perish, are not those which in any way affect their fertility. No domestic animal is more subject to disease than the sheep, yet it is remarkably prolific. The failure of animals to breed under confinement has been sometimes attributed exclusively to a failure in their sexual instincts: this may occasionally come into play, but there is no obvious reason why this instinct should be especially liable to be affected with perfectly tamed animals, except indeed indirectly through the reproductive system itself being disturbed. Moreover, numerous cases have been given of various animals which couple freely under confinement, but never conceive; or, if they conceive and produce young, these are fewer in number than is natural to the species. In the vegetable kingdom instinct of course can play no part; and we shall presently see that plants when removed from their natural conditions are affected in nearly the same manner as animals. Change of climate cannot be the cause of the loss of fertility, for, whilst many animals imported into Europe from extremely different climates breed freely, many others when confined in their native land are completely sterile. Change of food cannot be the chief cause; for ostriches, ducks, and many other animals, which must have undergone a great change in this respect, breed freely. Carnivorous birds when confined are extremely sterile; whilst most carnivorous mammals, except plantigrades, are moderately fertile. Nor can the amount of food be the cause; for a sufficient supply will certainly be given to valuable animals; and there is no reason to suppose that much more food would be given to them, than to our choice domestic productions which retain their full fertility. Lastly, we may infer from the case of the elephant, chetah, various hawks, and of many animals which are allowed to lead an almost free life in their native land, that want of exercise is not the sole cause.

It would appear that any change in the habits of life, whatever these habits may be, if great enough, tends to affect in an inexplicable manner the powers of reproduction. The result depends more on the constitution of the species than on the nature of the change; for certain whole groups are affected more than others; but exceptions always occur, for some species in the most fertile groups refuse to breed, and some in the most sterile groups breed freely. Those animals which usually breed freely under confinement, rarely breed, as I was assured, in the Zoological Gardens, within a year or two after their first importation. When an animal which is generally sterile under confinement happens to breed, the young apparently do not inherit this power; for had this been the case, various quadrupeds and birds, which are valuable for exhibition, would have become common. Dr. Broca even affirms[63] that many animals in the Jardin des Plantes, after having produced young for three or four successive generations, become sterile; but this may be the result of too close interbreeding. It is a remarkable circumstance that many mammals and birds have produced hybrids under confinement quite as readily as, or even more readily than, they have procreated their own kind. Of this fact many instances have been given;[64] and we are thus reminded of those plants which when cultivated refuse to be fertilised by their own pollen, but can easily be fertilised by that of a distinct species. Finally, we must conclude, limited as the conclusion is, that changed conditions of life have an especial power of acting injuriously on the reproductive system. The whole case is quite peculiar, for these organs, though not diseased, are thus rendered incapable of performing their proper functions, or perform them imperfectly.

Sterility of Domesticated Animals from changed conditions.—With respect to domesticated animals, as their domestication mainly depends on the accident of their breeding freely under captivity, we ought not to expect that their reproductive system would be affected by any moderate degree of change. Those orders of quadrupeds and birds, of which the wild species breed most readily in our menageries, have afforded us the greatest number of domesticated productions. Savages in most parts of the world are fond of taming animals;[65] and if any of these regularly produced

---

[63] 'Journal de Physiologie,' tom. ii. p. 347.
[64] For additional evidence on this subject, *see* F. Cuvier, in 'Annales du Muséum,' tom. xii. p 119.

[65] Numerous instances could be given. Thus Livingstone ('Travels,' p. 217) states that the King of the Barotse, an inland tribe which never had any communication with white men, was

young, and were at the same time useful, they would be at once domesticated. If, when their masters migrated into other countries, they were in addition found capable of withstanding various climates, they would be still more valuable; and it appears that the animals which breed readily in captivity can generally withstand different climates. Some few domesticated animals, such as the reindeer and camel, offer an exception to this rule. Many of our domesticated animals can bear with undiminished fertility the most unnatural conditions; for instance, rabbits, guinea-pigs, and ferrets breed in miserably confined hutches. Few European dogs of any kind withstand without degeneration the climate of India; but as long as they survive, they retain, as I hear from Dr. Falconer, their fertility; so it is, according to Dr. Daniell, with English dogs taken to Sierra Leone. The fowl, a native of the hot jungles of India, becomes more fertile than its parent-stock in every quarter of the world, until we advance as far north as Greenland and Northern Siberia, where this bird will not breed. Both fowls and pigeons, which I received during the autumn direct from Sierra Leone, were at once ready to couple.[66] I have, also, seen pigeons breeding as freely as the common kinds within a year after their importation from the Upper Nile. The guinea-fowl, an aboriginal of the hot and dry deserts of Africa, whilst living under our damp and cool climate, produces a large supply of eggs.

Nevertheless, our domesticated animals under new conditions occasionally show signs of lessened fertility. Roulin asserts that in the hot valleys of the equatorial Cordillera sheep are not fully fecund;[67] and according to Lord Somerville,[68] the merino-sheep which he imported from Spain were not at first perfectly fertile. It is said[69] that mares brought up on dry food in the stable, and turned out to grass, do not at first breed. The peahen, as we have seen, is said not to lay so many eggs in England as in India. It was long before the canary-bird was fully fertile, and even now first-rate breeding birds are not common.[70] In the hot and dry province of Delhi, the eggs of the turkey, as I hear from Dr. Falconer, though placed under a hen, are extremely liable to fail. According to Roulin, geese taken within a recent period to the lofty plateau of Bogota, at first laid seldom, and then only a few eggs; of these scarcely a fourth were hatched, and half the young birds died: in the second generation they were more fertile; and when Roulin wrote they were becoming as

---

extremely fond of taming animals, and every young antelope was brought to him. Mr. Galton informs me that the Damaras are likewise fond of keeping pets. The Indians of South America follow the same habit. Capt. Wilkes states that the Polynesians of the Samoan Islands tamed pigeons; and the New Zealanders, as Mr. Mantell informs me, kept various kinds of birds.

[66] For analogous cases with the fowl,

see Réaumur, 'Art de faire Eclorre,' &c., 1749, p. 243 ; and Col. Sykes, in 'Proc. Zoolog. Soc.,' 1832, &c. With respect to the fowl not breeding in northern regions, see Latham's 'Hist. of Birds,' vol. viii., 1823, p. 169.

[67] 'Mém. par divers Savans, Acad. des Sciences,' tom. vi., 1835, p. 347.

[68] Youatt on Sheep, p. 181.

[69] J. Mills, 'Treatise on Cattle,' 1776, p. 72.

[70] Bechstein, 'Stubenvögel,' s. 242.

fertile as our geese in Europe. In the Philippine Archipelago the goose, it is asserted, will not breed or even lay eggs.[71] A more curious case is that of the fowl, which, according to Roulin, when first introduced would not breed at Cusco in Bolivia, but subsequently became quite fertile; and the English Game fowl, lately introduced, had not as yet arrived at its full fertility, for to raise two or three chickens from a nest of eggs was thought fortunate. In Europe close confinement has a marked effect on the fertility of the fowl: it has been found in France that with fowls allowed considerable freedom only twenty per cent. of the eggs failed; when allowed less freedom forty per cent. failed; and in close confinement sixty out of the hundred were not hatched.[72] So we see that unnatural and changed conditions of life produce some effect on the fertility of our most thoroughly domesticated animals, in the same manner, though in a far less degree, as with captive wild animals.

It is by no means rare to find certain males and females which will not breed together, though both are known to be perfectly fertile with other males and females. We have no reason to suppose that this is caused by these animals having been subjected to any change in their habits of life; therefore such cases are hardly related to our present subject. The cause apparently lies in an innate sexual incompatibility of the pair which are matched. Several instances have been communicated to me by Mr. W. C. Spooner (well known for his essay on Cross-breeding), by Mr. Eyton of Eyton, by Mr. Wicksted and other breeders, and especially by Mr. Waring of Chelsfield, in relation to horses, cattle, pigs, foxhounds, other dogs, and pigeons.[73] In these cases, females, which either previously or subsequently were proved to be fertile, failed to breed with certain males, with whom it was particularly desired to match them. A change in the constitution of the female may sometimes have occurred before she was put to the second male; but in other cases this explanation is hardly tenable, for a female, known not to be barren, has been unsuccessfully paired seven or eight times with the same male likewise known to be perfectly fertile. With cart-mares, which sometimes will not breed with stallions of pure blood, but subsequently have bred with cart-stallions, Mr. Spooner is inclined to attribute the failure to the lesser sexual power of the race-horse. But I have heard from the greatest breeder of race-horses at the present day, through Mr. Waring, that " it frequently occurs with a mare to be put " several times during one or two seasons to a particular stallion of " acknowledged power, and yet prove barren; the mare afterwards " breeding at once with some other horse." These facts are worth recording, as they show, like so many previous facts, on what slight constitutional differences the fertility of an animal often depends.

[71] Crawfurd's 'Descriptive Dict. of the Indian Islands,' 1856, p. 145.
[72] 'Bull. de la Soc. Acclimat.,' tom. ix., 1862, pp. 380, 384.
[73] For pigeons, see Dr. Chapuis, 'Le Pigeon Voyageur Belge, 1865, p. 66.

Sterility of Plants from changed Conditions of Life, and from other causes.

In the vegetable kingdom cases of sterility frequently occur, analogous with those previously given in the animal kingdom. But the subject is obscured by several circumstances, presently to be discussed, namely, the contabescence of the anthers, as Gärtner has named a certain affection—monstrosities—doubleness of the flower—much-enlarged fruit—and long-continued or excessive propagation by buds.

It is notorious that many plants in our gardens and hot-houses, though preserved in the most perfect health, rarely or never produce seed. I do not allude to plants which run to leaves, from being kept too damp, or too warm, or too much manured; for these do not produce the reproductive individual or flower, and the case may be wholly different. Nor do I allude to fruit not ripening from want of heat, or rotting from too much moisture. But many exotic plants, with their ovules and pollen appearing perfectly sound, will not set any seed. The sterility in many cases, as I know from my own observation, is simply due to the absence of the proper insects for carrying the pollen to the stigma. But after excluding the several cases just specified, there are many plants in which the reproductive system has been seriously affected by the altered conditions of life to which they have been subjected.

It would be tedious to enter on many details. Linnæus long ago observed [74] that Alpine plants, although naturally loaded with seed, produce either few or none when cultivated in gardens. But exceptions often occur: the *Draba sylvestris*, one of our most thoroughly Alpine plants, multiplies itself by seed in Mr. H. C. Watson's garden, near London; and Kerner, who has particularly attended to the cultivation of Alpine plants, found that various kinds, when cultivated, spontaneously sowed themselves. [75] Many plants which naturally grow in peat-earth are entirely sterile in our gardens. I have noticed the same fact with several liliaceous plants, which nevertheless grew vigorously.

Too much manure renders some kinds utterly sterile, as I have myself observed. The tendency to sterility from this cause runs in families; thus, according to Gärtner, [76] it is hardly possible to give too much manure to most Gramineæ, Cruciferæ, and Leguminosæ, whilst succulent and bulbous-rooted plants are easily affected. Extreme poverty of soil is less

---

[74] 'Swedish Acts,' vol. i., 1739, p. 3. Pallas makes the same remark in his Travels (Eng. translat.), vol. i. p. 292.

[75] A. Kerner, 'Die Cultur der Alpenpflanzen,' 1864, s. 139; Watson's 'Cybele Britannica,' vol. i. p. 131; Mr.

D. Cameron, also, has written on the culture of Alpine plants in ' Gard. Chronicle,' 1848, pp. 253, 268, and mentions a few which seed.

[76] ' Beiträge zur Kenntniss der Befruchtung,' 1844, s. 333.

apt to induce sterility; but dwarfed plants of *Trifolium minus* and *repens*, growing on a lawn often mown and never manured, did not produce any seed. The temperature of the soil, and the season at which plants are watered, often have a marked effect on their fertility, as was observed by Kölreuter in the case of Mirabilis.[77]   Mr. Scott in the Botanic Gardens of Edinburgh observed that *Oncidium divaricatum* would not set seed when grown in a basket in which it throve, but was capable of fertilisation in a pot where it was a little damper. *Pelargonium fulgidum*, for many years after its introduction, seeded freely; it then became sterile; now it is fertile[78] if kept in a dry stove during the winter. Other varieties of pelargonium are sterile and others fertile without our being able to assign any cause. Very slight changes in the position of a plant, whether planted on a bank or at its base, sometimes make all the difference in its producing seed. Temperature apparently has a much more powerful influence on the fertility of plants than on that of animals. Nevertheless it is wonderful what changes some few plants will withstand with undiminished fertility: thus the *Zephyranthes candida*, a native of the moderately warm banks of the Plata, sows itself in the hot dry country near Lima, and in Yorkshire resists the severest frosts, and I have seen seeds gathered from pods which had been covered with snow during three weeks.[79] *Berberis Wallichii*, from the hot Khasia range in India, is uninjured by our sharpest frosts, and ripens its fruit under our cool summers. Nevertheless I presume we must attribute to change of climate the sterility of many foreign plants; thus the Persian and Chinese lilacs (*Syringa Persica* and *Chinensis*), though perfectly hardy, never here produce a seed; the common lilac (*S. vulgaris*) seeds with us moderately well, but in parts of Germany the capsules never contain seed.[80]

Some of the cases, given in the last chapter, of self-impotent plants, which are fertile both on the male and female side when united with distinct individuals or species, might have been here introduced; for as this peculiar form of sterility generally occurs with exotic plants or with endemic plants cultivated in pots, and as it disappeared in the *Passiflora alata* when grafted, we may conclude that in these cases it is the result of the treatment to which the plants or their parents have been exposed.

The liability of plants to be affected in their fertility by slightly changed conditions is the more remarkable, as the pollen when once in process of formation is not easily injured; a plant may be transplanted, or a branch with flower-buds be cut off and placed in water, and the pollen will be matured. Pollen, also, when once mature, may be kept for weeks or even months.[81] The female organs are more sensitive, for Gärtner[82] found that dicotyledonous plants, when carefully removed so that they did not in the least flag, could seldom be fertilised; this occurred even with potted

---

[77] 'Nova Acta Petrop.,' 1793, p. 391.

[78] 'Cottage Gardener,' 1856, pp. 44, 109.

[79] Dr. Herbert, 'Amaryllidaceæ,' p. 176.

[80] Gärtner, 'Beiträge zur Kenntniss,'
&c., s. 560, 564.

[81] 'Gardener's Chronicle,' 1844, p. 215; 1850, p. 470.

[82] 'Beiträge zur Kenntniss,' &c., s. 252, 333.

plants if the roots had grown out of the hole at the bottom. In some few cases, however, as with Digitalis, transplantation did not prevent fertilisation; and according to the testimony of Mawz, *Brassica rapa*, when pulled up by its roots and placed in water, ripened its seed. Flower-stems of several monocotyledonous plants when cut off and placed in water likewise produce seed. But in these cases I presume that the flowers had been already fertilised, for Herbert[83] found with the Crocus that the plants might be removed or mutilated after the act of fertilisation, and would still perfect their seeds; but that, if transplanted before being fertilised, the application of pollen was powerless.

Plants which have been long cultivated can generally endure with undiminished fertility various and great changes; but not in most cases so great a change of climate as domesticated animals. It is remarkable that many plants under these circumstances are so much affected that the proportions and the nature of their chemical ingredients are modified, yet their fertility is unimpaired. Thus, as Dr. Falconer informs me, there is a great difference in the character of the fibre in hemp, in the quantity of oil in the seed of the Linum, in the proportion of narcotin to morphine in the poppy, in gluten to starch in wheat, when these plants are cultivated on the plains and on the mountains of India; nevertheless, they all remain fully fertile.

Contabescence.—Gärtner has designated by this term a peculiar condition of the anthers in certain plants, in which they are shrivelled, or become brown and tough, and contain no good pollen. When in this state they exactly resemble the anthers of the most sterile hybrids. Gärtner,[84] in his discussion on this subject, has shown that plants of many orders are occasionally thus affected; but the Caryophyllaceæ and Liliaceæ suffer most, and to these orders, I think, the Ericaceæ may be added. Contabescence varies in degree, but on the same plant all the flowers are generally affected to nearly the same extent. The anthers are affected at a very early period in the flower-bud, and remain in the same state (with one recorded exception) during the life of the plant. The affection cannot be cured by any change of treatment, and is propagated by layers, cuttings, &c., and perhaps even by seed. In contabescent plants the female organs are seldom affected, or merely become precocious in their development. The cause of this affection is doubtful, and is different in different cases. Until I read Gärtner's discussion I attributed it, as apparently did Herbert, to the unnatural treatment of the plants; but its permanence under changed conditions, and the female organs not being affected, seem incompatible with this view. The fact of several endemic plants becoming contabescent in our gardens seems, at first sight, equally incompatible with this view; but Kölreuter believes that this is the result of their transplantation. The contabescent plants of Dianthus and Verbascum, found wild by Wiegmann, grew on a dry and sterile bank. The fact that exotic

---

[83] 'Journal of Hort. Soc.,' vol. ii. 1847, p. 83.

[84] 'Beiträge zur Kenntniss,' &c., s. 117 *et seq.*; Kölreuter, 'Zweite Fortsetz-ung,' s. 10, 121; 'Dritte Fortsetzung,' s. 57. Herbert, 'Amaryllidaceæ,' p. 355. Wiegmann, 'Ueber die Bastarderzeugung,' s. 27.

plants are eminently liable to this affection also seems to show that it is in some manner caused by their unnatural treatment. In some instances, as with Silene, Gärtner's view seems the most probable, namely, that it is caused by an inherent tendency in the species to become dioecious. I can add another cause, namely, the illegitimate unions of reciprocally dimorphic or trimorphic plants, for I have observed seedlings of three species of Primula and of *Lythrum salicaria*, which had been raised from plants illegitimately fertilised by their own-form pollen, with some or all their anthers in a contabescent state. There is perhaps an additional cause, namely, self-fertilisation; for many plants of Dianthus and Lobelia, which had been raised from self-fertilised seeds, had their anthers in this state; but these instances are not conclusive, as both genera are liable from other causes to this affection.

Cases of an opposite nature likewise occur, namely, plants with the female organs struck with sterility, whilst the male organs remain perfect. *Dianthus Japonicus*, a Passiflora, and Nicotiana, have been described by Gärtner[85] as being in this unusual condition.

Monstrosities as a cause of Sterility.—Great deviations of structure, even when the reproductive organs themselves are not seriously affected, some-times cause plants to become sterile. But in other cases plants may become monstrous to an extreme degree and yet retain their full fertility. Gallesio, who certainly had great experience,[86] often attributes sterility to this cause; but it may be suspected that in some of his cases sterility was the cause, and not the result, of the monstrous growths. The curious St. Valery apple, although it bears fruit, rarely produces seed. The wonderfully anomalous flowers of *Begonia frigida*, formerly described, though they appear fit for fructification, are sterile.[87] Species of Pri-mulæ, in which the calyx is brightly coloured, are said[88] to be often sterile, though I have known them to be fertile. On the other hand, Verlot gives several cases of proliferous flowers which can be propa-gated by seed. This was the case with a poppy, which had become monopetalous by the union of its petals.[89] Another extraordinary poppy, with the stamens replaced by numerous small supplementary capsules, likewise reproduces itself by seed. This has also occurred with a plant of *Saxifraga geum*, in which a series of adventitious carpels, bearing ovules on their margins, had been developed between the stamens and the normal carpels.[90] Lastly, with respect to peloric flowers, which depart wonder-fully from the natural structure,—those of *Linaria vulgaris* seem generally to be more or less sterile, whilst those before described of *Antirrhinum majus*, when artificially fertilised with their own pollen, are perfectly

---

[85] 'Bastarderzeugung,' s. 356.

[86] 'Teoria della Riproduzione,' 1816, p. 84; 'Traité du Citrus,' 1811, p. 67.

[87] Mr. C. W. Crocker, in 'Gard. Chronicle,' 1861, p. 1092.

[88] Verlot, 'Des Variétés,' 1865, p. 80.

[89] Verlot, idem, p. 88.

[90] Prof. Allman, Brit. Assoc., quoted

in the 'Phytologist,' vol. ii. p. 483. Prof. Harvey, on the authority of Mr. Andrews, who discovered the plant, informed me that this monstrosity could be propagated by seed. With respect to the poppy, *see* Prof. Goeppert, as quoted in 'Journal of Horticulture,' July 1st, 1863, p. 171.

fertile, though sterile when left to themselves, for bees are unable to crawl into the narrow tubular flower. The peloric flowers of *Corydalis solida*, according to Godron,[91] are barren; whilst those of Gloxinia are well known to yield plenty of seed. In our greenhouse Pelargoniums, the central flower of the truss is often peloric, and Mr. Masters informs me that he tried in vain during several years to get seed from these flowers. I likewise made many vain attempts, but sometimes succeeded in fertilising them with pollen from a normal flower of another variety; and conversely I several times fertilised ordinary flowers with peloric pollen. Only once I succeeded in raising a plant from a peloric flower fertilised by pollen from a peloric flower borne by another variety; but the plant, it may be added, presented nothing particular in its structure. Hence we may conclude that no general rule can be laid down; but any great deviation from the normal structure, even when the reproductive organs themselves are not seriously affected, certainly often leads to sexual impotence.

Double Flowers.—When the stamens are converted into petals, the plant becomes on the male side sterile; when both stamens and pistils are thus changed, the plant becomes completely barren. Symmetrical flowers having numerous stamens and petals are the most liable to become double, as perhaps follows from all multiple organs being the most subject to variability. But flowers furnished with only a few stamens, and others which are asymmetrical in structure, sometimes become double, as we see with the double gorse or Ulex, Petunia, and Antirrhinum. The Compositæ bear what are called double flowers by the abnormal development of the corolla of their central florets. Doubleness is sometimes connected with prolification,[92] or the continued growth of the axis of the flower. Doubleness is strongly inherited. No one has produced, as Lindley remarks,[93] double flowers by promoting the perfect health of the plant. On the contrary, unnatural conditions of life favour their production. There is some reason to believe that seeds kept during many years, and seeds believed to be imperfectly fertilised, yield double flowers more freely than fresh and perfectly fertilised seed.[94] Long-continued cultivation in rich soil seems to be the commonest exciting cause. A double narcissus and a double *Anthemis nobilis*, transplanted into very poor soil, have been observed to become single;[95] and I have seen a completely double white primrose rendered permanently single by being divided and transplanted whilst in full flower. It has been observed by Professor Morren that doubleness of the flowers and variegation of the leaves are antagonistic states; but so many exceptions to the rule have lately been recorded,[96] that, though general, it cannot be looked at as invariable.

[91] 'Comptes Rendus,' Dec. 19th, 1864, p. 1039.
[92] 'Gardener's Chronicle,' 1866, p. 681.
[93] 'Theory of Horticulture,' p. 333.
[94] Mr. Fairweather, in 'Transact. Hort. Soc.,' vol. iii. p. 406; Bosse, quoted by Bronn, 'Geschichte der Natur,' B. ii. s. 77. On the effects of the removal of the anthers, *see* Mr. Leitner, in Silliman's 'North American Journ. of Science,' vol. xxiii. p. 47; and Verlot, 'Des Variétés,' 1865, p. 84.
[95] Lindley's 'Theory of Horticulture,' p. 333.
[96] 'Gardener's Chronicle,' 1865, p. 626; 1866, pp. 290, 730; and Verlot, 'Des Variétés,' p. 75.

Variegation seems generally to result from a feeble or atrophied condition of the plant, and a large proportion of the seedlings raised from parents both of which are variegated usually perish at an early age; hence we may perhaps infer that doubleness, which is the antagonistic state, commonly arises from a plethoric condition. On the other hand, extremely poor soil sometimes, though rarely, appears to cause doubleness: I formerly described[97] some completely double, bud-like, flowers produced in large numbers by stunted wild plants of *Gentiana amarella* growing on a poor chalky bank. I have also noticed a distinct tendency to doubleness in the flowers of a Ranunculus, Horse-chesnut, and Bladder-nut (*Ranunculus repens, Œsculus pavia,* and *Staphylea*), growing under very unfavourable conditions. Professor Lehman[98] found several wild plants growing near a hot spring with double flowers. With respect to the cause of doubleness, which arises, as we see, under widely different circumstances, I shall presently attempt to show that the most probable view is that unnatural conditions first give a tendency to sterility, and that then, on the principle of compensation, as the reproductive organs do not perform their proper functions, they either become developed into petals, or additional petals are formed. This view has lately been supported by Mr. Laxton,[99] who advances the case of some common peas, which, after long-continued heavy rain, flowered a second time, and produced double flowers.

Seedless Fruit.—Many of our most valuable fruits, although consisting in a homological sense of widely different organs, are either quite sterile, or produce extremely few seeds. This is notoriously the case with our best pears, grapes, and figs, with the pine-apple, banana, bread-fruit, pomegranate, azarole, date-palms, and some members of the orange-tribe. Poorer varieties of these same fruits either habitually or occasionally yield seed.[100] Most horticulturists look at the great size and anomalous development of the fruit as the cause, and sterility as the result; but the opposite view, as we shall presently see, is more probable.

Sterility from the excessive development of the Organs of Growth or Vegetation.—Plants which from any cause grow too luxuriantly, and produce leaves, stems, runners, suckers, tubers, bulbs, &c., in excess, sometimes do not flower, or if they flower do not yield seed. To make European vegetables under the hot climate of India yield seed, it is necessary to check their growth; and, when one-third grown, they are taken up, and their stems and

---

[97] 'Gardener's Chronicle,' 1843, p. 628. In this article I suggested the following theory on the doubleness of flowers.

[98] Quoted by Gärtner, 'Bastarderzeugung,' s. 567.

[99] 'Gardener's Chronicle,' 1866, p. 901.

[100] Lindley, 'Theory of Horticulture,' p. 175-179; Godron, 'De l'Espèce,' tom. i. p. 106: Pickering, 'Races of Man;' Gallesio, 'Teoria della Riproduzione,

1816, p. 101-110. Meyen ('Reise um Erde,' Th. ii. s. 214) states that at Manilla one variety of the banana is full of seeds; and Chamisso (Hooker's 'Bot. Misc.,' vol. i. p. 310) describes a variety of the bread-fruit in the Mariana Islands with small fruit, containing seeds which are frequently perfect. Burnes, in his 'Travels in Bokhara,' remarks on the pomegranate seeding in Mazenderan, as a remarkable peculiarity.

tap-roots are cut or mutilated.[101] So it is with hybrids; for instance, Prof. Lecoq[102] had three plants of Mirabilis, which, though they grew luxuriantly and flowered, were quite sterile; but after beating one with a stick until a few branches alone were left, these at once yielded good seed. The sugar-cane, which grows vigorously and produces a large supply of succulent stems, never, according to various observers, bears seed in the West Indies, Malaga, India, Cochin China, or the Malay Archipelago.[103] Plants which produce a large number of tubers are apt to be sterile, as occurs, to a certain extent, with the common potato; and Mr. Fortune informs me that the sweet potato (*Convolvulus batatas*) in China never, as far as he has seen, yields seed. Dr. Royle remarks[104] that in India the *Agave vivipara*, when grown in rich soil, invariably produces bulbs, but no seeds; whilst a poor soil and dry climate leads to an opposite result. In China, according to Mr. Fortune, an extraordinary number of little bulbs are developed in the axils of the leaves of the yam, and this plant does not bear seed. Whether in these cases, as in those of double flowers and seedless fruit, sexual sterility from changed conditions of life is the primary cause which leads to the excessive development of the organs of vegetation, is doubtful; though some evidence might be advanced in favour of this view. It is perhaps a more probable view that plants which propagate themselves largely by one method, namely by buds, have not sufficient vital power or organised matter for the other method of sexual generation.

Several distinguished botanists and good practical judges believe that long-continued propagation by cuttings, runners, tubers, bulbs, &c., independently of any excessive development of these parts, is the cause of many plants failing to produce flowers and of others failing to produce fertile flowers,—it is as if they had lost the habit of sexual generation.[105] That many plants when thus propagated are sterile there can be no doubt, but whether the long continuance of this form of propagation is the actual cause of their sterility, I will not venture, from the want of sufficient evidence, to express an opinion.

That plants may be propagated for long periods by buds, without the aid of sexual generation, we may safely infer from this being the case with many plants which must have long survived in a state of nature. As I have had occasion before to allude to this subject, I will here give such cases as I have collected. Many alpine plants ascend mountains beyond the height at which they can produce seed.[106] Certain species of

---

[101] Ingledew, in 'Transact. of Agricult. and Hort. Soc. of India,' vol. ii.

[102] 'De la Fécondation,' 1862, p. 308.

[103] Hooker's 'Bot. Misc.,' vol. i. p. 99; Gallesio, 'Teoria della Riproduzione,' p. 110.

[104] 'Transact. Linn. Soc.,' vol. xvii. p. 563.

[105] Godron, 'De l'Espèce,' tom. ii. p. 106; Herbert on Crocus, in 'Journal

of Hort. Soc.,' vol. i., 1846, p. 254—Dr. Wight, from what he has seen in India, believes in this view; 'Madras Journal of Lit. and Science,' vol. iv., 1836, p. 61.

[106] Wahlenberg specifies eight species in this state on the Lapland Alps: *see* Appendix to Linnæus' 'Tour in Lapland,' translated by Sir J. E. Smith, vol. ii. pp. 274-280.

Poa and Festuca, when growing on mountain-pastures, propagate themselves, as I hear from Mr. Bentham, almost exclusively by bulblets. Kalm. gives a more curious instance[107] of several American trees, which grow so plentifully in marshes or in thick woods, that they are certainly well adapted for these stations, yet scarcely ever produce seeds; but when accidentally growing on the outside of the marsh or wood, are loaded with seed. The common ivy is found in Northern Sweden and Russia, but flowers and fruits only in the southern provinces. The *Acorus calamus* extends over a large portion of the globe, but so rarely perfects its fruit that this has been seen but by few botanists.[108] The *Hypericum calycinum*, which propagates itself so freely in our shrubberies by rhizomas, and is naturalised in Ireland, blossoms profusely, but sets no seed; nor did it set any when fertilised in my garden by pollen from plants growing at a distance. The *Lysimachia nummularia*, which is furnished with long runners, so seldom produces seed-capsules, that Prof. Decaisne,[109] who has especially attended to this plant, has never seen it in fruit. The *Carex rigida* often fails to perfect its seed in Scotland, Lapland, Greenland, Germany, and New Hampshire in the United States.[110] The periwinkle (*Vinca minor*), which spreads largely by runners, is said scarcely ever to produce fruit in England;[111] but this plant requires insect-aid for its fertilisation, and the proper insects may be absent or rare. The *Jussiœa grandiflora* has become naturalised in Southern France, and has spread by its rhizomas so extensively as to impede the navigation of the waters, but never produces fertile seed.[112] The horse-radish (*Cochlearia armoracia*) spreads pertinaciously and is naturalised in various parts of Europe; though it bears flowers, these rarely produce capsules: Professor Caspary also informs me that he has watched this plant since 1851, but has never seen its fruit; nor is this surprising, as he finds scarcely a grain of good pollen. The common little *Ranunculus ficaria* rarely, and some say never, bears seed in England, France, or Switzerland; but in 1863 I observed seeds on several plants growing near my house. According to M. Chatin, there are two forms of this Ranunculus; and it is the bulbiferous form which does not yield seed from producing no pollen.[113] Other cases analo-

---

[107] 'Travels in North America,' Eng. translat., vol. iii. p. 175.

[108] With respect to the ivy and Acorus, *see* Dr. Bromfield in the 'Phytologist,' vol. iii. p. 376. *See* also Lindley and Vaucher on the Acorus.

[109] 'Annal. des Sc. Nat.,' 3rd series, Zool., tom. iv. p. 280. Prof. Decaisne refers also to analogous cases with mosses and lichens near Paris.

[110] Mr. Tuckerman, in Silliman's 'American Journal of Science,' vol. xlv. p. 41.

[111] Sir J. E. Smith, 'English Flora,' vol. i. p. 339.

[112] G. Planchon, 'Flora de Montpellier,' 1864, p. 20.

[113] On the non-production of seeds in England *see* Mr. Crocker, in 'Gardener's Weekly Magazine,' 1852, p. 70; Vaucher, 'Hist. Phys. Plantes d'Europe,' tom. i. p. 33; Lecoq, 'Géograph. Bot. de l'Europe,' tom. iv. p. 466; Dr. D. Clos, in 'Annal. des Sc. Nat.,' 3rd series, Bot., tom. xvii., 1852, p. 129: this latter author refers to other analogous cases. On the non-production of pollen by this Ranunculus *see* Chatin, in 'Comptes Rendus,' June 11th, 1866.

gous with the foregoing could be given; for instance, some kinds of mosses and lichens have never been seen to fructify in France.

Some of these endemic and naturalised plants are probably rendered sterile from excessive multiplication by buds, and their consequent incapacity to produce and nourish seed. But the sterility of others more probably depends on the peculiar conditions under which they live, as in the case of the ivy in the northern parts of Europe, and of the trees in the swamps of the United States; yet these plants must be in some respects eminently well adapted for the stations which they occupy, for they hold their places against a host of competitors.

Finally, when we reflect on the sterility which accompanies the doubling of flowers,—the excessive development of fruit, —and a great increase in the organs of vegetation, we must bear in mind that the whole effect has seldom been caused at once. An incipient tendency is observed, and continued selection completes the work, as is known to be the case with our double flowers and best fruits. The view which seems the most probable, and which connects together all the foregoing facts and brings them within our present subject, is, that changed and unnatural conditions of life first give a tendency to sterility; and in consequence of this, the organs of reproduction being no longer able fully to perform their proper functions, a supply of organised matter, not required for the development of the seed, flows either into these same organs and renders them foliaceous, or into the fruit, stems, tubers, &c., increasing their size and succulency. But I am far from wishing to deny that there exists, independently of any incipient sterility, an antagonism between the two forms of reproduction, namely, by seed and by buds, when either is carried to an extreme degree. That incipient sterility plays an important part in the doubling of flowers, and in the other cases just specified, I infer chiefly from the following facts. When fertility is lost from a wholly different cause, namely, from hybridism, there is a strong tendency, as Gärtner[114] affirms, for flowers to become double, and this tendency is inherited. Moreover it is notorious that with hybrids the male organs become sterile before the female organs, and with double flowers the stamens first become foli-

[114] 'Bastarderzeugung,' s. 565. Kölreuter ('Dritte Fortsetzung,' s. 73, 87, 119) also shows that when two species, one single and the other double, are crossed, the hybrids are apt to be extremely double.

aceous. This latter fact is well shown by the male flowers of dioecious plants, which, according to Gallesio,[115] first become double. Again, Gärtner[116] often insists that the flowers of even utterly sterile hybrids, which do not produce any seed, generally yield perfect capsules or fruit,—a fact which has likewise been repeatedly observed by Naudin with the Cucurbitaceæ; so that the production of fruit by plants rendered sterile through any other and distinct cause is intelligible. Kölreuter has also expressed his unbounded astonishment at the size and development of the tubers in certain hybrids; and all experimentalists[117] have remarked on the strong tendency in hybrids to increase by roots, runners, and suckers. Seeing that hybrid plants, which from their nature are more or less sterile, thus tend to produce double flowers; that they have the parts including the seed, that is the fruit, perfectly developed, even when containing no seed; that they sometimes yield gigantic roots; that they almost invariably tend to increase largely by suckers and other such means;—seeing this, and knowing, from the many facts given in the earlier parts of this chapter, that almost all organic beings when exposed to unnatural conditions tend to become more or less sterile, it seems much the most probable view that with cultivated plants sterility is the exciting cause, and double flowers, rich seedless fruit, and in some cases largely-developed organs of vegetation, &c., are the indirect results—these results having been in most cases largely increased through continued selection by man.

[115] 'Teoria della Riproduzione Veg.,' 1816, p. 73.
[116] 'Bastarderzeugung,' s. 573.   [117] Ibid., s. 527.

CHAPTER XIX.

SUMMARY OF THE FOUR LAST CHAPTERS, WITH REMARKS ON HYBRIDISM.

ON THE EFFECTS OF CROSSING — THE INFLUENCE OF DOMESTICATION ON FERTILITY — CLOSE INTERBREEDING — GOOD AND EVIL RESULTS FROM CHANGED CONDITIONS OF LIFE — VARIETIES WHEN CROSSED NOT INVARIABLY FERTILE — ON THE DIFFERENCE IN FERTILITY BETWEEN CROSSED SPECIES AND VARIETIES — CONCLUSIONS WITH RESPECT TO HYBRIDISM — LIGHT THROWN ON HYBRIDISM BY THE ILLEGITIMATE PROGENY OF DIMORPHIC AND TRIMORPHIC PLANTS — STERILITY OF CROSSED SPECIES DUE TO DIFFERENCES CONFINED TO THE REPRODUCTIVE SYSTEM — NOT ACCUMULATED THROUGH NATURAL SELECTION — REASONS WHY DOMESTIC VARIETIES ARE NOT MUTUALLY STERILE — TOO MUCH STRESS HAS BEEN LAID ON THE DIFFERENCE IN FERTILITY BETWEEN CROSSED SPECIES AND CROSSED VARIETIES — CONCLUSION.

It was shown in the fifteenth chapter that when individuals of the same variety, or even of a distinct variety, are allowed freely to intercross, uniformity of character is ultimately acquired. Some few characters, however, are incapable of fusion, but these are unimportant, as they are almost always of a semi-monstrous nature, and have suddenly appeared. Hence, to preserve our domesticated breeds true, or to improve them by methodical selection, it is obviously necessary that they should be kept separate. Nevertheless, through unconscious selection, a whole body of individuals may be slowly modified, as we shall see in a future chapter, without separating them into distinct lots. Domestic races have often been intentionally modified by one or two crosses, made with some allied race, and occasionally even by repeated crosses with very distinct races; but in almost all such cases, long-continued and careful selection has been absolutely necessary, owing to the excessive variability of the crossed offspring, due to the principle of reversion. In a few instances, however, mongrels have retained a uniform character from their first production.

When two varieties are allowed to cross freely, and one is

much more numerous than the other, the former will ultimately absorb the latter. Should both varieties exist in nearly
equal numbers, it is probable that a considerable period would
elapse before the acquirement of a uniform character; and the
character ultimately acquired would largely depend on prepotency of transmission, and on the conditions of life; for the
nature of these conditions would generally favour one variety
more than another, so that a kind of natural selection would
come into play. Unless the crossed offspring were slaughtered
by man without the least discrimination, some degree of un-
methodical selection would likewise come into action. From
these several considerations we may infer, that when two or
more closely allied species first came into the possession of
the same tribe, their crossing will not have influenced, in so
great a degree as has often been supposed, the character of the
offspring in future times; although in some cases it probably
has had a considerable effect.

Domestication, as a general rule, increases the prolificness
of animals and plants. It eliminates the tendency to sterility
which is common to species when first taken from a state of
nature and crossed. On this latter head we have no direct
evidence; but as our races of dogs, cattle, pigs, &c., are almost
certainly descended from aboriginally distinct stocks, and as
these races are now fully fertile together, or at least incom-
parably more fertile than most species when crossed, we may
with much confidence accept this conclusion.

Abundant evidence has been given that crossing adds to the
size, vigour, and fertility of the offspring. This holds good when
there has been no previous close interbreeding. It applies to
the individuals of the same variety but belonging to different
families, to distinct varieties, sub-species, and partially even to
species. In the latter case, though size is often gained, fertility
is lost; but the increased size, vigour, and hardiness of many
hybrids cannot be accounted for solely on the principle of compensation from the inaction of the reproductive system. Certain
plants, both of pure and hybrid origin, though perfectly healthy,
have become self-impotent, apparently from the unnatural conditions to which they have been exposed; and such plants, as
well as others in their normal state, can be stimulated to fer-

tility only by crossing them with other individuals of the same species or even of a distinct species.

On the other hand, long-continued close interbreeding between the nearest relations diminishes the constitutional vigour, size, and fertility of the offspring; and occasionally leads to malformations, but not necessarily to general deterioration of form or structure. This failure of fertility shows that the evil results of interbreeding are independent of the augmentation of morbid tendencies common to both parents, though this augmentation no doubt is often highly injurious. Our belief that evil follows from close interbreeding rests to a large extent on the experience of practical breeders, especially of those who have reared many animals of the kinds which can be propagated quickly; but it likewise rests on several carefully recorded experiments. With some animals close interbreeding may be carried on for a long period with impunity by the selection of the most vigorous and healthy individuals; but sooner or later evil follows. The evil, however, comes on so slowly and gradually that it easily escapes observation, but can be recognised by the almost instantaneous manner in which size, constitutional vigour, and fertility are regained when animals that have long been interbred are crossed with a distinct family.

These two great classes of facts, namely, the good derived from crossing, and the evil from close interbreeding, with the consideration of the innumerable adaptations throughout nature for compelling, or favouring, or at least permitting, the occasional union of distinct individuals, taken together, lead to the conclusion that it is a law of nature that organic beings shall not fertilise themselves for perpetuity. This law was first plainly hinted at in 1799, with respect to plants, by Andrew Knight,[1] and, not long afterwards, that sagacious observer Kölreuter, after showing how well the Malvaceæ are adapted for

---

[1] 'Transactions Phil. Soc.,' 1799, p. 202. For Kölreuter, see 'Mém. de l'Acad. de St. Pétersbourg,' tom. iii., 1809 (published 1811), p. 197. In reading C. K. Sprengel's remarkable work, 'Das entdeckte Geheimniss,' &c., 1793, it is curious to observe how often this wonder- fully acute observer failed to understand the full meaning of the structure of the flowers which he has so well described, from not always having before his mind the key to the problem, namely, the good derived from the crossing of distinct individual plants.

crossing, asks, "an id aliquid in recessu habeat, quod hujus-
cemodi flores nunquam proprio suo pulvere, sed semper eo
aliarum suæ speciei impregnentur, merito quæritur? Certe
natura nil facit frustra." Although we may demur to Kölreuter's saying that nature does nothing in vain, seeing how
many organic beings retain rudimentary and useless organs,
yet undoubtedly the argument from the innumerable contrivances, which favour the crossing of distinct individuals of the
same species, is of the greatest weight. The most important
result of this law is that it leads to uniformity of character in
the individuals of the same species. In the case of certain
hermaphrodites, which probably intercross only at long intervals
of time, and with unisexual animals inhabiting somewhat separated localities, which can only occasionally come into contact
and pair, the greater vigour and fertility of the crossed offspring
will ultimately prevail in giving uniformity of character to the
individuals of the same species. But when we go beyond the
limits of the same species, free intercrossing is barred by the
law of sterility.

In searching for facts which might throw light on the cause of
the good effects from crossing, and of the evil effects from close
interbreeding, we have seen that, on the one hand, it is a widely
prevalent and ancient belief that animals and plants profit from
slight changes in their condition of life; and it would appear
that the germ, in a somewhat analogous manner, is more effectually stimulated by the male element, when taken from a distinct individual, and therefore slightly modified in nature, than
when taken from a male having the same identical constitution.
On the other hand, numerous facts have been given, showing
that when animals are first subjected to captivity, even in their
native land, and although allowed much liberty, their reproductive functions are often greatly impaired or quite annulled.
Some groups of animals are more affected than others, but with
apparently capricious exceptions in every group. Some animals
never or rarely couple: some couple freely, but never or rarely
conceive. The secondary male characters, the maternal functions and instincts, are occasionally affected. With plants,
when first subjected to cultivation, analogous facts have been
observed. We probably owe our double flowers, rich seedless

fruits, and in some cases greatly developed tubers, &c., to incipient sterility of the above nature combined with a copious supply of nutriment. Animals which have long been domesticated, and plants which have long been cultivated, can generally withstand with unimpaired fertility great changes in their conditions of life; though both are sometimes slightly affected. With animals the somewhat rare capacity of breeding freely under confinement has mainly determined, together with their utility, the kinds which have been domesticated.

We can in no case precisely say what is the cause of the diminished fertility of an animal when first captured, or of a plant when first cultivated; we can only infer that it is caused by a change of some kind in the natural conditions of life. The remarkable susceptibility of the reproductive system to such changes,—a susceptibility not common to any other organ,—apparently has an important bearing on Variability, as we shall see in a future chapter.

It is impossible not to be struck with the double parallelism between the two classes of facts just alluded to. On the one hand, slight changes in the conditions of life, and crosses between slightly modified forms or varieties, are beneficial as far as prolificness and constitutional vigour are concerned. On the other hand, changes in the conditions greater in degree, or of a different nature, and crosses between forms which have been slowly and greatly modified by natural means,—in other words, between species,—are highly injurious, as far as the reproductive system is concerned, and in some few instances as far as constitutional vigour is concerned. Can this parallelism be accidental ? Does it not rather indicate some real bond of connection ? As a fire goes out unless it be stirred up, so the vital forces are always tending, according to Mr. Herbert Spencer, to a state of equilibrium, unless disturbed and renovated through the action of other forces.

In some few cases varieties tend to keep distinct, by breeding at different periods, by great differences in size, or by sexual preference,—in this latter respect more especially resembling species in a state of nature. But the actual crossing of varieties, far from diminishing, generally adds to the fertility of both the first union and the mongrel offspring. Whether all the most widely distinct domestic varieties are invariably quite fertile when crossed, we do not positively know; much time and trouble would be requisite for the necessary experiments, and many difficulties occur, such as the descent of the various races from aboriginally distinct species, and the doubts whether certain forms ought to be ranked as species or varieties. Nevertheless, the wide experience of practical breeders proves that the great majority of varieties, even if some should hereafter prove not to be indefinitely fertile *inter se*, are far more fertile when crossed, than the vast majority of closely allied natural species. A few remarkable cases have, however, been given on the authority of excellent observers, showing that with plants certain forms, which undoubtedly must be ranked as varieties, yield fewer seeds when crossed than is natural to the parent-species. Other varieties have had their reproductive powers so far modified that they are either more or less fertile than are their parents, when crossed with a distinct species.

Nevertheless, the fact remains indisputable that domesticated varieties of animals and of plants, which differ greatly from each other in structure, but which are certainly descended from the same aboriginal species, such as the races of the fowl, pigeon, many vegetables, and a host of other productions, are extremely fertile when crossed; and this seems to make a broad and impassable barrier between domestic varieties and natural species. But, as I will now attempt to show, the distinction is not so great and overwhelmingly important as it at first appears.

On the Difference in Fertility between Varieties and Species when crossed.

This work is not the proper place for fully treating the subject of hybridism, and I have already given in my 'Origin of Species' a moderately full abstract. I will here merely enumerate the general conclusions which may be relied on, and which bear on our present point.

Firstly, the laws governing the production of hybrids are identical, or nearly identical, in the animal and vegetable kingdoms.

Secondly, the sterility of distinct species when first united, and that of their hybrid offspring, graduates, by an almost infinite number of steps, from zero, when the ovule is never impregnated and a seed-capsule is never formed, up to complete fertility. We can only escape the conclusion that some species are fully fertile when crossed, by determining to designate as varieties all the forms which are quite fertile. This high degree of fertility is, however, rare. Nevertheless plants, which have been exposed to unnatural conditions, sometimes become modified in so peculiar a manner, that they are much more fertile when crossed by a distinct species than when fertilised by their own pollen. Success in effecting a first union between two species, and the fertility of their hybrids, depends in an eminent degree on the conditions of life being favourable. The innate sterility of hybrids of the same parentage and raised from the same seed-capsule often differs much in degree.

Thirdly, the degree of sterility of a first cross between two species does not always run strictly parallel with that of their hybrid offspring. Many cases are known of species which can be crossed with ease, but yield hybrids excessively sterile; and conversely some which can be crossed with great difficulty, but produce fairly fertile hybrids. This is an inexplicable fact, on the view that species have been specially endowed with mutual sterility in order to keep them distinct.

Fourthly, the degree of sterility often differs greatly in two species when reciprocally crossed; for the first will readily fertilise the second; but the latter is incapable, after hundreds of trials, of fertilising the former. Hybrids produced from reciprocal crosses between the same two species, likewise sometimes differ in their degree of sterility. These cases also are utterly inexplicable on the view of sterility being a special endowment.

Fifthly, the degree of sterility of first crosses and of hybrids runs, to a certain extent, parallel with the general or systematic affinity of the forms which are united. For species belonging to distinct genera can rarely, and those belonging to distinct families can never, be crossed. The parallelism, however, is far from complete; for a multitude of closely allied species will not unite, or unite with extreme difficulty, whilst other species, widely different from each other, can be crossed with perfect facility. Nor does the difficulty depend on ordinary constitutional differences, for annual and perennial plants, deciduous and evergreen trees, plants flowering at different seasons, inhabiting different stations, and naturally living under the most opposite climates, can often be crossed with ease. The difficulty or facility apparently depends exclusively on the sexual constitution of the species which are crossed; or on their sexual elective affinity, *i. e. Wahlverwandtschaft* of Gärtner. As species rarely or never become modified in one character, without being at the same time modified in many, and as systematic affinity includes all visible resemblances and dissimilarities, any difference in sexual constitution between two species would naturally stand in more or less close relation with their systematic position.

Sixthly, the sterility of species when first crossed, and that of hybrids, may possibly depend to a certain extent on distinct causes. With pure species the reproductive organs are in a perfect condition, whilst with hybrids they are often plainly deteriorated. A hybrid embryo which partakes of the constitution of its father and mother is exposed to unnatural conditions, as long as it is nourished within the womb, or egg, or seed of the mother-form; and as we know that unnatural conditions often induce sterility, the reproductive organs of the hybrid might at this early age be permanently affected. But this cause has no bearing on the infertility of first unions. The diminished number of the offspring from first unions may often result, as is certainly sometimes the case, from the premature death of most of the hybrid embryos. But we shall immediately see that a law of an unknown nature apparently exists, which causes the offspring from unions, which are infertile, to be themselves more or less infertile; and this at present is all that can be said.

Seventhly, hybrids and mongrels present, with the one great exception of fertility, the most striking accordance in all other respects; namely, in the laws of their resemblance to their two parents, in their tendency to reversion, in their variability, and in being absorbed through repeated crosses by either parent-form.

Since arriving at the foregoing conclusions, condensed from my former work, I have been led to investigate a subject which throws considerable light on hybridism, namely, the fertility of

reciprocally dimorphic and trimorphic plants, when illegitimately united. I have had occasion several times to allude to
these plants, and I may here give a brief abstract[2] of my
observations. Several plants belonging to distinct orders present two forms, which exist in about equal numbers, and which
differ in no respect except in their reproductive organs; one
form having a long pistil with short stamens, the other a short
pistil with long stamens; both with differently sized pollen-
grains. With trimorphic plants there are three forms likewise
differing in the lengths of their pistils and stamens, in the size
and colour of the pollen-grains, and in some other respects; and
as in each of the three forms there are two sets of stamens,
there are altogether six sets of stamens and three kinds of
pistils. These organs are so proportioned in length to each other
that, in any two of the forms, half the stamens in each stand
on a level with the stigma of the third form. Now I have shown,
and the result has been confirmed by other observers, that, in
order to obtain full fertility with these plants, it is necessary
that the stigma of the one form should be fertilised by pollen
taken from the stamens of corresponding height in the other
form. So that with dimorphic species two unions, which may
be called legitimate, are fully fertile, and two, which may be
called illegitimate, are more or less infertile. With trimorphic
species six unions are legitimate or fully fertile, and twelve are
illegitimate or more or less infertile.

The infertility which may be observed in various dimorphic
and trimorphic plants, when they are illegitimately fertilised,
that is, by pollen taken from stamens not corresponding in height
with the pistil, differs much in degree, up to absolute and utter
sterility; just in the same manner as occurs in crossing distinct
species. As the degree of sterility in the latter case depends
in an eminent degree on the conditions of life being more or
less favourable, so I have found it with illegitimate unions. It
is well known that if pollen of a distinct species be placed on
the stigma of a flower, and its own pollen be afterwards, even

---

[2] This abstract was published in the
fourth edition (1866) of my 'Origin of
Species;' but as this edition will be in
the hands of but few persons, and as

my original observations on this point
have not as yet been published in
detail, I have ventured here to reprint
the abstract.

after a considerable interval of time, placed on the same stigma, its action is so strongly prepotent that it generally annihilates the effect of the foreign pollen; so it is with the pollen of the several forms of the same species, for legitimate pollen is strongly prepotent over illegitimate pollen, when both are placed on the same stigma. I ascertained this by fertilising several flowers, first illegitimately, and twenty-four hours afterwards legitimately, with pollen taken from a peculiarly coloured variety, and all the seedlings were similarly coloured; this shows that the legitimate pollen, though applied twenty-four hours subsequently, had wholly destroyed or prevented the action of the previously applied illegitimate pollen. Again, as, in making reciprocal crosses between the same two species, there is occasionally a great difference in the result, so something analogous occurs with dimorphic plants; for a short-styled cowslip (*P. veris*) yields more seed when fertilised by the long-styled form, and less seed when fertilised by its own form, compared with a long-styled cowslip when fertilised in the two corresponding methods.

In all these respects the forms of the same undoubted species, when illegitimately united, behave in exactly the same manner as do two distinct species when crossed. This led me carefully to observe during four years many seedlings, raised from several illegitimate unions. The chief result is that these illegitimate plants, as they may be called, are not fully fertile. It is possible to raise from dimorphic species, both long-styled and short-styled illegitimate plants, and from trimorphic plants all three illegitimate forms. These can then be properly united in a legitimate manner. When this is done, there is no apparent reason why they should not yield as many seeds as did their parents when legitimately fertilised. But such is not the case; they are all infertile, but in various degrees; some being so utterly and incurably sterile that they did not yield during four seasons a single seed or even seed-capsule. These illegitimate plants, which are so sterile, although united with each other in a legitimate manner, may be strictly compared with hybrids when crossed *inter se*, and it is well known how sterile these latter generally are. When, on the other hand, a hybrid is crossed with either pure parent-species, the sterility is usually much lessened: and so it is when an illegitimate plant is fertilised by a legitimate plant. In the same manner as the sterility of hybrids does not always run parallel with the difficulty of making the first cross between the two parent species, so the sterility of certain illegitimate plants was unusually great, whilst the sterility of the union from which they were derived was by no means great. With hybrids raised from the same seed-capsule the degree of sterility is innately variable, so it is in a marked manner with illegitimate plants. Lastly, many hybrids are profuse and persistent flowerers, whilst other and more sterile hybrids produce few flowers, and are weak, miserable dwarfs; exactly similar cases occur with the illegitimate offspring of various dimorphic and trimorphic plants.

Altogether there is the closest identity in character and behaviour between illegitimate plants and hybrids. It is hardly an exaggeration to maintain that the former are hybrids, but produced within the limits of the same species by the improper union of certain forms, whilst ordinary hybrids are produced from an improper union between so-called distinct species. We have already seen that there is the closest similarity in all respects between first illegitimate unions, and first crosses between distinct species. This will perhaps be made more fully apparent by an illustration: we may suppose that a botanist found two well-marked varieties (and such occur) of the long-styled form of the trimorphic *Lythrum salicaria*, and that he determined to try by crossing whether they were specifically distinct. He would find that they yielded only about one-fifth of the proper number of seed, and that they behaved in all the other above-specified respects as if they had been two distinct species. But to make the case sure, he would raise plants from his supposed hybridised seed, and he would find that the seedlings were miserably dwarfed and utterly sterile, and that they behaved in all other respects like ordinary hybrids. He might then maintain that he had actually proved, in accordance with the common view, that his two varieties were as good and as distinct species as any in the world; but he would be completely mistaken.

The facts now given on dimorphic and trimorphic plants are important, because they show us, firstly, that the physiological

test of lessened fertility, both in first crosses and in hybrids, is
no safe criterion of specific distinction; secondly, because we
may conclude that there must be some unknown law or bond
connecting the infertility of illegitimate unions with that of
their illegitimate offspring, and we are thus led to extend this
view to first crosses and hybrids; thirdly, because we find, and
this seems to me of especial importance, that with trimorphic
plants three forms of the same species exist, which when crossed
in a particular manner are infertile, and yet these forms differ
in no respect from each other, except in their reproductive
organs,—as in the relative length of the stamens and pistils,
in the size, form, and colour of the pollen-grains, in the structure of the stigma, and in the number and size of the seeds.
With these differences and no others, either in organisation or
constitution, we find that the illegitimate unions and the illegitimate progeny of these three forms are more or less sterile,
and closely resemble in a whole series of relations the first
unions and hybrid offspring of distinct species. From this we
may infer that the sterility of species when crossed and of their
hybrid progeny is likewise in all probability exclusively due
to differences confined to the reproductive system. We have
indeed been brought to a similar conclusion by observing that
the sterility of crossed species does not strictly coincide with
their systematic affinity, that is, with the sum of their external
resemblances; nor does it coincide with their similarity in
general constitution. But we are more especially led to this
same conclusion by considering reciprocal crosses, in which the
male of one species cannot be united, or can be united with
extreme difficulty, with the female of a second species, whilst
the converse cross can be effected with perfect facility; for
this difference in the facility of making reciprocal crosses, and
in the fertility of their offspring, must be attributed either to
the male or female element in the first species having been
differentiated with reference to the sexual element of the second
species in a higher degree than in the converse case. In so
complex a subject as Hybridism it is of considerable importance
thus to arrive at a definitive conclusion, namely, that the
sterility which almost invariably follows the union of distinct

species depends exclusively on differences in their sexual constitution.

On the principle which makes it necessary for man, whilst he is selecting and improving his domestic varieties, to keep them separate, it would clearly be advantageous to varieties in a state of nature, that is to incipient species, if they could be kept from blending, either through sexual aversion, or by becoming mutually sterile. Hence it at one time appeared to me probable, as it has to others, that this sterility might have been acquired through natural selection. On this view we must suppose that a shade of lessened fertility first spontaneously appeared, like any other modification, in certain individuals of a species when crossed with other individuals of the same species; and that successive slight degrees of infertility, from being advantageous, were slowly accumulated. This appears all the more probable, if we admit that the structural differences between the forms of dimorphic and trimorphic plants, as the length and curvature of the pistil, &c., have been co-adapted through natural selection; for if this be admitted, we can hardly avoid extending the same conclusion to their mutual infertility. Sterility moreover has been acquired through natural selection for other and widely different purposes, as with neuter insects in reference to their social economy. In the case of plants, the flowers on the circumference of the truss in the guelder-rose (*Viburnum opulus*) and those on the summit of the spike in the feather-hyacinth (*Muscari comosum*) have been rendered conspicuous, and apparently in consequence sterile, in order that insects might easily discover and visit the other flowers. But when we endeavour to apply the principle of natural selection to the acquirement by distinct species of mutual sterility, we meet with great difficulties. In the first place, it may be remarked that separate regions are often inhabited by groups of species or by single species, which when brought together and crossed are found to be more or less sterile; now it could clearly have been of no advantage to such separated species to have been rendered mutually sterile, and consequently this could not have been effected through natural selection; but it may perhaps be argued, that, if a species were rendered sterile with some one compatriot, sterility with other species would follow as a necessary consequence. In the second place, it is as much opposed to the theory of natural selection, as to the theory of special creation, that in reciprocal crosses the male element of one form should have been rendered utterly impotent on a second form, whilst at the same time the male element of this second form is enabled freely to fertilise the first form; for this peculiar state of the reproductive system could not possibly be advantageous to either species.

In considering the probability of natural selection having come into action in rendering species mutually sterile, one great difficulty will be found to lie in the existence of many graduated steps from slightly lessened fertility to absolute sterility. It may be admitted, on the principle above explained, that it would profit an incipient species if it were rendered in some slight degree sterile when crossed with its parent-form or with some other variety; for thus fewer bastardised and dete-riorated offspring would be produced to commingle their blood with the new species in process of formation. But he who will take the trouble to reflect on the steps by which this first degree of sterility could be increased through natural selec-tion to that higher degree which is common to so many species, and which is universal with species which have been dif-ferentiated to a generic or family rank, will find the subject extra-ordinarily complex. After mature reflection it seems to me that this could not have been effected through natural selection; for it could have been of no direct advantage to an individual animal to breed badly with another individual of a different variety, and thus leave few offspring; consequently such indi-viduals could not have been preserved or selected. Or take the case of two species which in their present state, when crossed, produce few and sterile offspring; now, what is there which could favour the survival of those individuals which happened to be endowed in a slightly higher degree with mutual infertility and which thus approached by one small step towards absolute sterility? yet an advance of this kind, if the theory of natural selection be brought to bear, must have incessantly occurred with many species, for a multitude are mutually quite barren. With sterile neuter insects we have reason to believe that modifications in their structure have been slowly accumulated by natural selection, from an advantage having been thus indirectly given to the community to which they belonged over other communities of the same species; but an individual animal, if rendered slightly sterile when crossed with some other variety, would not thus in itself gain any advantage, or indirectly give any advantage to its nearest relatives or to other individuals of the same variety, leading to their preservation. I infer from these considerations that, as far as animals are concerned, the various degrees of lessened fertility which occur with species when crossed cannot have been slowly accumulated by means of natural selection.

With plants, it is possible that the case may be somewhat different. With many kinds, insects constantly carry pollen from neighbouring plants to the stigmas of each flower; and with some species this is effected by the wind. Now, if the pollen of a variety, when deposited on the stigma of the same variety, should become by spontaneous variation in ever so slight a degree prepotent over the pollen of other varieties, this would certainly be an advantage to the variety; for its own pollen would thus obliterate the effects of the pollen of other varieties, and prevent deterioration of character. And the more prepotent the variety's own pollen could be rendered through natural selection, the greater the advantage would be. We know from the researches of Gärtner that, with species which are mutually sterile, the pollen of each is always prepotent on its own stigma over that of the other species; but we do not know whether this prepotency is a consequence of the mutual sterility, or the sterility a consequence of the prepotency. If the latter view be correct, as the prepotency became stronger through natural selection, from being advantageous to a species in process of formation, so the sterility consequent on prepotency would at the same time be augmented; and the final result would be various degrees of sterility, such as occurs with existing species. This view might be extended to animals, if the female before each birth received several males, so that the sexual element of the prepotent male of her own variety obliterated the effects of the access of previous males belonging to other varieties; but we have no reason to believe, at least with terrestrial animals, that this is the case; as most males and females pair for each birth, and some few for life.

On the whole we may conclude that with animals the sterility of crossed species has not been slowly augmented through natural selection; and as this sterility follows the same general laws in the vegetable as in the animal kingdom, it is improbable, though apparently possible, that with plants crossed species should have been rendered sterile by a different process. From this consideration, and remembering that species which have never co-existed in the same country, and which therefore could not have received any advantage from having been rendered mutually infertile, yet are generally sterile when crossed; and bearing in mind that in reciprocal crosses between the same two species there is sometimes the widest difference in their sterility, we must give up the belief that natural selection has come into play.

As species have not been rendered mutually infertile through the accumulative action of natural selection, and as we may safely conclude, from the previous as well as from other and more general considerations, that they have not been endowed through an act of creation with this quality, we must infer that it has arisen incidentally during their slow formation in connection with other and unknown changes in their organisation. By a quality arising incidentally, I refer to such cases as different species of animals and plants being differently affected by poisons to which they are not naturally exposed; and this difference in susceptibility is clearly incidental on other and unknown differences in their organisation. So again the capacity in different kinds of trees to be grafted on each other, or on a third species, differs much, and is of no advantage to these trees, but is incidental on structural or functional differences in their woody tissues. We need not feel surprise at sterility incidentally resulting from crosses between distinct species,—the modified descendants of a common progenitor,— when we bear in mind how easily the reproductive system is affected by various causes—often by extremely slight changes in the conditions of life, by too close interbreeding, and by other agencies. It is well to bear in mind such cases, as that of the *Passiflora alata*, which recovered its self-fertility from being grafted on a distinct species—the cases of plants which normally or abnormally are self-impotent, but can readily be fertilised by the pollen of a distinct species—and lastly the cases of individual domesticated animals which evince towards each other sexual incompatibility.

We now at last come to the immediate point under discussion: how is it that, with some few exceptions in the case of plants, domesticated varieties, such as those of the dog, fowl, pigeon, several fruit-trees, and culinary vegetables, which differ from each other in external characters more than many species, are perfectly fertile when crossed, or even fertile in excess, whilst closely allied species are almost invariably in some degree sterile? We can, to a certain extent, give a satisfactory answer to this question. Passing over the fact that the amount of external difference between two species is no sure guide to their degree of mutual sterility, so that similar differences in the case of varieties would be no sure guide, we know that with species the cause lies exclusively in differences in their sexual constitution. Now the conditions to which domesticated animals and cultivated plants have been subjected, have had so little tendency towards modifying the reproductive system in a manner leading to mutual sterility, that we have good grounds for admitting the directly opposite doctrine of Pallas, namely, that such conditions generally eliminate this tendency; so that the domesticated descendants of species, which in their natural state would have been in some degree sterile when crossed, become perfectly fertile together. With plants, so far is cultivation from giving a tendency towards mutual sterility, that in several well-authenticated cases, already often alluded to, certain species have been affected in a very different manner, for they have become self-impotent, whilst still retaining the capacity of fertilising, and being fertilised by, distinct species. If the Pallasian doctrine of the elimination of sterility through long-continued domestication be admitted, and it can hardly be rejected, it becomes in the highest degree improbable that similar circumstances should commonly both induce and eliminate the same tendency; though in certain cases, with species having a peculiar constitution, sterility might occasionally be thus induced. Thus, as I believe, we can understand why with domesticated animals varieties have not been produced which are mutually sterile; and why with plants only a few such cases have been observed, namely, by Gärtner, with certain varieties of maize and verbascum, by other experimentalists with varieties of the gourd and melon, and by Kölreuter with one kind of tobacco.

With respect to varieties which have originated in a state of nature, it is almost hopeless to expect to prove by direct evidence that they have been rendered mutually sterile; for if even a trace of sterility could be detected, such varieties would at once be raised by almost every naturalist to the rank of distinct species. If, for instance, Gärtner's statement were fully confirmed, that the blue and red-flowered forms of the pimpernel (*Anagallis arvensis*) are sterile when crossed, I presume that all the botanists who now maintain on various grounds that these two forms are merely fleeting varieties, would at once admit that they were specifically distinct.

The real difficulty in our present subject is not, as it appears to me, why domestic varieties have not become mutually infertile when crossed, but why this has so generally occurred with natural varieties as soon as they have been modified in a sufficient and permanent degree to take rank as species. We are far from precisely knowing the cause; nor is this surprising, seeing how profoundly ignorant we are in regard to the normal and abnormal action of the reproductive system. But we can see that species, owing to their struggle for life with numerous competitors, must have been exposed to more uniform conditions during long periods of time, than have been domestic varieties; and this may well make a wide difference in the result. For we know how commonly wild animals and plants, when taken from their natural conditions and subjected to captivity, are rendered sterile; and the reproductive functions of organic beings which have always lived and been slowly modified under natural conditions would probably in like manner be eminently sensitive to the influence of an unnatural cross. Domesticated productions, on the other hand, which, as shown by the mere fact of their domestication, were not originally highly sensitive to changes in their conditions of life, and which can now generally resist with undiminished fertility repeated changes of conditions, might be expected to produce varieties, which would be little liable to have their reproductive powers injuriously affected by the act of crossing with other varieties which had originated in a like manner.

Certain naturalists have recently laid too great stress, as it appears to me, on the difference in fertility between varieties and species when crossed. Some allied species of trees cannot be grafted on each other,—all varieties can be so grafted. Some allied animals are affected in a very different manner by the same poison, but with varieties no such case until recently was known, but now it has been proved that immunity from certain poisons stands in some cases in correlation with the colour of the hair. The period of gestation generally differs much with distinct species, but with varieties until lately no such difference had been observed. The time required for the germination of seeds differs in an analogous manner, and I am not aware that any difference in this respect has as yet been detected with varieties. Here we have various physiological differences, and no doubt others could be added, between one species and another of the same genus, which do not occur, or occur with extreme rarity, in the case of varieties; and these differences are apparently wholly or in chief part incidental on other constitutional differences, just in the same manner as the sterility of crossed species is incidental on differences confined to the sexual system. Why, then, should these latter differences, however serviceable they may indirectly be in keeping the inhabitants of the same country distinct, be thought of such paramount importance, in comparison with other incidental and functional differences? No sufficient answer to this question can be given. Hence the fact that the most distinct domestic varieties are, with rare exceptions, perfectly fertile when crossed, and produce fertile offspring, whilst closely allied species are, with rare exceptions, more or less sterile, is not nearly so formidable an objection as it appears at first to the theory of the common descent of allied species.

CHAPTER XX.

SELECTION BY MAN.

THE power of Selection, whether exercised by man, or brought into play under nature through the struggle for existence and the consequent survival of the fittest, absolutely depends on the variability of organic beings. Without variability nothing can be effected; slight individual differences, however, suffice for the work, and are probably the sole differences which are effective in the production of new species. Hence our discussion on the causes and laws of variability ought in strict order to have preceded our present subject, as well as the previous subjects of inheritance, crossing, &c.; but practically the present arrangement has been found the most convenient. Man does not attempt to cause variability; though he unintentionally effects this by exposing organisms to new conditions of life, and by crossing breeds already formed. But variability being granted, he works wonders. Unless some degree of selection be exercised, the free commingling of the individuals of the same variety soon obliterates, as we have previously seen, the slight differences which may arise, and gives to the whole body of individuals uniformity of character. In separated districts, long-continued exposure to different conditions of life may perhaps produce new races without the aid of selection; but to this difficult subject of the direct action of the conditions of life we shall in a future chapter recur.

When animals or plants are born with some conspicuous and firmly inherited new character, selection is reduced to the preservation of such individuals, and to the subsequent prevention of crosses ; so that nothing more need be said on the subject. But in the great majority of cases a new character, or some superiority in an old character, is at first faintly pronounced, and is not strongly inherited ; and then the full difficulty of selection is experienced. Indomitable patience, the finest powers of discrimination, and sound judgment must be exercised during many years. A clearly predetermined object must be kept steadily in view. Few men are endowed with all these qualities, especially with that of discriminating very slight differences ; judgment can be acquired only by long experience ; but if any of these qualities be wanting, the labour of a life may be thrown away. I have been astonished when celebrated breeders, whose skill and judgment have been proved by their success at exhibitions, have shown me their animals, which appeared all alike, and have assigned their reasons for matching this and that individual. The importance of the great principle of Selection mainly lies in this power of selecting scarcely appreciable differences, which nevertheless are found to be transmissible, and which can be accumulated until the result is made manifest to the eyes of every beholder.

The principle of selection may be conveniently divided into three kinds. *Methodical selection* is that which guides a man who systematically endeavours to modify a breed according to some predetermined standard. *Unconscious selection* is that which follows from men naturally preserving the most valued and destroying the less valued individuals, without any thought of altering the breed ; and undoubtedly this process slowly works great changes. Unconscious selection graduates into methodical, and only extreme cases can be distinctly separated ; for he who preserves a useful or perfect animal will generally breed from it with the hope of getting offspring of the same character ; but as long as he has not a predetermined purpose to improve the breed, he may be said to be selecting unconsciously.[1] Lastly, we have *Natural selection*, which implies that the individuals which are best fitted for the complex, and in the course of ages changing conditions to which they are exposed, generally survive and procreate their kind. With domestic productions, with which alone we are here strictly concerned, natural selection comes to a certain extent into action, independently of, and even in opposition to, the will of man.

Methodical Selection.—What man has effected within recent times in England by methodical selection is clearly shown by our exhibitions of improved quadrupeds and fancy birds. With respect to cattle, sheep, and pigs, we owe their great improvement to a long series of well-known names—Bakewell, Colling, Ellman, Bates, Jonas Webb, Lords Leicester and Western, Fisher Hobbs, and others. Agricultural writers are unanimous on the power of selection: any number of statements to this effect could be quoted; a few will suffice. Youatt, a sagacious and experienced observer, writes,[2] the principle of selection is " that which enables the agriculturist, not only to modify the character of his flock, but to change it altogether." A great breeder of shorthorns [3] says, " In the anatomy of the shoulder " modern breeders have made great improvements on the " Ketton shorthorns by correcting the defect in the knuckle or " shoulder-joint, and by laying the top of the shoulder more " snugly into the crop, and thereby filling up the hollow " behind it. . . . . The eye has its fashion at different periods: " at one time the eye high and outstanding from the head, and " at another time the sleepy eye sunk into the head; but these " extremes have merged into the medium of a full, clear, and " prominent eye with a placid look."

Again, hear what an excellent judge of pigs[4] says: " The legs

[1] The term *unconscious selection* has been objected to as a contradiction: but *see* some excellent observations on this head by Prof. Huxley ('Nat. Hist. Review,' Oct. 1864, p. 578), who remarks that when the wind heaps up sand-dunes it sifts and *unconsciously selects* from the gravel on the beach grains of sand of equal size.

[2] Sheep, 1838, p. 60.

[3] Mr. J. Wright on Shorthorn Cattle, in 'Journal of Royal Agricult. Soc.,' vol. vii. pp. 208, 209.

[4] H. D. Richardson on Pigs, 1847, p. 44.

" should be no longer than just to prevent the animal's belly
" from trailing on the ground. The leg is the least profitable
" portion of the hog, and we therefore require no more of it than
" is absolutely necessary for the support of the rest." Let any
one compare the wild-boar with any improved breed, and he will
see how effectually the legs have been shortened.

Few persons, except breeders, are aware of the systematic
care taken in selecting animals, and of the necessity of having a
clear and almost prophetic vision into futurity. Lord Spencer's
skill and judgment were well known; and he writes,[5] " It is
" therefore very desirable, before any man commences to breed
" either cattle or sheep, that he should make up his mind to the
" shape and qualities he wishes to obtain, and steadily pursue
" this object." Lord Somerville, in speaking of the marvellous
improvement of the New Leicester sheep, effected by Bakewell
and his successors, says, " It would seem as if they had first
drawn a perfect form, and then given it life." Youatt[6] urges
the necessity of annually drafting each flock, as many animals
will certainly degenerate " from the standard of excellence, which
the breeder has established in his own mind." Even with a
bird of such little importance as the canary, long ago (1780-
1790) rules were established, and a standard of perfection was
fixed, according to which the London fanciers tried to breed the
several sub-varieties.[7] A great winner of prizes at the Pigeon-
shows,[8] in describing the Short-faced Almond Tumbler, says,
" There are many first-rate fanciers who are particularly partial
" to what is called the goldfinch-beak, which is very beautiful;
" others say, take a full-size round cherry, then take a barley-
" corn, and judiciously placing and thrusting it into the cherry,
" form as it were your beak; and that is not all, for it will form
" a good head and beak, provided, as I said before, it is judi-
" ciously done; others take an oat; but as I think the gold-
" finch-beak the handsomest, I would advise the inexperienced
" fancier to get the head of a goldfinch, and keep it by him
" for his observation." Wonderfully different as is the beak
of the rock-pigeon and goldfinch, undoubtedly, as far as ex-

[5] 'Journal of R. Agricult. Soc.,' vol. i. p. 24.   [6] Sheep, pp. 520, 319.
[7] Loudon's ' Mag. of Nat. Hist.,' vol. viii., 1835, p. 618.
[8] 'A Treatise on the Art of Breeding the Almond Tumbler,' 1851, p. 9.

ternal shape and proportions are concerned, the end has been nearly gained.

Not only should our animals be examined with the greatest care whilst alive, but, as Anderson remarks,[9] their carcases should be scrutinised, " so as to breed from the descendants of such only as, in the language of the butcher, cut up well." The " grain of the meat " in cattle, and its being well marbled with fat,[10] and the greater or less accumulation of fat in the abdomen of our sheep, have been attended to with success. So with poultry, a writer,[11] speaking of Cochin-China fowls, which are said to differ much in the quality of their flesh, says, " the best " mode is to purchase two young brother-cocks, kill, dress, and " serve up one; if he be indifferent, similarly dispose of the " other, and try again ; if, however, he be fine and well-flavoured, " his brother will not be amiss for breeding purposes for the " table."

The great principle of the division of labour has been brought to bear on selection. In certain districts[12] " the breeding of " bulls is confined to a very limited number of persons, who by " devoting their whole attention to this department, are able " from year to year to furnish a class of bulls which are steadily " improving the general breed of the district." The rearing and letting of choice rams has long been, as is well known, a chief source of profit to several eminent breeders. In parts of Germany this principle is carried with merino sheep to an extreme point.[13] " So important is the proper selection of " breeding animals considered, that the best flock-masters do " not trust to their own judgment, or to that of their shepherds, " but employ persons called ' sheep-classifiers,' who make it their " special business to attend to this part of the management of " several flocks, and thus to preserve, or if possible to improve, " the best qualities of both parents in the lambs." In Saxony, " when the lambs are weaned, each in his turn is placed upon " a table that his wool and form may be minutely observed.

[9] 'Recreations in Agriculture,' vol. ii. p. 409.

[10] Youatt on Cattle, pp. 191, 227.

[11] Ferguson, ' Prize Poultry,' 1854, p. 208.

[12] Wilson, in ' Transact. Highland Agricult. Soc.,' quoted in ' Gard. Chronicle,' 1844, p. 29.

[13] Simmonds, quoted in ' Gard. Chronicle,' 1855, p. 637. And for the second quotation, see Youatt on Sheep, p. 171.

" The finest are selected for breeding and receive a first mark.
" When they are one year old, and prior to shearing them,
" another close examination of those previously marked takes
" place : those in which no defect can be found receive a second
" mark, and the rest are condemned.   A few months afterwards
" a third and last scrutiny is made; the prime rams and ewes
" receive a third and final mark, but the slightest blemish is
" sufficient to cause the rejection of the animal."   These sheep
are bred and valued almost exclusively for the fineness of their
wool; and the result corresponds with the labour bestowed on
their selection.   Instruments have been invented to measure
accurately the thickness of the fibres; and " an Austrian fleece
has been produced of which twelve hairs equalled in thickness
one from a Leicester sheep."

Throughout the world, wherever silk is produced, the greatest
care is bestowed on selecting the cocoons from which the moths
for breeding are to be reared.   A careful cultivator[14] likewise
examines the moths themselves, and destroys those that are not
perfect.   But what more immediately concerns us is that certain
families in France devote themselves to raising eggs for sale.[15]
In China, near Shanghai, the inhabitants of two small districts
have the privilege of raising eggs for the whole surrounding
country, and that they may give up their whole time to this
business, they are interdicted by law from producing silk.[16]

The care which successful breeders take in matching their
birds is surprising.   Sir John Sebright, whose fame is perpetuated
by the " Sebright Bantam," used to spend " two and three days
in examining, consulting, and disputing with a friend which
were the best of five or six birds."[17]   Mr. Bult, whose pouter-
pigeons won so many prizes and were exported to North
America under the charge of a man sent on purpose, told
me that he always deliberated for several days before he
matched each pair.   Hence we can understand the advice of an
eminent fancier, who writes,[18] " I would here particularly guard

---

[14] Robinet, ' Vers à Soie,' 1848, p. 271.

[15] Quatrefages, ' Les Maladies du
Ver à Soie,' 1859, p. 101.

[16] M. Simon, in ' Bull. de la Soc.
d'Acclimat.,' tom. ix., 1862, p. 221.

[17] ' The Poultry Chronicle,' vol. i.,
1854, p. 607.

[18] J. M. Eaton, ' A Treatise on Fancy
Pigeons,' 1852, p. xiv., and ' A Treatise
on the Almond Tumbler,' 1851, p. 11.

" you against having too great a variety of pigeons, otherwise
" you will know a little of all, but nothing about one as it
" ought to be known." Apparently it transcends the power of
the human intellect to breed all kinds: " it is possible that
" there may be a few fanciers that have a good general know-
" ledge of fancy pigeons; but there are many more who labour
" under the delusion of supposing they know what they do not."
The excellence of one sub-variety, the Almond Tumbler, lies in
the plumage, carriage, head, beak, and eye; but it is too presumptuous in the beginner to try for all these points. The
great judge above quoted says, " there are some young fanciers
" who are over-covetous, who go for all the above five properties
" at once; they have their reward by getting nothing." We
thus see that breeding even fancy pigeons is no simple art: we
may smile at the solemnity of these precepts, but he who laughs
will win no prizes.

What methodical selection has effected for our animals is
sufficiently proved, as already remarked, by our Exhibitions.
So greatly were the sheep belonging to some of the earlier
breeders, such as Bakewell and Lord Western, changed, that
many persons could not be persuaded that they had not been
crossed. Our pigs, as Mr. Corringham remarks,[19] during the
last twenty years have undergone, through rigorous selection
together with crossing, a complete metamorphosis. The first
exhibition for poultry was held in the Zoological Gardens in
1845; and the improvement effected since that time has been
great. As Mr. Baily, the great judge, remarked to me, it was
formerly ordered that the comb of the Spanish cock should be
upright, and in four or five years all good birds had upright
combs; it was ordered that the Polish cock should have no
comb or wattles, and now a bird thus furnished would be at
once disqualified; beards were ordered, and out of fifty-seven
pens lately (1860) exhibited at the Crystal Palace, all had
beards. So it has been in many other cases. But in all cases
the judges order only what is occasionally produced and what
can be improved and rendered constant by selection. The
steady increase of weight during the last few years in our

[19] 'Journal Royal Agricultural Soc.,' vol. vi. p. 22.

fowls, turkeys, ducks, and geese is notorious; "six-pound ducks are now common, whereas four pounds was formerly the average." As the actual time required to make a change has not often been recorded, it may be worth mentioning that it took Mr. Wicking thirteen years to put a clean white head on an almond tumbler's body, " a triumph," says another fancier, " of which he may be justly proud." [20]

Mr. Tollet, of Betley Hall, selected cows, and especially bulls, descended from good milkers, for the sole purpose of improving his cattle for the production of cheése; he steadily tested the milk with the lactometer, and in eight years he increased, as I was informed by him, the product in the proportion of four to three. Here is a curious case [21] of steady but slow progress, with the end not as yet fully attained: in 1784 a race of silk-worms was introduced into France, in which one hundred out of the thousand failed to produce white cocoons; but now, after careful selection during sixty-five generations, the proportion of yellow cocoons has been reduced to thirty-five in the thousand.

With plants selection has been followed with the same good results as with animals. But the process is simpler, for plants in the great majority of cases bear both sexes. Nevertheless, with most kinds it is necessary to take as much care to prevent crosses as with animals or unisexual plants; but with some plants, such as peas, this care does not seem to be necessary. With all improved plants, excepting of course those which are propagated by buds, cuttings, &c., it is almost indispensable to examine the seedlings and destroy those which depart from the proper type. This is called "roguing," and is, in fact, a form of selection, like the rejection of inferior animals. Experienced horticulturists and agriculturists incessantly urge every one to preserve the finest plants for the production of seed.

Although plants often present much more conspicuous variations than animals, yet the closest attention is generally requisite to detect each slight and favourable change. Mr. Masters relates [22] how " many a patient hour was devoted," whilst he was

---

[20] 'Poultry Chronicle,' vol. ii., 1855, p. 596.

[21] Isid. Geoffroy St. Hilaire, 'Hist. Nat. Gén.,' tom. iii. p. 254.

[22] 'Gardener's Chronicle,' 1850, p. 198.

young, to the detection of differences in peas intended for seed.
Mr. Barnet[23] remarks that the old scarlet American strawberry
was cultivated for more than a century without producing a
single variety; and another writer observes how singular it was
that when gardeners first began to attend to this fruit it began
to vary; the truth no doubt being that it had always varied,
but that, until slight varieties were selected and propagated by
seed, no conspicuous result was obtained. The finest shades of
difference in wheat have been discriminated and selected with
almost as much care, as we see in Colonel Le Couteur's works,
as in the case of the higher animals; but with our cereals the
process of selection has seldom or never been long continued.

It may be worth while to give a few examples of methodical
selection with plants; but in fact the great improvement of all
our anciently cultivated plants may be attributed to selection
long carried on, in part methodically, and in part unconsciously.
I have shown in a former chapter how the weight of the gooseberry has been increased by systematic selection and culture.
The flowers of the Heartsease have been similarly increased in
size and regularity of outline. With the Cineraria, Mr. Glenny[24]
" was bold enough, when the flowers were ragged and starry
" and ill defined in colour, to fix a standard which was then
" considered outrageously high and impossible, and which, even
" if reached, it was said, we should be no gainers by, as it would
" spoil the beauty of the flowers. He maintained that he was
" right; and the event has proved it to be so." The doubling
of flowers has several times been effected by careful selection:
the Rev. W. Williamson,[25] after sowing during several years
seed of *Anemone coronaria*, found a plant with one additional
petal; he sowed the seed of this, and by perseverance in the
same course obtained several varieties with six or seven rows of
petals. The single Scotch rose was doubled, and yielded eight
good varieties in nine or ten years.[26] The Canterbury bell
(*Campanula medium*) was doubled by careful selection in four
generations.[27] In four years Mr. Buckman,[28] by culture and

[23] 'Transact. Hort. Soc.,' vol. vi. p.
152.

[24] 'Journal of Horticulture,' 1862, p.
369.

[25] 'Transact. Hort. Soc.,' vol. iv. p. 381.

[26] 'Transact. Hort. Soc.,' vol. iv. p.
285.

[27] Rev. W. Bromehead, in 'Gard.
Chronicle,' 1857, p. 550.

[28] 'Gard. Chronicle,' 1862, p. 721.

careful selection, converted parsnips, raised from wild seed, into
a new and good variety. By selection during a long course of
years, the early maturity of peas has been hastened from ten to
twenty-one days.[29] A more curious case is offered by the beet-
plant, which, since its cultivation in France, has almost exactly
doubled its yield of sugar. This has been effected by the most
careful selection; the specific gravity of the roots being regularly tested, and the best roots saved for the production of
seed.[30]

Selection by Ancient and Semi-civilised People.

In attributing so much importance to the selection of animals
and plants, it may be objected that methodical selection would
not have been carried on during ancient times. A distinguished
naturalist considers it as absurd to suppose that semi-civilised
people should have practised selection of any kind. Undoubtedly the principle has been systematically acknowledged and
followed to a far greater extent within the last hundred years
than at any former period, and a corresponding result has
been gained; but it would be a great error to suppose, as we
shall immediately see, that its importance was not recognised
and acted on during the most ancient times, and by semi-
civilised people. I should premise that many facts now to be
given only show that care was taken in breeding; but when
this is the case, selection is almost sure to be practised to a
certain extent. We shall hereafter be enabled better to judge
how far selection, when only occasionally carried on, by a few
of the inhabitants of a country, will slowly produce a great
effect.

In a well-known passage in the thirtieth chapter of Genesis,
rules are given for influencing, as was then thought possible,
the colour of sheep; and speckled and dark breeds are spoken
of as being kept separate. By the time of David the fleece was
likened to snow. Youatt,[31] who has discussed all the passages
in relation to breeding in the Old Testament, concludes that

[29] Dr. Anderson, in 'The Bee,' vol.
vi. p. 96; Mr. Barnes, in 'Gard.
Chronicle,' 1844, p. 476.

[30] Godron, 'De l'Espèce,' 1859, tom.
ii. p. 69; 'Gard. Chronicle,' 1854, p. 258.
[31] On Sheep, p. 18.

at this early period " some of the best principles of breeding must have been steadily and long pursued." It was ordered, according to Moses, that "Thou shalt not let thy cattle gender with a diverse kind;" but mules were purchased,[32] so that at this early period other nations must have crossed the horse and ass. It is said[33] that Erichthonius, some generations before the Trojan war, had many brood-mares, "which by his care and judgment in the choice of stallions produced a breed of horses superior to any in the surrounding countries." Homer (Book v.) speaks of Æneas's horses as bred from mares which were put to the steeds of Laomedon. Plato, in his 'Republic,' says to Glaucus, "I see that you raise at your house a great many dogs for the chase. Do you take care about breeding and pairing them? Among animals of good blood, are there not always some which are superior to the rest?" To which Glaucus answers in the affirmative.[34] Alexander the Great selected the finest Indian cattle to send to Macedonia to improve the breed.[35] According to Pliny,[36] King Pyrrhus had an especially valuable breed of oxen; and he did not suffer the bulls and cows to come together till four years old, that the breed might not degenerate. Virgil, in his Georgics (lib. iii.), gives as strong advice as any modern agriculturist could do, carefully to select the breeding stock; "to note the tribe, the lineage, and the sire; whom to reserve for husband of the herd;"—to brand the progeny;—to select sheep of the purest white, and to examine if their tongues are swarthy. We have seen that the Romans kept pedigrees of their pigeons, and this would have been a senseless proceeding had not great care been taken in breeding them. Columella gives detailed instructions about breeding fowls: "Let the breeding hens therefore be of a choice colour, "a robust body, square-built, full-breasted, with large heads, "with upright and bright-red combs. Those are believed to be "the best bred which have five toes."[37] According to Tacitus, the Celts attended to the races of their domestic animals;

[32] Volz, 'Beiträge zur Kulturge-schichte,' 1852, s. 47.

[33] Mitford's 'History of Greece,' vol. i. p. 73.

[34] Dr. Dally, translated in 'Anthropo-logical Review,' May 1864, p. 101.

[35] Volz, 'Beiträge,' &c., 1852, s. 80.

[36] 'History of the World,' ch. 45.

[37] 'Gardener's Chronicle,' 1848, p. 323.

and Cæsar states that they paid high prices to merchants for fine imported horses.[38] In regard to plants, Virgil speaks of yearly culling the largest seeds; and Celsus says, "where the corn and crop is but small, we must pick out the best ears of corn, and of them lay up our seed separately by itself." [39]

Coming down the stream of time, we may be brief. At about the beginning of the ninth century Charlemagne expressly ordered his officers to take great care of his stallions; and if any proved bad or old, to forewarn him in good time before they were put to the mares.[40] Even in a country so little civilised as Ireland during the ninth century, it would appear from some ancient verses,[41] describing a ransom demanded by Cormac, that animals from particular places, or having a particular character, were valued. Thus it is said,—

> Two pigs of the pigs of Mac Lir,
> A ram and ewe both round and red,
> I brought with me from Aengus.
> I brought with me a stallion and a mare
> From the beautiful stud of Manannan,
> A bull and a white cow from Druim Cain.

Athelstan, in 930, received as a present from Germany, running-horses; and he prohibited the exportation of English horses. King John imported "one hundred chosen stallions from Flanders." [42] On June 16th, 1305, the Prince of Wales wrote to the Archbishop of Canterbury, begging for the loan of any choice stallion, and promising its return at the end of the season.[43] There are numerous records at ancient periods in English history of the importation of choice animals of various kinds, and of foolish laws against their exportation. In the reigns of Henry VII. and VIII. it was ordered that the magistrates, at Michaelmas, should scour the heaths and commons, and destroy all mares beneath a certain size.[44] Some of our earlier kings passed laws against the slaughtering rams of any good breed before they were seven years old, so that they

[38] Reynier, 'De l'Economie des Celtes,' 1818, pp. 487, 503.

[39] Le Couteur on Wheat, p. 15.

[40] Michel, 'Des Haras,' 1861, p. 84.

[41] Sir W. Wilde, an 'Essay on Un-manufactured Animal Remains,' &c.,

1860, p. 11.

[42] Col. Hamilton Smith, 'Nat. Library,' vol. xii., Horses, pp. 135, 140.

[43] Michel, 'Des Haras,' p. 90.

[44] Mr. Baker, 'History of the Horse,' Veterinary, vol. xiii. p. 423.

might have time to breed.   In Spain Cardinal Ximenes issued,
in 1509, regulations on the *selection* of good rams for breeding.[45]

The Emperor Akbar Khan before the year 1600 is said to
have "wonderfully improved" his pigeons by crossing the
breeds; and this necessarily implies careful selection. About
the same period the Dutch attended with the greatest care
to the breeding of these birds. Belon in 1555 says that good
managers in France examined the colour of their goslings in
order to get geese of a white colour and better kinds. Markham
in 1631 tells the breeder "to elect the largest and goodliest
conies," and enters into minute details. Even with respect
to seeds of plants for the flower-garden, Sir J. Hanmer writing
about the year 1660 [46] says, in "choosing seed, the best seed is
the most weighty, and is had from the lustiest and most vigorous stems;" and he then gives rules about leaving only a few
flowers on plants for seed; so that even such details were
attended to in our flower-gardens two hundred years ago. In
order to show that selection has been silently carried on in
places where it would not have been expected, I may add that
in the middle of the last century, in a remote part of North
America, Mr. Cooper improved by careful selection all his
vegetables, "so that they were greatly superior to those of any
" other person. When his radishes, for instance, are fit for use,
" he takes ten or twelve that he most approves, and plants
" them at least 100 yards from others that blossom at the same
" time. In the same manner he treats all his other plants,
" varying the circumstances according to their nature." [47]

In the great work on China published in the last century by
the Jesuits, and which is chiefly compiled from ancient Chinese
encyclopædias, it is said that with sheep " improving the breed
" consists in choosing with particular care the lambs which are
" destined for propagation, in nourishing them well, and in
" keeping the flocks separate." The same principles were
applied by the Chinese to various plants and fruit-trees.[48]   An

[45] M. l'Abbé Carlier, in 'Journal de
Physique,' vol. xxiv., 1784, p. 181 : this
memoir contains much information on
the ancient selection of sheep ; and is
my authority for rams not being killed
young in England.

[46] 'Gardener's Chronicle,' 1843, p. 389.
[47] Communications to Board of Agriculture, quoted in Dr. Darwin's 'Phyto-
logia,' 1800, p. 451.
[48] 'Mémoire sur les Chinois,' 1786,
tom. xi. p. 55 ; tom. v. p. 507.

imperial edict recommends the choice of seed of remarkable size; and selection was practised even by imperial hands, for it is said that the Ya-mi, or imperial rice, was noticed at an ancient period in a field by the Emperor Khang-hi, was saved and cultivated in his garden, and has since become valuable from being the only kind which will grow north of the Great Wall.[49] Even with flowers, the tree pæony (*P. moutan*) has been cultivated, according to Chinese traditions, for 1400 years; between 200 and 300 varieties have been raised, which are cherished like tulips formerly were by the Dutch.[50]

Turning now to semi-civilised people and to savages: it occurred to me, from what I had seen of several parts of South America, where fences do not exist, and where the animals are of little value, that there would be absolutely no care in breeding or selecting them; and this to a large extent is true. Roulin,[51] however, describes in Colombia a naked race of cattle, which are not allowed to increase, on account of their delicate constitution. According to Azara [52] horses are often born in Paraguay with curly hair; but, as the natives do not like them, they are destroyed. On the other hand, Azara states that a hornless bull, born in 1770, was preserved and propagated its race. I was informed of the existence in Banda Oriental of a breed with reversed hair; and the extraordinary niata cattle first appeared and have since been kept distinct in La Plata. Hence certain conspicuous variations have been preserved, and others have been habitually destroyed, in these countries, which are so little favourable for careful selection. We have also seen that the inhabitants sometimes introduce cattle on their estates to prevent the evil effects of close interbreeding. On the other hand, I have heard on reliable authority that the Gauchos of the Pampas never take any pains in selecting the best bulls or stallions for breeding; and this probably accounts for the cattle and horses being remarkably uniform in character throughout the immense range of the Argentine republic.

Looking to the Old World, in the Sahara Desert " The " Touareg is as careful in the selection of his breeding Mahari

[49] 'Recherches sur l'Agriculture des Chinois,' par L. D'Hervey-Saint-Denys, 1850, p. 229. With respect to Khang-hi, *see* Huc's ' Chinese Empire,' p. 311.

[50] Anderson, in ' Linn. Transact.,'

vol. xii. p. 253.

[51] ' Mém. de l'Acad.' (divers savans), tom. vi., 1835, p. 333.

[52] 'Des Quadrupèdes du Paraguay,' 1801, tom. ii. p. 333, 371.

" (a fine race of the dromedary) as the Arab is in that of his
" horse. The pedigrees are handed down, and many a dromedary
" can boast a genealogy far longer than the descendants of the
" Darley Arabian." [53]   According to Pallas the Mongolians
endeavour to breed the Yaks or horse-tailed buffaloes with
white tails, for these are sold to the Chinese mandarins as fly-
flappers; and Moorcroft, about seventy years after Pallas, found
that white-tailed animals were still selected for breeding.[54]

We have seen in the chapter on the Dog that savages in
different parts of North America and in Guiana cross their
dogs with wild Canidæ, as did the ancient Gauls, according
to Pliny. This was done to give their dogs strength and
vigour, in the same way as the keepers in large warrens
now sometimes cross their ferrets (as I have been informed by
Mr. Yarrell) with the wild polecat, "to give them more devil."
According to Varro, the wild ass was formerly caught and
crossed with the tame animal to improve the breed, in the
same manner as at the present day the natives of Java sometimes
drive their cattle into the forests to cross with the wild Banteng
(*Bos sondaicus*).[55]   In Northern Siberia, among the Ostyaks
the dogs vary in markings in different districts, but in each
place they are spotted black and white in a remarkably uniform
manner;[56] and from this fact alone we may infer careful
breeding, more especially as the dogs of one locality are famed
throughout the country for their superiority. I have heard of
certain tribes of Esquimaux who take pride in their teams of
dogs being uniformly coloured. In Guiana, as Sir R. Schomburgk informs me,[57] the dogs of the Turuma Indians are highly
valued and extensively bartered : the price of a good one is the
same as that given for a wife : they are kept in a sort of cage,
and the Indians "take great care when the female is in season
to prevent her uniting with a dog of an inferior description."
The Indians told Sir Robert that, if a dog proved bad or useless,

[53] 'The Great Sahara,' by the Rev.
H. B. Tristram, 1860, p. 238.

[54] Pallas, 'Act. Acad. St. Petersburg,'
1777, p. 249; Moorcroft and Trebeck,
'Travels in the Himalayan Provinces,'
1841.

[55] Quoted from Raffles, in the 'Indian

Field,' 1859, p. 196 : for Varro, *see*
Pallas, *ut supra*.

[56] Erman's 'Travels in Siberia,' Eng.
translat., vol. i. p. 453.

[57] *See* also 'Journal of R. Geograph.
Soc.,' vol. xiii. part i. p. 65.

he was not killed, but was left to die from sheer neglect. Hardly any nation is more barbarous than the Fuegians, but I hear from Mr. Bridges, the Catechist to the Mission, that, " when these savages have a large, strong, and active bitch, they " take care to put her to a fine dog, and even take care to feed " her well, that her young may be strong and well favoured."

In the interior of Africa, negroes, who have not associated with white men, show great anxiety to improve their animals : they "always choose the larger and stronger males for stock : " the Malakolo were much pleased at Livingstone's promise to send them a bull, and some Bakalolo carried a live cock all the way from Loanda into the interior.[58] Further south on the same continent, Andersson states that he has known a Damara give two fine oxen for a dog which struck his fancy. The Damaras take great delight in having whole droves of cattle of the same colour, and they prize their oxen in proportion to the size of their horns. " The Namaquas have a perfect mania for " a uniform team; and almost all the people of Southern Africa " value their cattle next to their women, and take a pride in " possessing animals that look high-bred." " They rarely or " never make use of a handsome animal as a beast of burden." [59] The power of discrimination which these savages possess is wonderful, and they can recognise to which tribe any cattle belong. Mr. Andersson further informs me that the natives frequently match a particular bull with a particular cow.

The most curious case of selection by semi-civilised people, or indeed by any people, which I have found recorded, is that given by Garcilazo de la Vega, a descendant of the Incas, as having been practised in Peru before the country was subjugated by the Spaniards.[60] The Incas annually held great hunts, when all the wild animals were driven from an immense circuit to a central point. The beasts of prey were first destroyed as injurious. The wild Guanacos and Vicunas were sheared; the old males and females killed, and the others set at liberty. The various kinds of deer were examined; the old males and females

[58] Livingstone's 'First Travels,' pp. 191, 439, 565 : see also 'Expedition to the Zambesi,' 1865, p. 495, for an analogous case respecting a good breed of goats.

[59] Andersson's 'Travels in South Africa,' pp. 232, 318, 319.

[60] Dr. Vavasseur, in 'Bull. de la Soc. d'Acclimat.,' tom. viii., 1861, p. 136.

were likewise killed; "but the young females, with a certain number of males, selected from the most beautiful and strong," were given their freedom. Here, then, we have selection by man aiding natural selection. So that the Incas followed exactly the reverse system of that which our Scottish sportsmen are accused of following, namely, of steadily killing the finest stags, thus causing the whole race to degenerate.[61] In regard to the domesticated llamas and alpacas, they were separated in the time of the Incas according to colour; and if by chance one in a flock was born of the wrong colour, it was eventually put into another flock.

In the genus Auchenia there are four forms,—the Guanaco and Vicuna, found wild and undoubtedly distinct species; the Llama and Alpaca, known only in a domesticated condition. These four animals appear so different, that most professed naturalists, especially those who have studied these animals in their native country, maintain that they are specifically distinct, notwithstanding that no one pretends to have seen a wild llama or alpaca. Mr. Ledger, however, who has closely studied these animals both in Peru and during their exportation to Australia, and who has made many experiments on their propagation, adduces arguments[62] which seem to me conclusive, that the llama is the domesticated descendant of the guanaco, and the alpaca of the vicuna. And now that we know that these animals many centuries ago were systematically bred and selected, there is nothing surprising in the great amount of change which they have undergone.

It appeared to me at one time probable that, though ancient and semi-civilised people might have attended to the improvement of their more useful animals in essential points, yet that they would have disregarded unimportant characters. But human nature is the same throughout the world: fashion everywhere reigns supreme, and man is apt to value whatever he may chance to possess. We have seen that in South America the niata cattle, which certainly are not made useful by their shortened faces and upturned nostrils, have been preserved. The Damaras of South Africa value their cattle for uniformity

[61] 'The Natural History of Dee Side,' 1855, p. 476.
[62] 'Bull. de la Soc. d'Acclimat.,' tom. vii., 1860, p. 457.

of colour and enormously long horns. The Mongolians value their yaks for their white tails. And I shall now show that there is hardly any peculiarity in our most useful animals which, from fashion, superstition, or some other motive, has not been valued, and consequently preserved. With respect to cattle, " an early record," according to Youatt,[63] speaks of a hundred " white cows with red ears being demanded as a compensation " by the princes of North and South Wales. If the cattle were " of a dark or black colour, 150 were to be presented." So that colour was attended to in Wales before its subjugation by England. In Central Africa, an ox that beats the ground with its tail is killed; and in South Africa some of the Damaras will not eat the flesh of a spotted ox. The Kaffirs value an animal with a musical voice; and " at a sale in British Kaffraria the " low of a heifer excited so much admiration that a sharp com- " petition sprung up for her possession, and she realised a " considerable price."[64] With respect to sheep, the Chinese prefer rams without horns; the Tartars prefer them with spirally wound horns, because the hornless are thought to lose courage.[65] Some of the Damaras will not eat the flesh of horn- less sheep. In regard to horses, at the end of the fifteenth century animals of the colour described as *liart pommé* were most valued in France. The Arabs have a proverb, " Never buy a horse with four white feet, for he carries his shroud with him ;"[66] the Arabs also, as we have seen, despise dun-coloured horses. So with dogs, Xenophon and others at an ancient period were prejudiced in favour of certain colours; and " white or slate-coloured hunting dogs were not esteemed."[67]

Turning to poultry, the old Roman gourmands thought that the liver of a white goose was the most savoury. In Paraguay black-skinned fowls are kept because they are thought to be more productive, and their flesh the most proper for invalids.[68] In Guiana, as I am informed by Sir R. Schomburgk, the aborigines will not eat the flesh or eggs of the fowl, but two

[63] 'Cattle,' p. 48.
[64] Livingstone's Travels, p. 576 ; Andersson, 'Lake Ngami,' 1856, p. 222. With respect to the sale in Kaffraria, *see* 'Quarterly Review,' 1860, p. 139.
[65] 'Mémoire sur les Chinois' (by the

Jesuits), 1786, tom. xi. p. 57.
[66] F. Michel, 'Des Haras,' pp. 47, 50.
[67] Col. Hamilton Smith, Dogs, in 'Nat. Lib.,' vol. x. p. 103.
[68] Azara, 'Quadrupèdes du Paraguay,' tom. ii. p. 324.

races are kept distinct merely for ornament. In the Philippines, no less than nine sub-varieties of the game cock are kept and named, so that they must be separately bred.

At the present time in Europe, the smallest peculiarities are carefully attended to in our most useful animals, either from fashion, or as a mark of purity of blood. Many examples could be given, two will suffice. "In the Western counties of England "the prejudice against a white pig is nearly as strong as against "a black one in Yorkshire." In one of the Berkshire sub-breeds, it is said, "the white should be confined to four white feet, "a white spot between the eyes, and a few white hairs behind "each shoulder." Mr. Saddler possessed "three hundred pigs, "every one of which was marked in this manner." [69] Marshall, towards the close of the last century, in speaking of a change in one of the Yorkshire breeds of cattle, says the horns have been considerably modified, as "a clean, small, sharp horn has been *fashionable* for the last twenty years." [70] In a part of Germany the cattle of the Race de Gfoehl are valued for many good qualities, but they must have horns of a particular curvature and tint, so much so that mechanical means are applied if they take a wrong direction; but the inhabitants "consider it "of the highest importance that the nostrils of the bull should "be flesh-coloured, and the eyelashes light; this is an indis- "pensable condition. A calf with blue nostrils would not be "purchased, or purchased at a very low price." [71] Therefore let no man say that any point or character is too trifling to be methodically attended to and selected by breeders.

Unconscious Selection.—By this term I mean, as already more than once explained, the preservation by man of the most valued, and the destruction of the least valued individuals, without any conscious intention on his part of altering the breed. It is difficult to offer direct proofs of the results which follow from this kind of selection; but the indirect evidence is abundant. In fact, except that in the one case man acts intentionally, and in the other unintentionally, there is little difference between

[69] Sidney's edit. of Youatt, 1860, pp. 24, 25.
[70] 'Rural Economy of Yorkshire' vol. ii. p. 182.
[71] Moll et Gayot, 'Du Bœuf,' 1860, p. 547.

methodical and unconscious selection. In both cases man preserves the animals which are most useful or pleasing to him,
and destroys or neglects the others. But no doubt a far more
rapid result follows from methodical than from unconscious
selection. The " roguing" of plants by gardeners, and the
destruction by law in Henry VIII.'s reign of all under-sized
mares, are instances of a process the reverse of selection in the
ordinary sense of the word, but leading to the same general
result. The influence of the destruction of individuals having
a particular character is well shown by the necessity of killing
every lamb with a trace of black about it, in order to keep the
flock white; or again, by the effects on the average height of
the men of France of the destructive wars of Napoleon, by
which many tall men were killed, the short ones being left to
be the fathers of families. This at least is the conclusion of
those who have closely studied the subject of the conscription;
and it is certain that since Napoleon's time the standard for
the army has been lowered two or three times.

Unconscious selection so blends into methodical that it is
scarcely possible to separate them. When a fancier long ago
first happened to notice a pigeon with an unusually short beak,
or one with the tail-feathers unusually developed, although
he bred from these birds with the distinct intention of propagating the variety, yet he could not have intended to make a
short-faced tumbler or a fantail, and was far from knowing that
he had made the first step towards this end. If he could have
seen the final result, he would have been struck with astonishment, but, from what we know of the habits of fanciers, probably
not with admiration. Our English carriers, barbs, and short-
faced tumblers have been greatly modified in the same manner,
as we may infer both from the historical evidence given in
the chapters on the Pigeon, and from the comparison of birds
brought from distant countries.

So it has been with dogs; our present fox-hounds differ from
the old English hound; our greyhounds have become lighter;
the wolf-dog, which belonged to the greyhound class, has become
extinct; the Scotch deer-hound has been modified, and is now
rare. Our bulldogs differ from those which were formerly used
for baiting bulls. Our pointers and Newfoundlands do not

closely resemble any native dog now found in the countries whence they were brought. These changes have been effected partly by crosses; but in every case the result has been governed by the strictest selection. Nevertheless there is no reason to suppose that man intentionally and methodically made the breeds exactly what they now are. As our horses became fleeter, and the country more cultivated and smoother, fleeter fox-hounds were desired and produced, but probably without any one distinctly foreseeing what they would become. Our pointers and setters, the latter almost certainly descended from large spaniels, have been greatly modified in accordance with fashion and the desire for increased speed. Wolves have become extinct, deer have become rarer, bulls are no longer baited, and the corresponding breeds of the dog have answered to the change. But we may feel almost sure that when, for instance, bulls were no longer baited, no man said to himself, I will now breed my dogs of smaller size, and thus create the present race. As circumstances changed, men unconsciously and slowly modified their course of selection.

With race-horses selection for swiftness has been followed methodically, and our horses can now easily beat their progenitors. The increased size and different appearance of the English race-horse led a good observer in India to ask, " Could any one in this year of 1856, looking at our race-horses, conceive that they were the result of the union of the Arab horse and the African mare ?"[72] This change has, it is probable, been largely effected through unconscious selection, that is, by the general wish to breed as fine horses as possible in each generation, combined with training and high feeding, but without any intention to give to them their present appearance. According to Youatt,[73] the introduction in Oliver Cromwell's time of three celebrated Eastern stallions speedily affected the English breed; "so that Lord Harleigh, one of the old school, complained that the great horse was fast disappearing." This is an excellent proof how carefully selection must have been attended to; for without such care, all traces of so small an infusion of Eastern blood would soon have been absorbed and

[72] ' The India Sporting Review,' vol. ii. p. 181; ' The Stud Farm,' by Cecil, p. 58.
[73] ' The Horse,' p. 22.

lost. Notwithstanding that the climate of England has never been esteemed particularly favourable to the horse, yet long-continued selection, both methodical and unconscious, together with that practised by the Arabs during a still longer and earlier period, has ended in giving us the best breed of horses in the world. Macaulay[74] remarks, "Two men whose authority on such " subjects was held in great esteem, the Duke of Newcastle and " Sir John Fenwick, pronounced that the meanest hack ever " imported from Tangier would produce a finer progeny than " could be expected from the best sire of our native breed. " They would not readily have believed that a time would " come when the princes and nobles of neighbouring lands " would be as eager to obtain horses from England as ever the " English had been to obtain horses from Barbary."

The London dray-horse, which differs so much in appearance from any natural species, and which from its size has so astonished many Eastern princes, was probably formed by the heaviest and most powerful animals having been selected during many generations in Flanders and England, but without the least intention or expectation of creating a horse such as we now see. If we go back to an early period of history, we behold in the antique Greek statues, as Schaaffhausen has remarked,[75] a horse equally unlike a race or dray horse, and differing from any existing breed.

The results of unconscious selection, in an early stage, are well shown in the difference between the flocks descended from the same stock, but separately reared by careful breeders. Youatt gives an excellent instance of this fact in the sheep belonging to Messrs. Buckley and Burgess, which " have been " purely bred from the original stock of Mr. Bakewell for " upwards of fifty years. There is not a suspicion existing in " the mind of any one at all acquainted with the subject that " the owner of either flock has deviated in any one instance " from the pure blood of Mr. Bakewell's flock ; yet the differ- " ence between the sheep possessed by these two gentlemen " is so great, that they have the appearance of being quite " different varieties."[76] I have seen several analogous and well-

[74] 'History of England,' vol. i. p. 316.

[75] 'Ueber Beständigkeit der Arten.'   [76] Youatt on Sheep, p. 315.

marked cases with pigeons: for instance, I had a family of
barbs, descended from those long bred by Sir J. Sebright, and
another family long bred by another fancier, and the two
families plainly differed from each other. Nathusius—and a
more competent witness could not be cited—observes that,
though the Shorthorns are remarkably uniform in appearance
(except in colouring), yet that the individual character and
wishes of each breeder become impressed on his cattle, so that
different herds differ slightly from each other.[77] The Hereford
cattle assumed their present well-marked character soon after
the year 1769, through careful selection by Mr. Tomkins,[78] and
the breed has lately split into two strains—one strain having
a white face, and differing slightly, it is said,[79] in some other
points; but there is no reason to believe that this split, the
origin of which is unknown, was intentionally made; it may with
much more probability be attributed to different breeders having
attended to different points. So again, the Berkshire breed of
swine in the year 1810 had greatly changed from what it had
been in 1780; and since 1810 at least two distinct sub-breeds
have borne this same name.[80] When we bear in mind how
rapidly all animals increase, and that some must be annually
slaughtered and some saved for breeding, then, if the same
breeder during a long course of years deliberately settles which
shall be saved and which shall be killed, it is almost inevitable
that his individual frame of mind will influence the character
of his stock, without his having had any intention to modify
the breed or form a new strain.

Unconscious selection in the strictest sense of the word, that
is, the saving of the more useful animals and the neglect or
slaughter of the less useful, without any thought of the future,
must have gone on occasionally from the remotest period and
amongst the most barbarous nations. Savages often suffer from
famines, and are sometimes expelled by war from their own
homes. In such cases it can hardly be doubted that they
would save their most useful animals. When the Fuegians

[77] 'Ueber Shorthorn Rindvieh,' 1857, s. 51.
[78] Low, 'Domesticated Animals,' 1845, p. 363.
[79] 'Quarterly Review,' 1849, p. 392.
[80] H. von Nathusius, 'Vorstudien .... Schweineschædel,' 1864, s. 140.

are hard pressed by want, they kill their old women for food rather than their dogs; for, as we were assured, "old women no use—dogs catch otters." The same sound sense would surely lead them to preserve their more useful dogs when still harder pressed by famine. Mr. Oldfield, who has seen so much of the aborigines of Australia, informs me that "they are all very glad to get a European kangaroo dog, and several instances have been known of the father killing his own infant that the mother might suckle the much-prized puppy." Different kinds of dogs would be useful to the Australian for hunting opossums and kangaroos, and to the Fuegian for catching fish and otters; and the occasional preservation in the two countries of the most useful animals would ultimately lead to the formation of two widely distinct breeds.

With plants, from the earliest dawn of civilisation, the best variety which at each period was known would generally have been cultivated and its seeds occasionally sown; so that there will have been some selection from an extremely remote period, but without any prefixed standard of excellence or thought of the future. We at the present day profit by a course of selection occasionally and unconsciously carried on during thousands of years. This is proved in an interesting manner by Oswald Heer's researches on the lake-inhabitants of Switzerland, as given in a former chapter; for he shows that the grain and seed of our present varieties of wheat, barley, oats, peas, beans, lentils, and poppy, exceed in size those which were cultivated in Switzerland during the Neolithic and Bronze periods. These ancient people, during the Neolithic period, possessed also a crab considerably larger than that now growing wild on the Jura.[81] The pears described by Pliny were evidently extremely inferior in quality to our present pears. We can realise the effects of long-continued selection and cultivation in another way, for would any one in his senses expect to raise a first-rate apple from the seed of a truly wild crab, or a luscious melting pear from the wild pear? Alphonse De Candolle informs me that he has lately seen on an ancient mosaic at Rome a representation of

[81] See also Dr. Christ, in 'Rütimeyer's Pfahlbauten,' 1861, s. 226.

the melon; and as the Romans, who were such gourmands, are silent on this fruit, he infers that the melon has been greatly ameliorated since the classical period.

Coming to later times, Buffon,[82] on comparing the flowers, fruit, and vegetables which were then cultivated, with some excellent drawings made a hundred and fifty years previously, was struck with surprise at the great improvement which had been effected; and remarks that these ancient flowers and vegetables would now be rejected, not only by a florist but by a village gardener. Since the time of Buffon the work of improvement has steadily and rapidly gone on. Every florist who compares our present flowers with those figured in books published not long since, is astonished at the change. A well-known amateur,[83] in speaking of the varieties of Pelargonium raised by Mr. Garth only twenty-two years before, remarks, " what a rage they excited: surely we had attained perfection, " it was said; and now not one of the flowers of those days " will be looked at. But none the less is the debt of gratitude " which we owe to those who saw what was to be done, and did " it." Mr. Paul, the well-known horticulturist, in writing of the same flower,[84] says he remembers when young being delighted with the portraits in Sweet's work; " but what are they in point " of beauty compared with the Pelargoniums of this day? Here " again nature did not advance by leaps; the improvement " was gradual, and, if we had neglected those very gradual " advances, we must have foregone the present grand results." How well this practical horticulturist appreciates and illustrates the gradual and accumulative force of selection! The Dahlia has advanced in beauty in a like manner; the line of improvement being guided by fashion, and by the successive modifications which the flower slowly underwent.[85] A steady and gradual change has been noticed in many other flowers: thus an old florist,[86] after describing the leading varieties of the Pink which were grown in 1813, adds, " the pinks of those days would now " be scarcely grown as border-flowers." The improvement of

[82] The passage is given 'Bull. Soc. d'Acclimat.,' 1858, p. 11.

[83] 'Journal of Horticulture,' 1862, p. 394.

[84] 'Gardener's Chronicle,' 1857, p. 85.

[85] See Mr. Wildman's address to the Floricult. Soc., in 'Gardener's Chronicle,' 1843, p. 86.

[86] 'Journal of Horticulture,' Oct. 24th, 1865, p. 230.

so many flowers and the number of the varieties which have been raised is all the more striking when we hear that the earliest known flower-garden in Europe, namely at Padua, dates only from the year 1545.[87]

Effects of Selection, as shown by the parts most valued by man presenting the greatest amount of Difference.—The power of long-continued selection, whether methodical or unconscious, or both combined, is well shown in a general way, namely, by the comparison of the differences between the varieties of distinct species, which are valued for different parts, such as for the leaves, or stems, or tubers, the seed, or fruit, or flowers. Whatever part man values most, that part will be found to present the greatest amount of difference. With trees cultivated for their fruit, Sageret remarks that the fruit is larger than in the parent-species, whilst with those cultivated for the seed, as with nuts, walnuts, almonds, chesnuts, &c., it is the seed itself which is larger; and he accounts for this fact by the fruit in the one case, and by the seed in the other, having been carefully attended to and selected during many ages. Gallesio has made the same observation. Godron insists on the diversity of the tuber in the potato, of the bulb in the onion, and of the fruit in the melon; and on the close similarity in these same plants of the other parts.[88]

In order to judge how far my own impression on this subject was correct, I cultivated numerous varieties of the same species close to each other. The comparison of the amount of difference between widely different organs is necessarily vague; I will therefore give the results in only a few cases. We have previously seen in the ninth chapter how greatly the varieties of the cabbage differ in their foliage and stems, which are the selected parts, and how closely they resemble each other in their flowers, capsules, and seeds. In seven varieties of the radish, the roots differed greatly in colour and shape, but no difference

[87] Prescott's 'Hist. of Mexico,' vol. ii. p. 61.

[88] Sageret, 'Pomologie Physiologique,' 1830, p. 47; Gallesio, 'Teoria della Riproduzione,' 1816, p. 88; Godron, 'De l'Espèce,' 1859, tom. ii. pp. 63, 67, 70. In my tenth and eleventh chapters I have given details on the potato; and I can confirm similar remarks with respect to the onion. I have also shown how far Naudin concurs in regard to the varieties of the melon.

whatever could be detected in their foliage, flowers, or seeds.
Now what a contrast is presented, if we compare the flowers of
the varieties of these two plants with those of any species cultivated in our flower-gardens for ornament; or if we compare
their seeds with those of the varieties of maize, peas, beans, &c.,
which are valued and cultivated for their seeds. In the ninth
chapter it was shown that the varieties of the pea differ but
little except in the tallness of the plant, moderately in the shape
of the pod, and greatly in the pea itself, and these are all selected
points. The varieties, however, of the *Pois sans parchemin*
differ much more in their pods, and these are eaten and valued.
I cultivated twelve varieties of the common bean; one alone,
the Dwarf Fan, differed considerably in general appearance;
two differed in the colour of their flowers, one being an albino,
and the other being wholly instead of partially purple; several
differed considerably in the shape and size of the pod, but far
more in the bean itself, and this is the valued and selected part.
Toker's bean, for instance, is twice-and-a-half as long and broad
as the horse-bean, and is much thinner and of a different shape.

The varieties of the gooseberry, as formerly described, differ
much in their fruit, but hardly perceptibly in their flowers or
organs of vegetation. With the plum, the differences likewise
appear to be greater in the fruit than in the flowers or leaves.
On the other hand, the seed of the strawberry, which corresponds with the fruit of the plum, differs hardly at all; whilst
every one knows how greatly the fruit—that is, the enlarged
receptacle—differs in the several varieties. In apples, pears,
and peaches the flowers and leaves differ considerably, but not,
as far as I can judge, in proportion with the fruit. The Chinese
double-flowering peaches, on the other hand, show that varieties
of this tree have been formed, which differ more in the flower
than in fruit. If, as is highly probable, the peach is the modified descendant of the almond, a surprising amount of change
has been effected in the same species, in the fleshy covering of
the former and in the kernels of the latter.

When parts stand in such close relation to each other as the
fleshy covering of the fruit (whatever its homological nature may
be) and the seed, when one part is modified, so generally is the
other, but by no means necessarily in the same degree. With

the plum-tree, for instance, some varieties produce plums which are nearly alike, but include stones extremely dissimilar in shape; whilst conversely other varieties produce dissimilar fruit with barely distinguishable stones; and generally the stones, though they have never been subjected to selection, differ greatly in the several varieties of the plum. In other cases organs which are not manifestly related, through some unknown bond vary together, and are consequently liable, without any intention on man's part, to be simultaneously acted on by selection. Thus the varieties of the stock (Matthiola) have been selected solely for the beauty of their flowers, but the seeds differ greatly in colour and somewhat in size. Varieties of the lettuce have been selected solely on account of their leaves, yet produce seeds which likewise differ in colour. Generally, through the law of correlation, when a variety differs greatly from its fellow-varieties in any one character, it differs to a certain extent in several other characters. I observed this fact when I cultivated together many varieties of the same species, for I used first to make a list of the varieties which differed most from each other in their foliage and manner of growth, afterwards of those that differed most in their flowers, then in their seed-capsules, and lastly in their mature seed; and I found that the same names generally occurred in two, three, or four of the successive lists. Nevertheless the greatest amount of difference between the varieties was always exhibited, as far as I could judge, by that part or organ for which the plant was cultivated.

When we bear in mind that each plant was at first cultivated because useful to man, and that its variation was a subsequent, often a long subsequent, event, we cannot explain the greater amount of diversity in the valuable parts by supposing that species endowed with an especial tendency to vary in any particular manner, were originally chosen. We must attribute the result to the variations in these parts having been successively preserved, and thus continually augmented; whilst other variations, excepting such as inevitably appeared through correlation, were neglected and lost. Hence we may infer that most plants might be made, through long-continued selection, to yield races as different from each other in any character as they now are in those parts for which they are valued and cultivated.

With animals we see something of the same kind; but they have not been domesticated in sufficient number or yielded sufficient varieties for a fair comparison. Sheep are valued for their wool, and the wool differs much more in the several races than the hair in cattle. Neither sheep, goats, European cattle, nor pigs are valued for their fleetness or strength; and we do not possess breeds differing in these respects like the race-horse and dray-horse. But fleetness and strength are valued in camels and dogs; and we have with the former the swift dromedary and heavy camel; with the latter the greyhound and mastiff. But dogs are valued even in a higher degree for their mental qualities and senses; and every one knows how greatly the races differ in these respects. On the other hand, where the dog is valued solely to serve for food, as in the Polynesian islands and China, it is described as an extremely stupid animal.[89] Blumenbach remarks that "many dogs, such as the badger- " dog, have a build so marked and so appropriate for particular " purposes, that I should find it very difficult to persuade myself " that this astonishing figure was an accidental consequence of " degeneration."[90] But had Blumenbach reflected on the great principle of selection, he would not have used the term degeneration, and he would not have been astonished that dogs and other animals should become excellently adapted for the service of man.

On the whole we may conclude that whatever part or character is most valued—whether the leaves, stems, tubers, bulbs, flowers, fruit, or seed of plants, or the size, strength, fleetness, hairy covering, or intellect of animals—that character will almost invariably be found to present the greatest amount of difference both in kind and degree. And this result may be safely attributed to man having preserved during a long course of generations the variations which were useful to him, and neglected the others.

I will conclude this chapter by some remarks on an important subject. With animals such as the giraffe, of which

[89] Godron, 'De l'Espèce,' tom. ii. p. 27.
[90] 'The Anthropological Treatises of Blumenbach,' 1865, p. 292.

the whole structure is admirably co-ordinated for certain purposes, it has been supposed that all the parts must have been
simultaneously modified; and it has been argued that, on the
principle of natural selection, this is scarcely possible. But in
thus arguing, it has been tacitly assumed that the variations must
have been abrupt and great. No doubt, if the neck of a ruminant
were suddenly to become greatly elongated, the fore limbs and
back would have to be simultaneously strengthened and modified;
but it cannot be denied that an animal might have its neck, or
head, or tongue, or fore-limbs elongated a very little without
any corresponding modification in other parts of the body; and
animals thus slightly modified would, during a dearth, have a
slight advantage, and be enabled to browse on higher twigs, and
thus survive. A few mouthfuls more or less every day would
make all the difference between life and death. By the repeti-
tion of the same process, and by the occasional intercrossing
of the survivors, there would be some progress, slow and fluctuating though it would be, towards the admirably co-ordinated
structure of the giraffe. If the short-faced tumbler-pigeon, with
its small conical beak, globular head, rounded body, short wings,
and small feet—characters which appear all in harmony—had
been a natural species, its whole structure would have been
viewed as well fitted for its life; but in this case we know that
inexperienced breeders are urged to attend to point after point,
and not to attempt improving the whole structure at the same
time. Look at the greyhound, that perfect image of grace,
symmetry, and vigour; no natural species can boast of a more
admirably co-ordinated structure, with its tapering head, slim
body, deep chest, tucked-up abdomen, rat-like tail, and long
muscular limbs, all adapted for extreme fleetness, and for
running down weak prey. Now, from what we see of the
variability of animals, and from what we know of the method
which different men follow in improving their stock—some chiefly
attending to one point, others to another point, others again
correcting defects by crosses, and so forth—we may feel assured
that if we could see the long line of ancestors of a first-rate
greyhound, up to its wild wolf-like progenitor, we should behold
an infinite number of the finest gradations, sometimes in one
character and sometimes in another, but all leading towards our

present perfect type. By small and doubtful steps such as these, nature, as we may confidently believe, has progressed on her grand march of improvement and development.

A similar line of reasoning is as applicable to separate organs as to the whole organisation. A writer[91] has recently maintained that " it is probably no exaggeration to suppose that, in order to "improve such an organ as the eye at all, it must be improved " in ten different ways at once. And the improbability of any "complex organ being produced and brought to perfection in " any such way is an improbability of the same kind and degree " as that of producing a poem or a mathematical demonstration " by throwing letters at random on a table." If the eye were abruptly and greatly modified, no doubt many parts would have to be simultaneously altered, in order that the organ should remain serviceable.

But is this the case with smaller changes? There are persons who can see distinctly only in a dull light, and this condition depends, I believe, on the abnormal sensitiveness of the retina, and is known to be inherited. Now, if a bird, for instance, received some great advantage from seeing well in the twilight, all the individuals with the most sensitive retina would succeed best and be the most likely to survive; and why should not all those which happened to have the eye itself a little larger, or the pupil capable of greater dilatation, be likewise preserved, whether or not these modifications were strictly simultaneous? These individuals would subsequently intercross and blend their respective advantages. By such slight successive changes, the eye of a diurnal bird would be brought into the condition of that of an owl, which has often been advanced as an excellent instance of adaptation. Short-sight, which is often inherited, permits a person to see distinctly a minute object at so near a distance that it would be indistinct to ordinary eyes; and here we have a capacity which might be serviceable under certain conditions, abruptly gained. The Fuegians on board the

[91] Mr. J. J. Murphy in his opening address to the Belfast Nat. Hist. Soc., as given in the Belfast Northern Whig, Nov. 19, 1866. Mr. Murphy here follows the line of argument against my views previously and more cautiously given by the Rev. C. Pritchard, Pres. Royal Astronomical Soc., in his sermon (Appendix, p. 33) preached before the British Association at Nottingham, 1866.

Beagle could certainly see distant objects more distinctly than our sailors with all their long practice; I do not know whether this depends on nervous sensitiveness or on the power of adjustment in the focus; but this capacity for distant vision might, it is probable, be slightly augmented by successive modifications of either kind. Amphibious animals, which are enabled to see both in the water and in the air, require and possess, as M. Plateau has shown,[92] eyes constructed on the following plan : " the cornea is always flat, or at least much flattened in front of " the crystalline and over a space equal to the diameter of that " lens, whilst the lateral portions may be much curved." The crystalline is very nearly a sphere, and the humours have nearly the same density as water. Now, as a terrestrial animal slowly became more and more aquatic in its habits, very slight changes, first in the curvature of the cornea or crystalline, and then in the density of the humours, or conversely, might successively occur, and would be advantageous to the animal whilst under water, without serious detriment to its power of vision in the air. It is of course impossible to conjecture by what steps the fundamental structure of the eye in the Vertebrata was originally acquired, for we know absolutely nothing about this organ in the first progenitors of the class. With respect to the lowest animals in the scale, the transitional states through which the eye at first probably passed, can by the aid of analogy be indicated, as I have attempted to show in my 'Origin of Species.'[93]

[92] On the Vision of Fishes and Amphibia, translated in 'Annals and Mag. of Nat. Hist.,' vol. xviii., 1866, p. 469.　　　[93] Fourth edition, 1866, p. 215.

CHAPTER XXI.

SELECTION, *continued.*

NATURAL SELECTION AS AFFECTING DOMESTIC PRODUCTIONS — CHARACTERS WHICH
APPEAR OF TRIFLING VALUE OFTEN OF REAL IMPORTANCE — CIRCUMSTANCES
FAVOURABLE TO SELECTION BY MAN — FACILITY IN PREVENTING CROSSES, AND
THE NATURE OF THE CONDITIONS — CLOSE ATTENTION AND PERSEVERANCE INDIS-
PENSABLE — THE PRODUCTION OF A LARGE NUMBER OF INDIVIDUALS ESPECIALLY
FAVOURABLE — WHEN NO SELECTION IS APPLIED, DISTINCT RACES ARE NOT FORMED
— HIGHLY-BRED ANIMALS LIABLE TO DEGENERATION — TENDENCY IN MAN TO
CARRY THE SELECTION OF EACH CHARACTER TO AN EXTREME POINT, LEADING TO
DIVERGENCE OF CHARACTER, RARELY TO CONVERGENCE — CHARACTERS CONTINUING
TO VARY IN THE SAME DIRECTION IN WHICH THEY HAVE ALREADY VARIED —
DIVERGENCE OF CHARACTER, WITH THE EXTINCTION OF INTERMEDIATE VARIETIES,
LEADS TO DISTINCTNESS IN OUR DOMESTIC RACES — LIMIT TO THE POWER OF
SELECTION — LAPSE OF TIME IMPORTANT — MANNER IN WHICH DOMESTIC RACES
HAVE ORIGINATED — SUMMARY.

*Natural Selection, or the Survival of the Fittest, as affecting
domestic productions.*—WE know little on this head. But as
animals kept by savages have to provide their own food,
either entirely or to a large extent, throughout the year, it
can hardly be doubted that, in different countries, varieties differing in constitution and in various characters would succeed
best, and so be naturally selected. Hence perhaps it is that the
few domesticated animals kept by savages partake, as has been
remarked by more than one writer, of the wild appearance of
their masters, and likewise resemble natural species. Even in
long-civilised countries, at least in the wilder parts, natural
selection must act on our domestic races. It is obvious that
varieties, having very different habits, constitution, and structure, would succeed best on mountains and on rich lowland
pastures. For example, the improved Leicester sheep were
formerly taken to the Lammermuir Hills; but an intelligent
sheep-master reported that " our coarse lean pastures were
" unequal to the task of supporting such heavy-bodied sheep;
" and they gradually dwindled away into less and less bulk;

" each generation was inferior to the preceding one; and when
" the spring was severe, seldom more than two-thirds of the
" lambs survived the ravages of the storms."[1] So with the
mountain cattle of North Wales and the Hebrides, it has been
found that they could not withstand being crossed with the larger
and more delicate lowland breeds. Two French naturalists, in
describing the horses of Circassia, remark that, subjected as
they are to extreme vicissitudes of climate, having to search
for scanty pasture, and exposed to constant danger from wolves,
the strongest and most vigorous alone survive.[2]

Every one must have been struck with the surpassing grace,
strength, and vigour of the Game-cock, with its bold and confident air, its long, yet firm neck, compact body, powerful and
closely pressed wings, muscular thighs, strong beak massive at
the base, dense and sharp spurs set low on the legs for
delivering the fatal blow, and its compact, glossy, and mail-like
plumage serving as a defence. Now the English game-cock has
not only been improved during many years by man's careful
selection, but in addition, as Mr. Tegetmeier has remarked,[3] by
a kind of natural selection, for the strongest, most active and
courageous birds have stricken down their antagonists in the
cockpit, generation after generation, and have subsequently
served as the progenitors of their kind.

In Great Britain, in former times, almost every district had
its own breed of cattle and sheep; " they were indigenous to
" the soil, climate, and pasturage of the locality on which they
" grazed: they seemed to have been formed for it and by it."[4]
But in this case we are quite unable to disentangle the effects
of the direct action of the conditions of life,—of use or habit—of
natural selection—and of that kind of selection which we have
seen is occasionally and unconsciously followed by man even
during the rudest periods of history.

Let us now look to the action of natural selection on special
characters. Although nature is difficult to resist, yet man often
strives against her power, and sometimes, as we shall see, with

[1] Quoted by Youatt on Sheep, p.
325. *See* also Youatt on Cattle, pp. 62,
69.

[2] MM. Lherbette and De Quatre-

fages, in ' Bull. Soc. Acclimat.,' tom.
viii., 1861, p. 311.

[3] ' The Poultry Book,' 1866, p. 123.

[4] Youatt on Sheep, p. 312.

success. From the facts to be given, it will also be seen that natural selection would powerfully affect many of our domestic productions if left unprotected. This is a point of much interest, for we thus learn that differences apparently of very slight importance would certainly determine the survival of a form when forced to struggle for its own existence. It may have occurred to some naturalists, as it formerly did to me, that, though selection acting under natural conditions would determine the structure of all important organs, yet that it could not affect characters which are esteemed by us of little importance; but this is an error to which we are eminently liable, from our ignorance of what characters are of real value to each living creature.

When man attempts to breed an animal with some serious defect in structure, or in the mutual relation of parts, he will either partially or completely fail, or encounter much difficulty; and this is in fact a form of natural selection. We have seen that the attempt was once made in Yorkshire to breed cattle with enormous buttocks, but the cows perished so often in bringing forth their calves, that the attempt had to be given up. In rearing short-faced tumblers, Mr. Eaton says,[5] "I am "convinced that better head and beak birds have perished in "the shell than ever were hatched; the reason being that the "amazingly short-faced bird cannot reach and break the shell "with its beak, and so perishes." Here is a more curious case, in which natural selection comes into play only at long intervals of time: during ordinary seasons the Niata cattle can graze as well as others, but occasionally, as from 1827 to 1830, the plains of La Plata suffer from long-continued droughts and the pasture is burnt up; at such times common cattle and horses perish by the thousand, but many survive by browsing on twigs, reeds, &c.; this the Niata cattle cannot so well effect from their upturned jaws and the shape of their lips; consequently, if not attended to, they perish before the other cattle. In Colombia, according to Roulin, there is a breed of nearly hairless cattle, called Pelones; these succeed in their native hot district, but are found too tender for the Cordillera; in this case, natural selection

---

[5] 'Treatise on the Almond Tumbler,' 1851, p. 33.

determines only the range of the variety. It is obvious that a
host of artificial races could never survive in a state of nature;
—such as Italian greyhounds,—hairless and almost toothless
Turkish dogs,—fantail pigeons, which cannot fly well against a
strong wind,—barbs with their vision impeded by their eye-
wattle,—Polish fowls with their vision impeded by their great
topknots,—hornless bulls and rams which consequently cannot
cope with other males, and thus have a poor chance of leaving
offspring,—seedless plants, and many other such cases.

Colour is generally esteemed by the systematic naturalist as
unimportant: let us, therefore, see how far it indirectly affects
our domestic productions, and how far it would affect them if
they were left exposed to the full force of natural selection.   In
a future chapter I shall have to show that constitutional peculiarities of the strangest kind, entailing liability to the action
of certain poisons, are correlated with the colour of the skin.
I will here give a single case, on the high authority of Professor
Wyman; he informs me that, being surprised at all the pigs in
a part of Virginia being black, he made inquiries, and ascertained that these animals feed on the roots of the *Lachnanthes
tinctoria*, which colours their bones pink, and, excepting in the
case of the black varieties, causes the hoofs to drop off.   Hence,
as one of the squatters remarked, " we select the black members
of the litter for raising, as they alone have a good chance of
living."   So that here we have artificial and natural selection
working hand in hand. I may add that in the Tarentino the
inhabitants keep black sheep alone, because the *Hypericum
crispum* abounds there; and this plant does not injure black
sheep, but kills the white ones in about a fortnight's time.[6]

Complexion, and liability to certain diseases, are believed
to run together in man and the lower animals. Thus white
terriers suffer more than terriers of any other colour from the
fatal Distemper.[7] In North America plum-trees are liable
to a disease which Downing[8] believes is not caused by insects;
the kinds bearing purple fruit are most affected, "and we have
" never known the green or yellow fruited varieties infected

[6] Dr. Heusinger, ' Wochenschrift für
die Heilkunde,' Berlin, 1846, s. 279.

[7] Youatt on the Dog, p. 232.

[8] ' The Fruit-trees of America,' 1845,
p. 270 : for peaches, p. 466.

" until the other sorts had first become filled with the knots."
On the other hand, peaches in North America suffer much
from a disease called the *yellows*, which seems to be peculiar
to that continent, and " more than nine-tenths of the victims,
" when the disease first appeared, were the yellow-fleshed
" peaches. The white-fleshed kinds are much more rarely
" attacked; in some parts of the country never." In Mauritius,
the white sugar-canes have of late years been so severely
attacked by a disease, that many planters have been compelled
to give up growing this variety (although fresh plants were
imported from China for trial), and cultivate only red canes.[9]
Now, if these plants had been forced to struggle with other
competing plants and enemies, there cannot be a doubt that the
colour of the flesh or skin of the fruit, unimportant as these
characters are considered, would have rigorously determined their
existence.

Liability to the attacks of parasites is also connected with
colour. It appears that white chickens are certainly more subject than dark-coloured chickens to the *gapes*, which is caused
by a parasitic worm in the trachea.[10] On the other hand,
experience has shown that in France the caterpillars which
produce white cocoons resist the deadly fungus better than
those producing yellow cocoons.[11] Analogous facts have been
observed with plants: a new and beautiful white onion, imported
from France, though planted close to other kinds, was alone
attacked by a parasitic fungus.[12] White verbenas are especially
liable to mildew.[13] Near Malaga, during an early period of the
vine-disease, the green sorts suffered most; " and red and black
grapes, even when interwoven with the sick plants, suffered not
at all." In France whole groups of varieties were comparatively free, and others, such as the Chasselas, did not afford a
single fortunate exception; but I do not know whether any
correlation between colour and liability to disease was here
observed.[14] In a former chapter it was shown how curiously
liable one variety of the strawberry is to mildew.

[9] 'Proc. Royal Soc. of Arts and Sciences of Mauritius,' 1852, p. cxxxv.
[10] 'Gardener's Chronicle,' 1856, p. 379.
[11] Quatrefages, 'Maladies Actuelles du Ver à Soie,' 1850, pp. 12, 214.
[12] 'Gardener's Chronicle,' 1851, p. 595.
[13] 'Journal of Horticulture,' 1862, p. 476.
[14] 'Gardener's Chronicle,' 1852, pp. 435, 691.

It is certain that insects regulate in many cases the range and even the existence of the higher animals, whilst living under their natural conditions. Under domestication light-coloured animals suffer most: in Thuringia [15] the inhabitants do not like grey, white, or pale cattle, because they are much more troubled by various kinds of flies than the brown, red, or black cattle. An Albino negro, it has been remarked,[16] was peculiarly sensitive to the bites of insects. In the West Indies [17] it is said that " the only horned cattle fit for work are those which " have a good deal of black in them. The white are terribly " tormented by the insects; and they are weak and sluggish in " proportion to the white."

In Devonshire there is a prejudice against white pigs, because it is believed that the sun blisters them when turned out; [18] and I knew a man who would not keep white pigs in Kent, for the same reason. The scorching of flowers by the sun seems like-wise to depend much on colour; thus, dark pelargoniums suffer most; and from various accounts it is clear that the cloth-of-gold variety will not withstand a degree of exposure to sunshine which other varieties enjoy. Another amateur asserts that not only all dark-coloured verbenas, but likewise scarlets, suffer from the sun; " the paler kinds stand better, and pale blue is perhaps the best of all." So again with the heartsease (*Viola tricolor*); hot weather suits the blotched sorts, whilst it destroys the beautiful markings of some other kinds.[19] During one extremely cold season in Holland all red-flowered hyacinths were observed to be very inferior in quality. It is believed by many agriculturists that red wheat is hardier in northern climates than white wheat.[20]

With animals, white varieties from being conspicuous are the most liable to be attacked by beasts and birds of prey. In parts of France and Germany where hawks abound, persons are advised not to keep white pigeons; for, as Parmentier says, " it

[15] Bechstein, 'Naturgesch. Deutsch-lands,' 1801, B. i. s. 310.

[16] Prichard, 'Phys. Hist. of Mankind,' 1851, vol. i. p. 224.

[17] G. Lewis's 'Journal of Residence in West Indies,' · 'Home and Col. Library,' p. 100.

[18] Sidney's edit. of Youatt on the Pig, p. 24.

[19] 'Journal of Horticulture,' 1862, pp. 476, 498; 1865, p. 460. With respect to the heartsease, ' Gardener's Chronicle,' 1863, p. 628.

[20] 'Des Jacinthes, de leur Culture,' 1768, p. 53 : on wheat, 'Gardener's Chronicle,' 1846, p. 653.

is certain that in a flock the white always first fall victims to
the kite." In Belgium, where so many societies have been established for the flight of carrier-pigeons, white is the one colour
which for the same reason is disliked.[21] On the other hand, it
is said that the sea-eagle (*Falco ossifragus*, Linn.) on the west
coast of Ireland picks out the black fowls, so that "the vil-
lagers avoid as much as possible rearing birds of that colour."
M. Daudin,[22] speaking of white rabbits kept in warrens in
Russia, remarks that their colour is a great disadvantage, as
they are thus more exposed to attack, and can be seen during
bright nights from a distance. A gentleman in Kent, who failed
to stock his woods with a nearly white and hardy kind of rabbit,
accounted in the same manner for their early disappearance.
Any one who will watch a white cat prowling after her prey
will soon perceive under what a disadvantage she lies.

The white Tartarian cherry, "owing either to its colour being
so much like that of the leaves, or to the fruit always appearing
from a distance unripe," is not so readily attacked by birds
as other sorts. The yellow-fruited raspberry, which generally
comes nearly true by seed, "is very little molested by birds,
who evidently are not fond of it; so that nets may be dispensed
with in places where nothing else will protect the red fruit."[23]
This immunity, though a benefit to the gardener, would be a
disadvantage in a state of nature both to the cherry and
raspberry, as their dissemination depends on birds. I noticed
during several winters that some trees of the yellow-berried
holly, which were raised from seed from a wild tree found by
my father, remained covered with fruit, whilst not a scarlet
berry could be seen on the adjoining trees of the common kind.
A friend informs me that a mountain-ash (*Pyrus aucuparia*)
growing in his garden bears berries which, though not differently
coloured, are always devoured by birds before those on the other
trees. This variety of the mountain-ash would thus be more
freely disseminated, and the yellow-berried variety of the holly
less freely, than the common varieties of these two trees.

[21] W. B. Tegetmeier, 'The Field,'
Feb. 25, 1865. With respect to black
fowls, *see* a quotation in Thompson's
'Nat. Hist. of Ireland,' 1849, vol. i. p. 22.
[22] 'Bull. de la Soc. d'Acclimat.,' tom.

vii. 1860, p. 359.
[23] 'Transact. Hort. Soc.,' vol. i. 2nd
series, 1835, p. 275. For raspberries,
see 'Gard. Chronicle,' 1855, p. 154, and
1863, p. 245.

Independently of colour, other trifling differences are sometimes found to be of importance to plants under cultivation, and
would be of paramount importance if they had to fight their
own battle and to struggle with many competitors. The thin-
shelled peas, called *pois sans parchemin*, are attacked by
birds[24] much more than common peas. On the other hand, the
purple-podded pea, which has a hard shell, escaped the attacks of
tomtits (*Parus major*) in my garden far better than any other
kind. The thin-shelled walnut likewise suffers greatly from
the tomtit.[25] These same birds have been observed to pass over
and thus favour the filbert, destroying only the other kinds of
nuts which grew in the same orchard.[26]

Certain varieties of the pear have soft bark, and these suffer
severely from boring wood-beetles; whilst other varieties are
known to resist their attacks much better.[27] In North America
the smoothness, or absence of down on the fruit, makes a great
difference in the attacks of the weevil, " which is the uncom-
promising foe of all smooth stone-fruits;" and the cultivator
" has the frequent mortification of seeing nearly all, or indeed
often the whole crop, fall from the trees when half or two-thirds
grown." Hence the nectarine suffers more than the peach. A particular variety of the Morello cherry, raised in North America, is
without any assignable cause more liable to be injured by this
same insect than other cherry-trees.[28] From some unknown cause,
the Winter Majetin apple enjoys the great advantage of not being
infested by the coccus. On the other hand, a particular case
has been recorded in which aphides confined themselves to the
Winter Nelis pear, and touched no other kind in an extensive
orchard.[29] The existence of minute glands on the leaves of
peaches, nectarines, and apricots, would not be esteemed by
botanists as a character of the least importance, for they are
present or absent in closely related sub-varieties, descended from
the same parent-tree; yet there is good evidence[30] that the

[24] 'Gardener's Chronicle,' 1843, p. 806.

[25] Ibid., 1850, p. 732.

[26] Ibid., 1860, p. 956.

[27] J. De Jonghe, in 'Gard. Chronicle,' 1860, p. 120.

[28] Downing, 'Fruit-trees of North America,' pp. 266, 501 : in regard to the cherry, p. 198.

[29] 'Gardener's Chronicle,' 1849, p. 755.

[30] 'Journal of Horticulture,' Sept. 26th, 1865, p. 254; *see* other references given in chap. x.

absence of glands leads to mildew, which is highly injurious to these trees.

A difference either in flavour or in the amount of nutriment in certain varieties causes them to be more eagerly attacked by various enemies than other varieties of the same species. Bullfinches (*Pyrrhula vulgaris*) injure our fruit-trees by devouring the flower-buds, and a pair of these birds have been seen "to denude a large plum-tree in a couple of days of almost every bud;" but certain varieties [31] of the apple and thorn (*Cratægus oxycantha*) are more especially liable to be attacked. A striking instance of this was observed in Mr. Rivers's garden, in which two rows of a particular variety of plum [32] had to be carefully protected, as·they were usually stripped of all their buds during the winter, whilst other sorts growing near them escaped. The root (or enlarged stem) of Laing's Swedish turnip is preferred by hares, and therefore suffers more than other varieties. Hares and rabbits eat down common rye before St. John's-day-rye, when both grow together. [33] In the South of France, when an orchard of almond-trees is formed, the nuts of the bitter variety are sown, "in order that they may not be devoured by field-mice;" [34] so we see the use of the bitter principle in almonds.

Other slight differences, which would be thought quite unimportant, are no doubt sometimes of great service both to plants and animals. The Whitesmith's gooseberry, as formerly stated, produces its leaves later than other varieties, and, as the flowers are thus left unprotected, the fruit often fails. In one variety of the cherry, according to Mr. Rivers, [35] the petals are much curled backwards, and in consequence of this the stigmas were observed to be killed by a severe frost; whilst at the same time, in another variety with incurved petals, the stigmas were not in the least injured. The straw of the Fenton wheat is remarkably unequal in height; and a competent observer believes that this variety is highly productive, partly because the ears, from being distributed at various heights above the ground,

[31] Mr. Selby, in 'Mag. of Zoology and Botany,' Edinburgh, vol. ii., 1838, p. 393.

[32] The Reine Claude de Bavay, 'Journal of Horticulture,' Dec. 27, 1864, p. 511.

[33] Mr. Pusey, in 'Journal of R. Agricult. Soc.,' vol. vi. p. 179. For Swedish turnips, *see* 'Gard. Chron.,' 1847, p. 91.

[34] Godron, 'De l'Espèce,' tom. ii. p. 98.

[35] 'Gardener's Chron.,' 1866, p. 732.

are less crowded together. The same observer maintains that in the upright varieties the divergent awns are serviceable by breaking the shocks when the ears are dashed together by the wind.[36] If several varieties of a plant are grown together, and the seed is indiscriminately harvested, it is clear that the hardier and more productive kinds will, by a sort of natural selection, gradually prevail over the others; this takes place, as Colonel Le Couteur believes,[37] in our wheat-fields, for, as formerly shown, no variety is quite uniform in character. The same thing, as I am assured by nurserymen, would take place in our flower-gardens, if the seed of the different varieties were not separately saved. When the eggs of the wild and tame duck are hatched together, the young wild ducks almost invariably perish, from being of smaller size and not getting their fair share of food.[38]

Facts in sufficient number have now been given showing that natural selection often checks, but occasionally favours, man's power of selection. These facts teach us, in addition, a valuable lesson, namely, that we ought to be extremely cautious in judging what characters are of importance in a state of nature to animals and plants, which have to struggle from the hour of their birth to that of their death for existence,—their existence depending on conditions, about which we are profoundly ignorant.

Circumstances favourable to Selection by Man.

The possibility of selection rests on variability, and this, as we shall see in the following chapters, mainly depends on changed conditions of life, but is governed by infinitely complex, and, to a great extent, unknown laws. Domestication, even when long continued, occasionally causes but a small amount of variability, as in the case of the goose and turkey. The slight differences, however, which characterise each individual animal and plant would in most, probably in all cases, suffice for the production of distinct races through careful and prolonged selection. We see what selection, though acting on mere individual differences, can effect when families of cattle, sheep,

[36] 'Gardener's Chronicle,' 1862, pp. 820, 821.
[37] 'On the Varieties of Wheat,' p. 59.
[38] Mr. Hewitt and others, in 'Journal of Hort.,' 1862, p. 773.

pigeons, &c., of the same race, have been separately bred during a number of years by different men without any wish on their part to modify the breed. We see the same fact in the difference between hounds bred for hunting in different districts,[39] and in many other such cases.

In order that selection should produce any result, it is manifest that the crossing of distinct races must be prevented; hence facility in pairing, as with the pigeon, is highly favourable for the work; and difficulty in pairing, as with cats, prevents the formation of distinct breeds. On nearly the same principle the cattle of the small island of Jersey have been improved in their milking qualities "with a rapidity that could not have been obtained in a widely extended country like France."[40] Although free crossing is a danger on the one side which every one can see, too close interbreeding is a hidden danger on the other side. Unfavourable conditions of life overrule the power of selection. Our improved heavy breeds of cattle and sheep could not have been formed on mountainous pastures; nor could dray-horses have been raised on a barren and inhospitable land, such as the Falkland islands, where even the light horses of La Plata rapidly decrease in size. Nor could the wool of sheep have been much increased in length within the Tropics; yet selection has kept Merino sheep nearly true under diversified and unfavourable conditions of life. The power of selection is so great, that breeds of the dog, sheep, and poultry, of the largest and least size, long and short beaked pigeons, and other breeds with opposite characters, have had their characteristic qualities augmented, though treated in every way alike, being exposed to the same climate and fed on the same food. Selection, however, is either checked or favoured by the effects of use or habit. Our wonderfully-improved pigs could never have been formed if they had been forced to search for their own food; the English racehorse and greyhound could not have been improved up to their present high standard of excellence without constant training.

As conspicuous deviations of structure occur rarely, the improvement of each breed is generally the result, as already

---

[39] 'Encyclop. of Rural Sports,' p. 405.
[40] Col. Le Couteur, 'Journal Roy. Agricult. Soc.,' vol. iv. p. 43.

remarked, of the selection of slight individual differences. Hence the closest attention, the sharpest powers of observation, and indomitable perseverance, are indispensable. It is, also, highly important that many individuals of the breed which is to be improved should be raised; for thus there will be a better chance of the appearance of variations in the right direction, and individuals varying in an unfavourable manner may be freely rejected or destroyed. But that a large number of individuals should be raised, it is necessary that the conditions of life should favour the propagation of the species. Had the peacock been bred as easily as the fowl, we should probably ere this have had many distinct races. We see the importance of a large number of plants, from the fact of nursery gardeners almost always beating amateurs in the exhibition of new varieties. In 1845 it was estimated [41] that between 4000 and 5000 pelargoniums were annually raised from seed in England, yet a decidedly improved variety is rarely obtained. At Messrs. Carter's grounds, in Essex, where such flowers as the Lobelia, Nemophila, Mignonette, &c., are grown by the acre for seed, "scarcely a season passes without some new kinds being raised, or some improvement effected on old kinds." [42] At Kew, as Mr. Beaton remarks, where many seedlings of common plants are raised, "you see new forms of Laburnums, Spiræas, and other shrubs." [43] So with animals: Marshall, [44] in speaking of the sheep in one part of Yorkshire, remarks, "as they belong to poor people, and are mostly in small lots, they never can be improved." Lord Rivers, when asked how he succeeded in always having first-rate greyhounds, answered, "I breed many, and hang many." This, as another man remarks, "was the secret of his success; and the same will be found in exhibiting fowls, —successful competitors breed largely, and keep the best." [45]

It follows from this that the capacity of breeding at an early age and at short successive intervals, as with pigeons, rabbits, &c., facilitates selection; for the result is thus soon made visible, and perseverance in the work is encouraged. It can hardly be

[41] 'Gardener's Chronicle,' 1845, p. 273.

[42] 'Journal of Horticulture,' 1862, p. 157.

[43] 'Cottage Gardener,' 1860, p. 368.

[44] 'A Review of Reports,' 1808, p. 406.

[45] 'Gardener's Chronicle,' 1853, p. 45.

accidental that the great majority of the culinary and agricultural plants which have yielded numerous races are annuals or biennials, which therefore are capable of rapid propagation and thus of improvement. Sea-kale, asparagus, common and Jerusalem artichokes, potatoes, and onions, alone are perennials. Onions are propagated like annuals, and of the other plants just specified, none, with the exception of the potato, have yielded more than one or two varieties. No doubt fruit-trees, which cannot be propagated quickly by seed, have yielded a host of varieties, though not permanent races; but these, judging from pre-historic remains, were produced at a later and more civilised epoch than the races of culinary and agricultural plants.

A species may be highly variable, but distinct races will not be formed, if from any cause selection be not applied. The carp is highly variable, but it would be extremely difficult to select slight variations in fishes whilst living in their natural state, and distinct races have not been formed;[46] on the other hand, a closely allied species, the gold-fish, from being reared in glass or open vessels, and from having been carefully attended to by the Chinese, has yielded many races. Neither the bee, which has been semi-domesticated from an extremely remote period, nor the cochineal insect, which was cultivated by the aboriginal Mexicans, has yielded races; and it would be impossible to match the queen-bee with any particular drone, and most difficult to match cochineal insects. Silk-moths, on the other hand, have been subjected to rigorous selection, and have produced a host of races. Cats, which from their nocturnal habits cannot be selected for breeding, do not, as formerly remarked, yield distinct races in the same country. The ass in England varies much in colour and size; but it is an animal of little value, bred by poor people; consequently there has been no selection, and distinct races have not been formed. We must not attribute the inferiority of our asses to climate, for in India they are of even smaller size than in Europe. But when selection is brought to bear on the ass, all is changed. Near Cordova, as I am informed (Feb. 1860) by Mr. W. E. Webb, C.E., they are carefully bred, as much as 200l. having been paid for a stallion ass,

---

[46] Isidore Geoffroy St. Hilaire, 'Hist. Nat. Gén.,' tom. iii. p. 49. On the Cochineal Insect, p. 46.

and they have been immensely improved. In Kentucky, asses
have been imported (for breeding mules) from Spain, Malta,
and France; these "seldom averaged more than fourteen hands
"high; but the Kentuckians, by great care, have raised them
"up to fifteen hands, and sometimes even to sixteen. The prices
"paid for these splendid animals, for such they really are, will
"prove how much they are in request. One male, of great
"celebrity, was sold for upwards of one thousand pounds sterling."
These choice asses are sent to cattle-shows, one day being given
to their exhibition.[47]

Analogous facts have been observed with plants: the nutmeg-
tree in the Malay archipelago is highly variable, but there has
been no selection, and there are no distinct races.[48] The common
mignonette (*Reseda odorata*), from bearing inconspicuous flowers,
valued solely for their fragrance, "remains in the same unim-
"proved condition as when first introduced."[49] Our common
forest-trees are very variable, as may be seen in every extensive
nursery-ground; but as they are not valued like fruit-trees,
and as they seed late in life, no selection has been applied to
them; consequently, as Mr. Patrick Matthews remarks,[50] they
have not yielded distinct races, leafing at different periods,
growing to different sizes, and producing timber fit for different
purposes. We have gained only some fanciful and semi-
monstrous varieties, which no doubt appeared suddenly as we
now see them.

Some botanists have argued that plants cannot have so strong
a tendency to vary as is generally supposed, because many
species long grown in botanic gardens, or unintentionally cultivated year after year mingled with our corn crops, have not produced distinct races; but this is accounted for by slight variations not having been selected and propagated. Let a plant
which is now grown in a botanic garden, or any common weed,
be cultivated on a large scale, and let a sharp-sighted gardener
look out for each slight variety and sow the seed, and then, if
distinct races are not produced, the argument will be valid.

---

[47] Capt. Marryat, quoted by Blyth
in 'Journ. Asiatic Soc. of Bengal,' vol.
xxviii. p. 229.

[48] Mr. Oxley, 'Journal of the Indian
Archipelago,' vol. ii., 1848, p. 645.

[49] Mr. Abbey, in 'Journal of Horticulture,' Dec. 1, 1863, p. 430.

[50] 'On Naval Timber,' 1831, p. 107.

238 SELECTION. Chap. XXI.

The importance of selection is likewise shown by considering special characters. For instance, with most breeds of fowls the form of the comb and the colour of the plumage have been attended to, and are eminently characteristic of each race; but in Dorkings, fashion has never demanded uniformity of comb or colour; and the utmost diversity in these respects prevails. Rose-combs, double-combs, cup-combs, &c., and colours of all kinds, may be seen in purely-bred and closely related Dorking fowls, whilst other points, such as the general form of body, and the presence of an additional toe, have been attended to, and are invariably present. It has also been ascertained that colour can be fixed in this breed, as well as in any other.[51]

During the formation or improvement of a breed, its members will always be found to vary much in those characters to which especial attention is directed, and of which each slight improvement is eagerly sought and selected. Thus with short-faced tumbler-pigeons, the shortness of the beak, shape of head and plumage,—with carriers, the length of the beak and wattle,—with fantails, the tail and carriage,—with Spanish fowls, the white face and comb,—with long-eared rabbits, the length of ear, are all points which are eminently variable. So it is in every case, and the large price paid for first-rate animals proves the difficulty of breeding them up to the highest standard of excellence. This subject has been discussed by fanciers,[52] and the greater prizes given for highly improved breeds, in comparison with those given for old breeds which are not now undergoing rapid improvement, has been fully justified. Nathusius makes[53] a similar remark when discussing the less uniform character of improved Shorthorn cattle and of the English horse, in comparison, for example, with the unennobled cattle of Hungary, or with the horses of the Asiatic steppes. This want of uniformity in the parts which at the time are undergoing selection, chiefly depends on the strength of the principle of reversion: but it likewise depends to a certain extent on the continued

[51] Mr. Baily, in 'The Poultry Chronicle,' vol. ii., 1854, p. 150. Also vol. i. p. 342; vol. iii. p. 245.
[52] 'Cottage Gardener,' 1855, December, p. 171; 1856, January, pp. 248, 323.
[53] 'Ueber Shorthorn Rindvieh,' 1857, s. 51.

variability of the parts which have recently varied. That the same parts do continue varying in the same manner we must admit, for, if it were not so, there could be no improvement beyond an early standard of excellence, and we know that such improvement is not only possible, but is of general occurrence.

As a consequence of continued variability, and more especially of reversion, all highly improved races, if neglected or not subjected to incessant selection, soon degenerate. Youatt gives a curious instance of this in some cattle formerly kept in Glamorganshire; but in this case the cattle were not fed with sufficient care. Mr. Baker, in his memoir on the Horse, sums up: "It must have "been observed in the preceding pages that, whenever there has "been neglect, the breed has proportionally deteriorated." [54] If a considerable number of improved cattle, sheep, or other animals of the same race, were allowed to breed freely together, with no selection, but with no change in their condition of life, there can be no doubt that after a score or hundred generations they would be very far from excellent of their kind; but, from what we see of the many common races of dogs, cattle, fowls, pigeons, &c., which without any particular care have long retained nearly the same character, we have no grounds for believing that they would altogether depart from their type.

It is a general belief amongst breeders that characters of all kinds become fixed by long-continued inheritance. But I have attempted to show in the fourteenth chapter that this belief apparently resolves itself into the following proposition, namely, that all characters whatever, whether recently acquired or ancient, tend to be transmitted, but that those which have already long withstood all counteracting influences, will, as a general rule, continue to withstand them, and consequently be faithfully transmitted.

Tendency in Man to carry the practice of Selection to an extreme point.

It is an important principle that in the process of selection man almost invariably wishes to go to an extreme point. Thus, in useful qualities, there is no limit to his desire to breed certain

[54] 'The Veterinary,' vol. xiii. p. 720. For the Glamorganshire cattle, see Youatt on Cattle, p. 51.

horses and dogs as fleet as possible, and others as strong as possible; certain kinds of sheep for extreme fineness, and others for extreme length of wool; and he wishes to produce fruit, grain, tubers, and other useful parts of plants, as large and excellent as possible. With animals bred for amusement, the same principle is even more powerful; for fashion, as we see even in our dress, always runs to extremes. This view has been expressly admitted by fanciers. Instances were given in the chapters on the pigeon, but here is another: Mr. Eaton, after describing a comparatively new variety, namely, the Archangel, remarks, "What "fanciers intend doing with this bird I am at a loss to know, "whether they intend to breed it down to the tumbler's head "and beak, or carry it out to the carrier's head and beak; leaving "it as they found it, is not progressing." Ferguson, speaking of fowls, says, "their peculiarities, whatever they may be, must "necessarily be fully developed: a little peculiarity forms nought "but ugliness, seeing it violates the existing laws of symmetry." So Mr. Brent, in discussing the merits of the sub-varieties of the Belgian canary-bird, remarks, "Fanciers always go to extremes; "they do not admire indefinite properties."[55]

This principle, which necessarily leads to divergence of character, explains the present state of various domestic races. We can thus see how it is that race-horses and dray-horses, greyhounds and mastiffs, which are opposed to each other in every character,—how varieties so distinct as Cochin-China fowls and bantams, or carrier-pigeons with very long beaks, and tumblers with excessively short beaks, have been derived from the same stock. As each breed is slowly improved, the inferior varieties are first neglected and finally lost. In a few cases, by the aid of old records, or from intermediate varieties still existing in countries where other fashions have prevailed, we are enabled partially to trace the graduated changes through which certain breeds have passed. Selection, whether methodical or unconscious, always tending towards an extreme point, together with the neglect and slow extinction of the intermediate and less-valued forms, is the key which unlocks the mystery how man has produced such wonderful results.

[55] J. M. Eaton, 'A Treatise on Fancy Pigeons,' p. 82; Ferguson, on 'Rare and Prize Poultry,' p. 162; Mr. Brent, in 'Cottage Gardener,' Oct. 1860, p. 13.

In a few instances selection, guided by utility for a single purpose, has led to convergence of character. All the improved and different races of the pig, as Nathusius has well shown,[56] closely approach each other in character, in their shortened legs and muzzles, their almost hairless, large, rounded bodies, and small tusks. We see some degree of convergence in the similar outline of the body in well-bred cattle belonging to distinct races.[57] I know of no other such cases.

Continued divergence of character depends on, and is indeed a clear proof, as previously remarked, of the same parts continuing to vary in the same direction. The tendency to mere general variability or plasticity of organisation can certainly be inherited, even from one parent, as has been shown by Gärtner and Kölreuter, in the production of varying hybrids from two species, of which one alone was variable. It is in itself probable that, when an organ has varied in any manner, it will again vary in the same manner, if the conditions which first caused the being to vary remain, as far as can be judged, the same. This is either tacitly or expressly admitted by all horticulturists: if a gardener observes one or two additional petals in a flower, he feels confident that in a few generations he will be able to raise a double flower, crowded with petals. Some of the seedlings from the weeping Moccas oak were so prostrate that they only crawled along the ground. A seedling from the fastigate or upright Irish yew is described as differing greatly from the parent-form "by the exaggeration of the fastigate habit of its branches."[58] Mr. Sheriff, who has been more successful than any other man in raising new kinds of wheat, remarks, "A good variety may safely be regarded as the forerunner of a better one."[59] A great rose-grower, Mr. Rivers, has made the same remark with respect to roses. Sageret,[60] who had large experience, in speaking of the future progress of fruit-trees, observes that the most important principle is "that the more plants have departed from their original type, the more they tend to depart from it." There is apparently much truth in this

---

[56] 'Die Racen des Schweines,' 1860, s. 48.

[57] *See* some good remarks on this head by M. de Quatrefages, 'Unité de l'Espèce Humaine,' 1861, p. 119.

[58] Verlot, 'Des Variétés,' 1865, p. 94.

[59] Mr. Patrick Sheriff, in 'Gard. Chronicle,' 1858, p. 771.

[60] 'Pomologie Physiolog.,' 1830, p. 106.

remark; for we can in no other way understand the surprising
amount of difference between varieties in the parts or qualities
which are valued, whilst other parts retain nearly their original
character.

The foregoing discussion naturally leads to the question, what
is the limit to the possible amount of variation in any part
or quality, and, consequently, is there any limit to what selection can effect? Will a race-horse ever be reared fleeter than
Eclipse? Can our prize-cattle and sheep be still further improved? Will a gooseberry ever weigh more than that produced by "London" in 1852? Will the beet-root in France
yield a greater percentage of sugar? Will future varieties of
wheat and other grain produce heavier crops than our present
varieties? These questions cannot be positively answered; but
it is certain that we ought to be cautious in answering by a
negative. In some lines of variation the limit has probably been
reached. Youatt believes that the reduction of bone in some
of our sheep has already been carried so far that it entails great
delicacy of constitution.[61] But seeing the great improvement
within recent times in our cattle and sheep, and especially in
our pigs; seeing the wonderful increase in weight in our poultry
of all kinds during the last few years; he would be a bold man
who would assert that perfection has been reached. Eclipse
perhaps may never be beaten until all our race-horses have
been rendered swifter, through the selection of the best horses
during many generations; and then the old Eclipse may possibly
be eclipsed; but, as Mr. Wallace has remarked, there must be an
ultimate limit to the fleetness of every animal, whether under
nature or domestication; and with the horse this limit has perhaps
been reached. Until our fields are better manured, it may be
impossible for a new variety of wheat to yield a heavier crop. But
in many cases those who are best qualified to judge do not believe
that the extreme point has as yet been reached even with respect
to characters which have already been carried to a high standard
of perfection. For instance, the short-faced tumbler-pigeon has
been greatly modified; nevertheless, according to Mr. Eaton,[62]
"the field is still as open for fresh competitors as it was one
hundred years ago." Over and over again it has been said that

---

[61] Youatt on Sheep, p. 521.   [62] 'A Treatise on the Almond Tumbler,' p. i.

perfection had been attained with our flowers, but a higher standard has soon been reached. Hardly any fruit has been more improved than the strawberry, yet a great authority remarks,[63] " it must not be concealed that we are far from the extreme limits at which we may arrive."

Time is an important element in the formation of our domestic races, as it permitts innumerable individuals to be born, and these when exposed to diversified conditions are rendered variable. Methodical selection has been occasionally practised from an ancient period to the present day, even by semi-civilised people, and during former times will have produced some effect. Unconscious selection will have been still more effective ; for during a lengthened period the more valuable individual animals will occasionally have been saved, and the less valuable neglected. In the course, also, of time, different varieties, especially in the less civilised countries, will have been more or less modified through natural selection. It is generally believed, though on this head we have little or no evidence, that new characters in time become fixed ; and after having long remained fixed it seems possible that under new conditions they might again be rendered variable.

How great the lapse of time has been since man first domesticated animals and cultivated plants, we begin dimly to see. When the lake-buildings of Switzerland were inhabited during the Neolithic period, several animals were already domesticated and various plants cultivated. If we may judge from what we now see of the habits of savages, it is probable that the men of the earlier Stone period—when many great quadrupeds were living which are now extinct, and when the face of the country was widely different from what it now is—possessed at least some few domesticated animals, although their remains have not as yet been discovered. If the science of language can be trusted, the art of ploughing and sowing the land was followed, and the chief animals had been already domesticated, at an epoch so immensely remote, that the Sanskrit, Greek, Latin, Gothic, Celtic, and Sclavonic languages had not as yet diverged from their common parent-tongue.[64]

[63] M. J. de Jonghe, in ' Gard. Chron.,' 1858, p. 173.
[64] Max. Müller, ' Science of Language,' 1861, p. 223.

It is scarcely possible to overrate the effects of selection occasionally carried on in various ways and places during thousands of generations. All that we know, and, in a still stronger degree, all that we do not know,[65] of the history of the great majority of our breeds, even of our more modern breeds, agrees with the view that their production, through the action of unconscious and methodical selection, has been almost insensibly slow. When a man attends rather more closely than is usual to the breeding of his animals, he is almost sure to improve them to a slight extent. They are in consequence valued in his immediate neighbourhood, and are bred by others; and their characteristic features, whatever these may be, will then slowly but steadily be increased, sometimes by methodical and almost always by unconscious selection. At last a strain, deserving to be called a sub-variety, becomes a little more widely known, receives a local name, and spreads. The spreading will have been extremely slow during ancient and less civilised times, but now is rapid. By the time that the new breed had assumed a somewhat distinct character, its history, hardly noticed at the time, will have been completely forgotten; for, as Low remarks,[66] "we know how quickly the memory of such events is effaced."

As soon as a new breed is thus formed, it is liable through the same process to break up into new strains and subvarieties. For different varieties are suited for, and are valued under, different circumstances. Fashion changes, but, should a fashion last for even a moderate length of time, so strong is the principle of inheritance, that some effect will probably be impressed on the breed. Thus varieties go on increasing in number, and history shows us how wonderfully they have increased since the earliest records.[67] As each new variety is produced, the earlier, intermediate, and less valuable forms will be neglected, and perish. When a breed, from not being valued, is kept in small numbers, its extinction almost inevitably follows sooner or later, either from accidental causes of destruction or from close interbreeding; and this is an event which, in the case of well-marked breeds, excites attention. The birth or production of a new domestic race is so slow a process that it

---

[65] Youatt on Cattle, pp. 116, 128.   [66] 'Domesticated Animals,' p. 188.
[67] Volz, 'Beiträge zur Kulturgeschichte,' 1852, s. 99 et passim.

escapes notice ; its death or destruction is comparatively sudden, is often recorded, and when too late sometimes regretted.

Several authors have drawn a wide distinction between arti-ficial and natural races. The latter are more uniform in cha-racter, possessing in a high degree the character of natural species, and are of ancient origin. They are generally found in less civilised countries, and have probably been largely modi-fied by natural selection, and only to a small extent by man's unconscious and methodical selection. They have, also, during a long period, been directly acted on by the physical conditions of the countries which they inhabit. The so-called artificial races, on the other hand, are not so uniform in character; some have a semi-monstrous character, such as "the wry-legged terriers so useful in rabbit-shooting,"[68] turnspit dogs, ancon sheep, niata oxen, Polish fowls, fantail-pigeons, &c.; their charac-teristic features have generally been acquired suddenly, though subsequently increased in many cases by careful selection. Other races, which certainly must be called artificial, for they have been largely modified by methodical selection and by crossing, as the English race-horse, terrier-dogs, the English game-cock, Antwerp carrier-pigeons, &c., nevertheless cannot be said to have an unnatural appearance; and no distinct line, as it seems to me, can be drawn between natural and artificial races.

It is not surprising that domestic races should generally present a different aspect from natural species. Man selects and propagates modifications solely for his own use or fancy, and not for the creature's own good. His attention is struck by strongly marked modifications, which have appeared suddenly, due to some great disturbing cause in the organisation. He attends almost exclusively to external characters; and when he succeeds in modifying internal organs,—when for instance he reduces the bones and offal, or loads the viscera with fat, or gives early maturity, &c.,—the chances are strong that he will at the same time weaken the constitution. On the other hand, when an animal has to struggle throughout its life with many com-petitors and enemies, under circumstances inconceivably complex and liable to change, modifications of the most varied nature—in the internal organs as well as in external characters, in the

<hr>

[68] Blaine, 'Encyclop. of Rural Sports,' p. 213.

functions and mutual relations of parts—will be rigorously tested, preserved, or rejected. Natural selection often checks man's comparatively feeble and capricious attempts at improvement; and if this were not so, the result of his work, and of nature's work, would be even still more different. Nevertheless, we must not overrate the amount of difference between natural species and domestic races; the most experienced naturalists have often disputed whether the latter are descended from one or from several aboriginal stocks, and this clearly shows that there is no palpable difference between species and races.

Domestic races propagate their kind far more truly, and endure for much longer periods, than most naturalists are willing to admit. Breeders feel no doubt on this head; ask a man who has long reared Shorthorn or Hereford cattle, Leicester or Southdown sheep, Spanish or Game poultry, tumbler or carrier-pigeons, whether these races may not have been derived from common progenitors, and he will probably laugh you to scorn. The breeder admits that he may hope to produce sheep with finer or longer wool and with better carcases, or handsomer fowls, or carrier-pigeons with beaks just perceptibly longer to the practised eye, and thus be successful at an exhibition. Thus far he will go, but no farther. He does not reflect on what follows from adding up during a long course of time many, slight, successive modifications; nor does he reflect on the former existence of numerous varieties, connecting the links in each divergent line of descent. He concludes, as was shown in the earlier chapters, that all the chief breeds to which he has long attended are aboriginal productions. The systematic naturalist, on the other hand, who generally knows nothing of the art of breeding, who does not pretend to know how and when the several domestic races were formed, who cannot have seen the intermediate gradations, for they do not now exist, nevertheless feels no doubt that these races are sprung from a single source. But ask him whether the closely allied natural species which he has studied may not have descended from a common progenitor, and he in his turn will perhaps reject the notion with scorn. Thus the naturalist and breeder may mutually learn a useful lesson from each other.

Summary on Selection by Man.—There can be no doubt that

methodical selection has effected and will effect wonderful results. It was occasionally practised in ancient times, and is still practised by semi-civilised people. Characters of the highest importance, and others of trifling value, have been attended to, and modified. I need not here repeat what has been so often said on the part which unconscious selection has played : we see its power in the difference between flocks which have been separately bred, and in the slow changes, as circumstances have slowly changed, which many animals have undergone in the same country, or when transported into a foreign land. We see the combined effects of methodical and unconscious selection in the great amount of difference between varieties in those parts or qualities which are valued by man, in comparison with those which are not valued, and consequently have not been attended to. Natural selection often determines man's power of selection. We sometimes err in imagining that characters, which are considered as unimportant by the systematic naturalist, could not be affected by the struggle for existence, and therefore be acted on by natural selection ; but striking cases have been given, showing how great an error this is.

The possibility of selection coming into action rests on variability ; and this is mainly caused, as we shall hereafter see, by changes in the conditions of life. Selection is sometimes rendered difficult, or even impossible, by the conditions being opposed to the desired character or quality. It is sometimes checked by the lessened fertility and weakened constitution which follow from long-continued close interbreeding. That methodical selection may be successful, the closest attention and discernment, combined with unwearied patience, are absolutely necessary ; and these same qualities, though not indispensable, are highly serviceable in the case of unconscious selection. It is almost necessary that a large number of individuals should be reared ; for thus there will be a fair chance of variations of the desired nature arising, and every individual with the slightest blemish or in any degree inferior may be freely rejected. Hence length of time is an important element of success. Thus, also, propagation at an early age and at short intervals favours the work. Facility in pairing animals, or their inhabiting a confined area, is advantageous as a check to free crossing. Whenever and wherever selection is not practised, distinct races are not formed. When any one part of the body or quality is not attended to, it remains either unchanged or varies in a fluctuating manner, whilst at the same time other parts and other qualities may become permanently and greatly modified. But from the tendency to reversion and to continued variability, those parts or organs which are now undergoing rapid improvement through selection, are likewise found to vary much. Consequently highly-bred animals, when neglected, soon degenerate; but we have no reason to believe that the effects of long-continued selection would, if the conditions of life remained the same, be soon and completely lost.

Man always tends to go to an extreme point in the selection, whether methodical or unconscious, of all useful and pleasing qualities. This is an important principle, as it leads to continued divergence, and in some rare cases to convergence of character. The possibility of continued divergence rests on the tendency in each part or organ to go on varying in the same manner in which it has already varied; and that this occurs, is proved by the steady and gradual improvement of many animals and plants during lengthened periods. The principle of divergence of character, combined with the neglect and final extinction of all previous, less-valued, and intermediate varieties, explains the amount of difference and the distinctness of our several races. Although we may have reached the utmost limit to which certain characters can be modified, yet we are far from having reached, as we have good reason to believe, the limit in the majority of cases. Finally, from the difference between selection as carried on by man and by nature, we can understand how it is that domestic races often, though by no means always, differ in general aspect from closely allied natural species.

Throughout this chapter and elsewhere I have spoken of selection as the paramount power, yet its action absolutely depends on what we in our ignorance call spontaneous or accidental variability. Let an architect be compelled to build an edifice with uncut stones, fallen from a precipice. The shape of each fragment may be called accidental; yet the shape of each has been determined by the force of gravity, the nature

of the rock, and the slope of the precipice,—events and circumstances, all of which depend on natural laws; but there is no
relation between these laws and the purpose for which each
fragment is used by the builder. In the same manner the
variations of each creature are determined by fixed and im-
mutable laws; but these bear no relation to the living structure which is slowly built up through the power of selection,
whether this be natural or artificial selection.

If our architect succeeded in rearing a noble edifice, using the
rough wedge-shaped fragments for the arches, the longer stones
for the lintels, and so forth, we should admire his skill even in
a higher degree than if he had used stones shaped for the
purpose. So it is with selection, whether applied by man or by
nature; for though variability is indispensably necessary, yet,
when we look at some highly complex and excellently adapted
organism, variability sinks to a quite subordinate position in
importance in comparison with selection, in the same manner as
the shape of each fragment used by our supposed architect is
unimportant in comparison with his skill.

CHAPTER XXII.

CAUSES OF VARIABILITY.

WE will now consider, as far as we can, the causes of the almost universal variability of our domesticated productions. The subject is an obscure one; but it may be useful to probe our ignorance. Some authors, for instance Dr. Prosper Lucas, look at variability as a necessary contingent on reproduction, and as much an aboriginal law, as growth or inheritance. Others have of late encouraged, perhaps unintentionally, this view by speaking of inheritance and variability as equal and antagonistic principles. Pallas maintained, and he has had some followers, that variability depends exclusively on the crossing of primordially distinct forms. Other authors attribute the tendency to variability to an excess of food, and with animals to an excess relatively to the amount of exercise taken, or again to the effects of a more genial climate. That these causes are all effective is highly probable. But we must, I think, take a broader view, and conclude that organic beings, when subjected during several generations to any change whatever in their conditions, tend to vary; the kind of variation which ensues depending in a far higher degree on the nature or constitution of the being, than on the nature of the changed conditions.

Those authors who believe that it is a law of nature that each individual should differ in some slight degree from every other, may maintain, apparently with truth, that this is the fact, not only with all domesticated animals and cultivated plants, but likewise with all organic beings in a state of nature. The Laplander by long practice knows and gives a name to each reindeer, though, as Linnæus remarks, " to distinguish one from another among such multitudes was beyond my comprehension, for they were like ants on an ant-hill." In Germany shepherds have won wagers by recognising each sheep in a flock of a hundred, which they had never seen until the previous fortnight. This power of discrimination, however, is as nothing compared to that which some florists have acquired. Verlot mentions a gardener who could distinguish 150 kinds of camellia, when not in flower; and it has been positively asserted that the famous old Dutch florist Voorhelm, who kept above 1200 varieties of the hyacinth, was hardly ever deceived in knowing each variety by the bulb alone. Hence we must conclude that the bulbs of the hyacinth and the branches and leaves of the camellia, though appearing to an unpractised eye absolutely undistinguishable, yet really differ.[1]

As Linnæus has compared the reindeer in number to ants, I may add that each ant knows its fellow of the same community. Several times I carried ants of the same species (*Formica rufa*) from one ant-hill to another, inhabited apparently by tens of thousands of ants; but the strangers were instantly detected and killed. I then put some ants taken from a very large nest into a bottle strongly perfumed with assafœtida, and after an interval of twenty-four hours returned them to their home; they were at first threatened by their fellows, but were soon recognised and allowed to pass. Hence each ant certainly recognises, independently of odour, its fellow; and if all the ants of the same community have not some countersign or watchword, they must present to each other's senses some distinguishable character.

[1] 'Des Jacinthes,' &c., Amsterdam, 1768, p. 43; Verlot, 'Des Variétés,' &c., p. 86. On the reindeer, *see* Linnæus, 'Tour in Lapland,' translated by Sir J. E. Smith, vol. i. p. 314. The statement in regard to German shepherds is given on the authority of Dr. Weinland.

The dissimilarity of brothers or sisters of the same family, and of seedlings from the same capsule, may be in part accounted for by the unequal blending of the characters of the two parents, and by the more or less complete recovery through reversion of ancestral characters on either side; but we thus only push the difficulty further back in time, for what made the parents or their progenitors different? Hence the belief[2] that an innate tendency to vary exists, independently of external conditions, seems at first sight probable But even the seeds nurtured in the same capsule are not subjected to absolutely uniform conditions, as they draw their nourishment from different points; and we shall see in a future chapter that this difference sometimes suffices greatly to affect the character of the future plant. The less close similarity of the successive children of the same family in comparison with human twins, which often resemble each other in external appearance, mental disposition, and constitution, in so extraordinary a manner, apparently proves that the state of the parents at the exact period of conception, or the nature of the subsequent embryonic development, has a direct and powerful influence on the character of the offspring. Nevertheless, when we reflect on the

[2] Müller's ' Physiology,' Eng. translation, vol. ii. p. 1662. With respect to the similarity of twins in constitution, Dr. William Ogle has given me the following extract from Professor Trousseau's Lectures (' Clinique Médicale,' tom. i. p. 523), in which a curious case is recorded:—" J'ai donné mes soins à deux frères jumeaux, tous deux si extraordinairement ressemblants qu'il m'était impossible de les reconnaître, à moins de les voir l'un à côté de l'autre. Cette ressemblance physique s'étendait plus loin: ils avaient, permettez-moi l'expression, une similitude pathologique plus remarquable encore. Ainsi l'un d'eux que je voyais aux néothermes à Paris malade d'une ophthalmie rhumatismale me disait, ' En ce moment mon frère doit avoir une ophthalmie comme la mienne;' et comme je m'étais récrié, il me montrait quelques jours après une lettre qu'il venait de recevoir de ce frère alors à Vienne, et qui lui écrivait en effet— ' J'ai mon ophthalmie, tu dois avoir la tienne.' Quelque singulier que ceci puisse paraître, le fait non est pas moins exact: on ne me l'a pas raconté, je l'ai vu, et j'en ai vu d'autres analogues dans ma pratique. Ces deux jumeaux étaient aussi tous deux asthmatiques, et asthmatiques à un effroyable degré. Originaires de Marseille, ils n'ont jamais pu demeurer dans cette ville, où leurs intérêts les appelaient souvent, sans être pris de leurs accès ; jamais ils n'en éprouvaient à Paris. Bien mieux, il leur suffisait de gagner Toulon pour être guéris de leurs attaques de Marseille. Voyageant sans cesse et dans tous pays pour leurs affaires, ils avaient remarqué que certaines localités leur étaient funestes, que dans d'autres ils étaient exempts de tout phénomène d'oppression."

individual differences between organic beings in a state of nature, as shown by every wild animal knowing its mate; and when we reflect on the infinite diversity of the many varieties of our domesticated productions, we may well be inclined to exclaim, though falsely as I believe, that Variability must be looked at as an ultimate fact, necessarily contingent on reproduction.

Those authors who adopt this latter view would probably deny that each separate variation has its own proper exciting cause. Although we can seldom trace the precise relation between cause and effect, yet the considerations presently to be given lead to the conclusion that each modification must have its own distinct cause. When we hear of an infant born, for instance, with a crooked finger, a misplaced tooth, or other slight deviation of structure, it is difficult to bring the conviction home to the mind that such abnormal cases are the result of fixed laws, and not of what we blindly call accident. Under this point of view the following case, which has been carefully examined and communicated to me by Dr. William Ogle, is highly instructive. Two girls, born as twins, and in all respects extremely alike, had their little fingers on both hands crooked; and in both children the second bicuspid tooth in the upper jaw, of the second dentition, was misplaced; for these teeth, instead of standing in a line with the others, grew from the roof of the mouth behind the first bicuspids. Neither the parents nor any other member of the family had exhibited any similar peculiarity. Now, as both these children were affected in exactly the same manner by both deviations of structure, the idea of accident is at once excluded; and we are compelled to admit that there must have existed some precise and sufficient cause which, if it had occurred a hundred times, would have affected a hundred children.

We will now consider the general arguments, which appear to me to have great weight, in favour of the view that variations of all kinds and degrees are directly or indirectly caused by the conditions of life to which each being, and more especially its ancestors, have been exposed.

No one doubts that domesticated productions are more variable than organic beings which have never been removed from their natural conditions. Monstrosities graduate so insensibly into mere variations that it is impossible to separate them; and all those who have studied monstrosities believe that they are far commoner with domesticated than with wild animals and plants;[3] and in the case of plants, monstrosities would be equally noticeable in the natural as in the cultivated state. Under nature, the individuals of the same species are exposed to nearly uniform conditions, for they are rigorously kept to their proper places by a host of competing animals and plants; they have, also, long been habituated to their conditions of life; but it cannot be said that they are subject to quite uniform conditions, and they are liable to a certain amount of variation. The circumstances under which our domestic productions are reared are widely different: they are protected from competition; they have not only been removed from their natural conditions and often from their native land, but they are frequently carried from district to district, where they are treated differently, so that they never remain during a considerable length of time exposed to closely similar conditions. In conformity with this, all our domesticated productions, with the rarest exceptions, vary far more than natural species. The hive-bee, which feeds itself and follows in most respects its natural habits of life, is the least variable of all domesticated animals, and probably the goose is the next least variable; but even the goose varies more than almost any wild bird, so that it cannot be affiliated with perfect certainty to any natural species. Hardly a single plant can be named, which has long been cultivated and propagated by seed, that is not highly variable; common rye (*Secale cereale*) has afforded fewer and less marked varieties than almost any other cultivated plant;[4] but it may be doubted whether the variations of this, the least valuable of all our cereals, have been closely observed.

Bud-variation, which was fully discussed in a former chapter, shows us that variability may be quite independent of seminal reproduction, and likewise of reversion to long-lost ancestral characters. No one will maintain that the sudden appearance

---

[3] Isid. Geoffroy St. Hilaire, 'Hist. des Anomalies,' tom. iii. p. 352; Moquin-Tandon, 'Tératologie Végétale,' 1841, p. 115.

[4] Metzger, 'Die Getreidearten,' 1841, s. 39.

of a moss-rose on a Provence-rose is a return to a former state, for mossiness of the calyx has been observed in no natural species; the same argument is applicable to variegated and laciniated leaves; nor can the appearance of nectarines on peach-trees be accounted for with any probability on the principle of reversion. But bud-variations more immediately concern us, as they occur far more frequently on plants which have been highly cultivated during a length of time, than on other and less highly cultivated plants; and very few well-marked instances have been observed with plants growing under strictly natural conditions. I have given one instance of an ash-tree growing in a gentleman's pleasure-grounds; and occasionally there may be seen, on beech and other trees, twigs leafing at a different period from the other branches. But our forest trees in England can hardly be considered as living under strictly natural conditions; the seedlings are raised and protected in nursery-grounds, and must often be transplanted into places where wild trees of the kind would not naturally grow. It would be esteemed a prodigy if a dog-rose growing in a hedge produced by bud-variation a moss-rose, or a wild bullace or wild cherry-tree yielded a branch bearing fruit of a different shape and colour from the ordinary fruit. The prodigy would be enhanced if these varying branches were found capable of propagation, not only by grafts, but sometimes by seed; yet analogous cases have occurred with many of our highly cultivated trees and herbs.

These several considerations alone render it probable that variability of every kind is directly or indirectly caused by changed conditions of life. Or, to put the case under another point of view, if it were possible to expose all the individuals of a species during many generations to absolutely uniform conditions of life, there would be no variability.

On the Nature of the Changes in the Conditions of Life which induce Variability.

From a remote period to the present day, under climates and circumstances as different as it is possible to conceive, organic beings of all kinds, when domesticated or cultivated, have varied. We see this with the many domestic races of quadrupeds and birds belonging to different orders, with gold-fish and silk-worms, with plants of many kinds, raised in various quarters of the world. In the deserts of northern Africa the date-palm has yielded thirty-eight varieties; in the fertile plains of India it is notorious how many varieties of rice and of a host of other plants exist; in a single Polynesian island, twenty-four varieties of the bread-fruit, the same number of the banana, and twenty-two varieties of the arum, are cultivated by the natives; the mulberry-tree in India and Europe has yielded many varieties serving as food for the silkworm; and in China sixty-three varieties of the bamboo are used for various domestic purposes.[5] These facts alone, and innumerable others could be added, indicate that a change of almost any kind in the conditions of life suffices to cause variability—different changes acting on different organisms.

Andrew Knight[6] attributed the variation of both animals and plants to a more abundant supply of nourishment, or to a more favourable climate, than that natural to the species. A more genial climate, however, is far from necessary; the kidney-bean, which is often injured by our spring frosts, and peaches, which require the protection of a wall, have varied much in England, as has the orange-tree in northern Italy, where it is barely able to exist.[7] Nor can we overlook the fact, though not immediately connected with our present subject, that the plants and shells of the arctic regions are eminently variable.[8] More-over, it does not appear that a change of climate, whether more or less genial, is one of the most potent causes of variability; for in regard to plants Alph. De Candolle, in his ' Géographie

---

[5] On the date-palm, *see* Vogel, 'Annals and Mag. of Nat. Hist.,' 1854, p. 460. On Indian varieties, Dr. F. Hamilton, ' Transact. Linn. Soc.,' vol. xiv. p. 296. On the varieties cultivated in Tahiti, *see* Dr. Bennett, in London's ' Mag. of N. Hist.,' vol. v., 1832, p. 484. Also Ellis, ' Polynesian Researches,' vol. i. pp. 375, 370. On twenty varieties of the Pandanus and other trees in the Marianne Island, *see* 'Hooker's Miscellany,' vol. i. p. 308. On the bamboo in China, *see* Huc's

' Chinese Empire,' vol. ii. p. 307.

[6] ' Treatise on the Culture of the Apple,' &c., p. 3.

[7] Gallesio, ' Teoria della Riproduzione Veg.,' p. 125.

[8] *See* Dr. Hooker's Memoir on Arctic Plants in ' Linn. Transact.,' vol. xxiii. part ii. Mr. Woodward, and a higher authority cannot be quoted, speaks of the Arctic mollusca (in his ' Rudimentary Treatise,' 1856, p. 355) as remarkably subject to variation.

Botanique,' repeatedly shows that the native country of a plant, where in most cases it has been longest cultivated, is that where it has yielded the greatest number of varieties.

It is doubtful whether a change in the nature of the food is a potent cause of variability. Scarcely any domesticated animal has varied more than the pigeon or the fowl, but their food, especially that of highly-bred pigeons, is generally the same. Nor can our cattle and sheep have been subjected to any great change in this respect. But in all these cases the food probably is much less varied in kind than that which was consumed by the species in its natural state.[9]

Of all the causes which induce variability, excess of food, whether or not changed in nature, is probably the most powerful. This view was held with regard to plants by Andrew Knight, and is now held by Schleiden, more especially in reference to the inorganic elements of the food.[10] In order to give a plant more food it suffices in most cases to grow it separately, and thus prevent other plants robbing its roots. It is surprising, as I have often seen, how vigorously our common wild plants flourish when planted by themselves, though not in highly manured land. Growing plants separately is, in fact, the first step in cultivation. We see the converse of the belief that excess of food induces variability in the following statement by a great raiser of seeds of all kinds.[11] "It is a rule invariably with us, when " we desire to keep a true stock of any one kind of seed, to " grow it on poor land without dung; but when we grow for " quantity, we act contrary, and sometimes have dearly to " repent of it."

In the case of animals the want of a proper amount of exercise, as Bechstein has remarked, has perhaps played, independently of the direct effects of the disuse of any particular organ, an important part in causing variability. We can see in a vague manner that, when the organised and nutrient fluids of the body are not used during growth, or by the wear and tear of the tissues,

---

[9] Bechstein, in his ' Naturgeschichte der Stubenvögel,' 1840, s. 238, has some good remarks on this subject. He states that his canary-birds varied in colour, though kept on uniform food.

[10] 'The Plant,' by Schleiden, trans-

lated by Henfrey, 1848, p. 169. *See* also Alex. Braun, in ' Bot. Memoirs,' Ray. Soc., 1853, p. 313.

[11] Messrs. Hardy and Son, of Maldon, in ' Gard. Chronicle,' 1856, p. 458.

they will be in excess; and as growth, nutrition, and reproduction are intimately allied processes, this superfluity might disturb the due and proper action of the reproductive organs, and consequently affect the character of the future offspring. But it may be argued that neither an excess of food nor a superfluity in the organised fluids of the body necessarily induces variability. The goose and the turkey have been well fed for many generations, yet have varied very little. Our fruit-trees and culinary plants, which are so variable, have been cultivated from an ancient period, and, though they probably still receive more nutriment than in their natural state, yet they must have received during many generations nearly the same amount; and it might be thought that they would have become habituated to the excess. Nevertheless, on the whole, Knight's view, that excess of food is one of the most potent causes of variability, appears, as far as I can judge, probable.

Whether or not our various cultivated plants have received nutriment in excess, all have been exposed to changes of various kinds. Fruit-trees are grafted on different stocks, and grown in various soils. The seeds of culinary and agricultural plants are carried from place to place; and during the last century the rotation of our crops and the manures used have been greatly changed.

Slight changes of treatment often suffice to induce variability. The simple fact of almost all our cultivated plants and domesticated animals having varied in all places and at all times, leads to this conclusion. Seeds taken from common English forest-trees, grown under their native climate, not highly manured or otherwise artificially treated, yield seedlings which vary much, as may be seen in every extensive seed-bed. I have shown in a former chapter what a number of well marked and singular varieties the thorn (*Cratægus oxycantha*) has produced; yet this tree has been subjected to hardly any cultivation. In Staffordshire I carefully examined a large number of two British plants, namely, *Geranium phœum* and *Pyrenaicum*, which have never been highly cultivated. These plants had spread spontaneously by seed from a common garden into an open plantation; and the seedlings varied in almost every single character, both in their flowers and foliage, to a degree which

I have never seen exceeded; yet they could not have been
exposed to any great change in their conditions.

With respect to animals, Azara has remarked with much surprise,[12] that, whilst the feral horses on the Pampas are always of
one of three colours, and the cattle always of a uniform colour,
yet these animals, when bred on the unenclosed estancias, though
kept in a state which can hardly be called domesticated, and
apparently exposed to almost identically the same conditions as
when they are feral, nevertheless display a great diversity of colour.
So again in India several species of fresh-water fish are only so
far treated artificially, that they are reared in great tanks; but
this small change is sufficient to induce much variability.[13]

Some facts on the effects of grafting, in regard to the variability
of trees, deserve attention. Cabanis asserts that when certain
pears are grafted on the quince, their seeds yield more varieties
than do the seeds of the same variety of pear when grafted on
the wild pear.[14] But as the pear and quince are distinct species,
though so closely related that the one can be readily grafted
and succeeds admirably on the other, the fact of variability being
thus caused is not surprising; we are, however, here enabled
to see the cause, namely, the different nature of the stock with
its roots and the rest of the tree. Several North American
varieties of the plum and peach are well known to reproduce
themselves truly by seed; but Downing asserts,[15] " that when a
" graft is taken from one of these trees and placed upon another
" stock, this grafted tree is found to lose its singular property of
" producing the same variety by seed, and becomes like all other
" worked trees ;"—that is, its seedlings become highly variable.
Another case is worth giving: the Lalande variety of the
walnut-tree leafs between April 20th and May 15th, and its
seedlings invariably inherit the same habit; whilst several
other varieties of the walnut leaf in June. Now, if seedlings are
raised from the May-leafing Lalande variety, grafted on another
May-leafing variety, though both stock and graft have the same
early habit of leafing, yet the seedlings leaf at various times,

[12] 'Quadrupèdes du Paraguay,' 1801,
tom. ii. p. 319.

[13] M'Clelland on Indian Cyprinidæ,
'Asiatic Researches,' vol. xix. part ii.,
1839, pp. 266, 268, 313.

[14] Quoted by Sageret, 'Pom. Phys.,'
1830, p. 43.

[15] 'The Fruits of America,' 1845,
p. 5.

even as late as the 5th of June.[16] Such facts as these are well
fitted to show, on what obscure and slight causes variability
rests.

I may here just allude to the appearance of new and valuable varieties
of fruit-trees and of wheat in woods and waste places, which at first sight
seems a most anomalous circumstance. In France a considerable number
of the best pears have been discovered in woods; and this has occurred
so frequently, that Poiteau asserts that "improved varieties of our cultivated fruits rarely originate with nurserymen.[17] In England, on the
other hand, no instance of a good pear having been found wild has been
recorded; and Mr. Rivers informs me that he knows of only one instance
with apples, namely, the Bess Poole, which was discovered in a wood in
Nottinghamshire. This difference between the two countries may be
in part accounted for by the more favourable climate of France, but
chiefly from the great number of seedlings which spring up there in the
woods. I infer that this is the case from a remark made by a French
gardener,[18] who regards it as a national calamity that such a number of
pear-trees are periodically cut down for firewood, before they have borne
fruit. The new varieties which thus spring up in the woods, though they
cannot have received any excess of nutriment, will have been exposed to
abruptly changed conditions, but whether this is the cause of their production is very doubtful. These varieties, however, are probably all
descended [19] from old cultivated kinds growing in adjoining orchards,—a
circumstance which will account for their variability; and out of a vast
number of varying trees there will always be a good chance of the appearance of a valuable kind. In North America, where fruit-trees frequently
spring up in waste places, the Washington pear was found in a hedge,
and the Emperor peach in a wood.[20]

With respect to wheat, some writers have spoken [21] as if it were an ordinary event for new varieties to be found in waste places; the Fenton wheat
was certainly discovered growing on a pile of basaltic detritus in a quarry,
but in such a situation the plant would probably receive a sufficient amount

---

[16] M. Cardan, in 'Comptes Rendus,'
Dec. 1848, quoted in 'Gard. Chronicle,'
1849, p. 101.

[17] M. Alexis Jordan mentions four
excellent pears found in woods in
France, and alludes to others ('Mém.
Acad. de Lyon,' tom. ii. 1852, p. 159).
Poiteau's remark is quoted in 'Gardener's Mag.,' vol. iv., 1828, p. 385.
See 'Gard. Chronicle,' 1862, p. 335,
for another case of a new variety of the
pear found in a hedge in France. Also
for another case, see Loudon's 'Encyclop. of Gardening,' p. 901. Mr. Rivers
has given me similar information.

[18] Duval, 'Hist. du Poirier,' 1849,
p. 2.

[19] I infer that this is the fact from
Van Mons' statement ('Arbres Fruitiers,' 1835, tom. i. p. 446) that he finds
in the woods seedlings resembling all
the chief cultivated races of both the
pear and apple. Van Mons, however,
looked at these wild varieties as aboriginal species.

[20] Downing, 'Fruit-trees of North
America,' p. 422; Foley, in 'Transact.
Hort. Soc.,' vol. vi. p. 412.

[21] 'Gard. Chronicle,' 1847, p. 244.

of nutriment. The Chidham wheat was raised from an ear found *on* a hedge; and Hunter's wheat was discovered *by* the roadside in Scotland, but it is not said that this latter variety grew where it was found.[22]

Whether our domestic productions would ever become so completely habituated to the conditions under which they now live, as to cease varying, we have no sufficient means for judging. But, in fact, our domestic productions are never exposed for a great length of time to uniform conditions, and it is certain that our most anciently cultivated plants, as well as animals, still go on varying, for all have recently undergone marked improvement. In some few cases, however, plants have become habituated to new conditions. Thus Metzger, who cultivated in Germany during many years numerous varieties of wheat, brought from different countries,[23] states that some kinds were at first extremely variable, but gradually, in one instance after an interval of twenty-five years, became constant; and it does not appear that this resulted from the selection of the more constant forms.

On the Accumulative Action of changed Conditions of Life.— We have good grounds for believing that the influence of changed conditions accumulates, so that no effect is produced on a species until it has been exposed during several generations to continued cultivation or domestication. Universal experience shows us that when new flowers are first introduced into our gardens they do not vary; but ultimately all, with the rarest exceptions, vary to a greater or less extent. In a few cases the requisite number of generations, as well as the successive steps in the progress of variation, have been recorded, as in the often-quoted instance of the Dahlia.[24] After several years' culture the Zinnia has only lately (1860) begun to vary in any great degree. " In the first seven or "eight years of high cultivation the Swan River daisy (*Brachycome* "*iberidifolia*) kept to its original colour; it then varied into lilac "and purple and other minor shades."[25] Analogous facts have been recorded with the Scotch rose. In discussing the variability of plants several experienced horticulturists have spoken to the

[22] Gardener's Chronicle,' 1841, p. 383; 1850, p. 700; 1854, p. 650.

[23] 'Die Getreidearten,' 1843, s. 66, 116, 117.

[24] Sabine, in 'Hort. Transact.,' vol.

iii. p. 225; Bronn, 'Geschichte der Natur,' b. ii. s. 119.

[25] 'Journal of Horticulture,' 1861, p. 112; on Zinnia, 'Gardener's Chronicle,' 1860, p. 852.

same general effect. Mr. Salter[26] remarks, " Every one knows
" that the chief difficulty is in breaking through the original form
" and colour of the species, and every one will be on the look-out
" for any natural sport, either from seed or branch ; that being
" once obtained, however trifling the change may be, the result
" depends upon himself." M. de Jonghe, who has had so much
success in raising new varieties of pears and strawberries,[27] remarks with respect to the former, "There is another principle,
" namely, that the more a type has entered into a state of variation,
" the greater is its tendency to continue doing so ; and the more
" it has varied from the original type, the more it is disposed to
" vary still farther." We have, indeed, already discussed this
latter point when treating of the power which man possesses,
through selection, of continually augmenting in the same direction each modification ; for this power depends on continued
variability of the same general kind. The most celebrated horticulturist in France, namely, Vilmorin,[28] even maintains that,
when any particular variation is desired, the first step is to
get the plant to vary in any manner whatever, and to go on
selecting the most variable individuals, even though they vary
in the wrong direction ; for the fixed character of the species
being once broken, the desired variation will sooner or later
appear.

As nearly all our animals were domesticated at an extremely
remote epoch, we cannot, of course, say whether they varied
quickly or slowly when first subjected to new conditions. But
Dr. Bachman[29] states that he has seen turkeys raised from
the eggs of the wild species lose their metallic tints and become
spotted with white in the third generation. Mr. Yarrell many
years ago informed me that the wild ducks bred on the ponds
in St. James's Park, which had never been crossed, as it is believed, with domestic ducks, lost their true plumage after a
few generations. An excellent observer,[30] who has often reared
birds from the eggs of the wild duck, and who took precautions

[26] 'The Chrysanthemum, its History,
&c.,' 1865, p. 3.
    [27] 'Gardener's Chron.,' 1855, p. 54;
'Journal of Horticulture,' May 9, 1865,
p. 363.
    [28] Quoted by Verlot, 'Des Variétés,'

&c., 1865, p. 28.
    [29] 'Examination of the Characteris-
tics of Genera and Species :' Charleston,
1855, p. 14.
    [30] Mr. Hewitt, 'Journal of Hort.,'
1863, p. 39.

that there should be no crossing with domestic breeds, has given, as previously stated, full details on the changes which they gradually undergo. He found that he could not breed these wild ducks true for more than five or six generations, " as they then " proved so much less beautiful. The white collar round the neck " of the mallard became much broader and more irregular, and " white feathers appeared in the ducklings' wings." They increased also in size of body; their legs became less fine, and they lost their elegant carriage. Fresh eggs were then procured from wild birds; but again the same result followed. In these cases of the duck and turkey we see that animals, like plants, do not depart from their primitive type until they have been subjected during several generations to domestication. On the other hand, Mr. Yarrell informed me that the Australian dingos, bred in the Zoological Gardens, almost invariably produced in the first generation puppies marked with white and other colours; but these introduced dingos had probably been procured from the natives, who keep them in a semi-domesticated state. It is certainly a remarkable fact that changed conditions should at first produce, as far as we can see, absolutely no effect; but that they should subsequently cause the character of the species to change. In the chapter on pangenesis I shall attempt to throw a little light on this fact.

Returning now to the causes which are supposed to induce variability. Some authors[31] believe that close interbreeding gives this tendency, and leads to the production of monstrosities. In the seventeenth chapter some few facts were advanced, showing that monstrosities are, as it appears, occasionally thus caused; and there can be no doubt that close interbreeding induces lessened fertility and a weakened constitution; hence it may lead to variability: but I have not sufficient evidence on this head. On the other hand, close interbreeding, if not carried to an injurious extreme, far from causing variability, tends to fix the character of each breed.

It was formerly a common belief, still held by some persons, that the imagination of the mother affects the child in

[31] Devay, 'Mariages Consanguins,' pp. 97, 125. In conversation I have found two or three naturalists of the same opinion.

the womb.[32] This view is evidently not applicable to the lower
animals, which lay unimpregnated eggs, or to plants. Dr. William
Hunter, in the last century, told my father that during many
years every woman in a large London Lying-in Hospital was
asked before her confinement whether anything had specially
affected her mind, and the answer was written down; and it so
happened that in no one instance could a coincidence be detected between the woman's answer and any abnormal structure;
but when she knew the nature of the structure, she frequently
suggested some fresh cause. The belief in the power of the
mother's imagination may perhaps have arisen from the children
of a second marriage resembling the previous father, as certainly
sometimes occurs, in accordance with the facts given in the
eleventh chapter.

Crossing as a Cause of Variability.—In an early part of this
chapter it was stated that Pallas[33] and a few other naturalists
maintain that variability is wholly due to crossing. If this means
that new characters never spontaneously appear in our domestic
races, but that they are all directly derived from certain aboriginal
species, the doctrine is little less than absurd; for it implies
that animals like Italian greyhounds, pug-dogs, bull-dogs, pouter
and fantail pigeons, &c., were able to exist in a state of nature.
But the doctrine may mean something widely different, namely,
that the crossing of distinct species is the sole cause of the first
appearance of new characters, and that without this aid man could
not have formed his various breeds. As, however, new characters
have appeared in certain cases by bud-variation, we may conclude
with certainty that crossing is not necessary for variability. It is,
moreover, almost certain that the breeds of various animals, such as
of the rabbit, pigeon, duck, &c., and the varieties of several plants,
are the modified descendants of a single wild species. Nevertheless, it is probable that the crossing of two forms, when one or
both have long been domesticated or cultivated, adds to the variability of the offspring, independently of the commingling of the
characters derived from the two parent-forms; and this implies

---

[32] Müller has conclusively argued
against this belief, 'Elements of
Phys.,' Eng. translat., vol. ii., 1842,
p. 1405.

[33] 'Act. Acad. St. Petersburg,' 1780,
part ii. p. 84, &c.

that new characters actually arise. But we must not forget the facts advanced in the thirteenth chapter, which clearly prove that the act of crossing often leads to the reappearance or reversion of long-lost characters; and in most cases it would be impossible to distinguish between the reappearance of ancient characters and the first appearance of new characters. Practically, whether new or old, they would be new to the breed in which they reappeared.

Gärtner declares,[34] and his experience is of the highest value on such a point, that, when he crossed native plants which had not been cultivated, he never once saw in the offspring any new character; but that from the odd manner in which the characters derived from the parents were combined, they sometimes appeared as if new. When, on the other hand, he crossed cultivated plants, he admits that new characters occasionally appeared, but he is strongly inclined to attribute their appearance to ordinary variability, not in any way to the cross. An opposite conclusion, however, appears to me the more probable. According to Kölreuter, hybrids in the genus Mirabilis vary almost infinitely, and he describes new and singular characters in the form of the seeds, in the colour of the anthers, in the cotyledons being of immense size, in new and highly peculiar odours, in the flowers expanding early in the season, and in their closing at night. With respect to one lot of these hybrids, he remarks that they presented characters exactly the reverse of what might have been expected from their parentage.[35]

Prof. Lecoq[36] speaks strongly to the same effect in regard to this same genus, and asserts that many of the hybrids from *Mirabilis jalapa* and *multiflora* might easily be mistaken for distinct species, and adds that they differed in a greater degree, than the other species of the genus, from *M. jalapa*. Herbert, also, has described[37] the offspring from a hybrid Rhododendron as being " as *unlike all others* in foliage, as if they had been a " separate species." The common experience of floriculturists proves that the crossing and recrossing of distinct but allied plants, such as the species of Petunia, Calceolaria, Fuchsia, Verbena, &c., induces excessive variability; hence the appearance of quite new characters is probable. M. Carrière[38] has lately discussed this subject: he states that *Erythrina cristagalli* had been multiplied by seed for many years, but had not yielded any varieties: it was then crossed with the allied *E. herbacea*, and " the " resistance was now overcome, and varieties were produced with flowers " of extremely different size, form, and colour."

From the general and apparently well-founded belief that the crossing

[34] 'Bastarderzeugung,' s. 249, 255, 295.

[35] 'Nova Acta, St. Petersburg,' 1794, p. 378; 1795, pp. 307, 313, 316; 1787, p. 407.

[36] 'De la Fécondation,' 1862, p. 311.

[37] 'Amaryllidaceæ,' 1837, p. 362.

[38] Abstracted in ' Gard. Chronicle, 1860, p. 1081.

of distinct species, besides commingling their characters, adds greatly to their variability, it has probably arisen that some botanists have gone so far as to maintain [39] that, when a genus includes only a single species, this when cultivated never varies. The proposition made so broadly cannot be admitted; but it is probably true that the variability of cultivated monotypic genera is much less than that of genera including numerous species, and this quite independently of the effects of crossing. I have stated in my 'Origin of Species,' and in a future work shall more fully show, that the species belonging to small genera generally yield a less number of varieties in a state of nature than those belonging to large genera. Hence the species of small genera would, it is probable, produce fewer varieties under cultivation than the already variable species of larger genera.

Although we have not at present sufficient evidence that the crossing of species, which have never been cultivated, leads to the appearance of new characters, this apparently does occur with species which have been already rendered in some degree variable through cultivation. Hence crossing, like any other change in the conditions of life, seems to be an element, probably a potent one, in causing variability. But we seldom have the means of distinguishing, as previously remarked, between the appearance of really new characters and the reappearance of long-lost characters, evoked through the act of crossing. I will give an instance of the difficulty in distinguishing such cases. The species of Datura may be divided into two sections, those having white flowers with green stems, and those having purple flowers with brown stems: now Naudin [40] crossed *Datura lævis* and *ferox*, both of which belong to the white section, and raised from them 205 hybrids. Of these hybrids, every one had brown stems and bore purple flowers; so that they resembled the species of the other section of the genus, and not their own two parents. Naudin was so much astonished at this fact, that he was led carefully to observe both parent-species, and he discovered that the pure seedlings of *D. ferox*, immediately after germination, had dark purple stems, extending from the young roots up to the cotyledons, and that this tint remained ever afterwards as a ring round the base of the stem of the plant when old. Now I have shown in the thirteenth chapter that the retention or exaggeration of an early character is so intimately related to reversion, that it evidently comes under the same principle. Hence probably we ought to look at the purple flowers and brown stems of these hybrids, not as new characters due to variability, but as a return to the former state of some ancient progenitor.

Independently of the appearance of new characters from crossing, a few words may be added to what has been said in former chapters on the unequal combination and transmission of the characters proper to the two parent-forms. When two species or races are crossed, the offspring of

[39] This was the opinion of the elder De Candolle, as quoted in 'Dic. Class. d'Hist. Nat.,' tom. viii. p. 405. Puvis, in his work, 'De la Dégénération,' 1837, p. 37, has discussed this same point.

[40] 'Comptes Rendus,' Novembre 21, 1864, p. 838.

the first generation are generally uniform, but subsequently they display an almost infinite diversity of character. He who wishes, says Kölreuter,[41] to obtain an endless number of varieties from hybrids should cross and recross them. There is also much variability when hybrids or mongrels are reduced or absorbed by repeated crosses with either pure parent-form; and a still higher degree of variability when three distinct species, and most of all when four species, are blended together by successive crosses. Beyond this point Gärtner,[42] on whose authority the foregoing statements are made, never succeeded in effecting a union; but Max Wichura[43] united six distinct species of willows into a single hybrid. The sex of the parent-species affects in an inexplicable manner the degree of variability of hybrids; for Gärtner[44] repeatedly found that when a hybrid was used as the father, and either one of the pure parent-species, or a third species, was used as the mother, the offspring were more variable than when the same hybrid was used as the mother, and either pure parent or the same third species as the father: thus seedlings from *Dianthus barbatus* crossed by the hybrid *D. chinensi-barbatus* were more variable than those raised from this latter hybrid fertilised by the pure *D. barbatus*. Max Wichura[45] insists strongly on an analogous result with his hybrid willows. Again Gärtner[46] asserts that the degree of variability sometimes differs in hybrids raised from reciprocal crosses between the same two species; and here the sole difference is, that the one species is first used as the father and then as the mother. On the whole we see that, independently of the appearance of new characters, the variability of successive crossed generations is extremely complex, partly from the offspring partaking unequally of the characters of the two parent-forms, and more especially from their unequal tendency to revert to these same characters or to those of more ancient progenitors.

On the Manner and on the Period of Action of the Causes which induce Variability.—This is an extremely obscure subject, and we need here only briefly consider, firstly, whether inherited variations are caused by the organisation being directly acted on, or indirectly through the reproductive system ; and secondly, at what period of life or growth they are primarily caused. We shall see in the two following chapters that various agencies, such as an abundant supply of food, exposure to a different climate, increased use or disuse of parts, &c., prolonged during several generations, certainly modify either the whole organisation or certain organs. This direct action of changed conditions perhaps comes into play much more frequently than can be proved, and it is at least clear that in all cases of bud-

---

[41] 'Nova Acta, St. Petersburg,' 1794, p. 391.

[42] 'Bastarderzeugung,' s. 507, 516, 572.

[43] 'Die Bastardbefruchtung.' &c.,

1865, s. 24.

[44] 'Bastarderzeugung,' s. 452, 507.

[45] 'Die Bastardbefruchtung,' s. 56.

[46] 'Bastarderzeugung,' s. 423.

variation the action cannot have been through the reproductive system.

With respect to the part which the reproductive system takes in causing variability, we have seen in the eighteenth chapter that even slight changes in the conditions of life have a remarkable power in causing a greater or less degree of sterility. Hence it seems not improbable that beings generated through a system so easily affected should themselves be affected, or should fail to inherit, or inherit in excess, characters proper to their parents. We know that certain groups of organic beings, but with exceptions in each group, have their reproductive systems much more easily affected by changed conditions than other groups; for instance, carnivorous birds more readily than carnivorous mammals, and parrots more readily than pigeons; and this fact harmonizes with the apparently capricious manner and degree in which various groups of animals and plants vary under domestication.

Kölreuter [47] was struck with the parallelism between the excessive variability of hybrids when crossed and recrossed in various ways,— these hybrids having their reproductive powers more or less affected,— and the variability of anciently cultivated plants. Max Wichura [48] has gone one step farther, and shows that with many of our highly cultivated plants, such as the hyacinth, tulip, auricula, snapdragon, potato, cabbage, &c., which there is no reason to believe have been hybridized, the anthers contain many irregular pollen-grains, in the same state as in hybrids. He finds also in certain wild forms, the same coincidence between the state of the pollen and a high degree of variability, as in many species of Rubus; but in *R. cæsius* and *idæus*, which are not highly variable species, the pollen is sound. It is also notorious that many cultivated plants, such as the banana, pine-apple, breadfruit, and others previously mentioned, have their reproductive organs so seriously affected as to be generally quite sterile; and when they do yield seed, the seedlings, judging from the large number of cultivated races which exist, must be variable in an extreme degree. These facts indicate that there is some relation between the state of the reproductive organs and a tendency to variability; but we must not conclude that the relation is strict. Although many of our highly cultivated plants may have their pollen in a deteriorated condition, yet, as we have previously seen, they yield more seed, and our anciently domesticated animals are more prolific, than the corresponding species in a state of nature. The peacock is almost the only bird which is believed to be less fertile under domestication than in its native state, and it has varied in a remarkably small degree. From these considerations it would seem that changes in the conditions of life lead either to sterility or to variability, or to both; and not that sterility induces variability. On the whole it is probable that any cause affecting the organs of reproduction would likewise affect their product,—that is, the offspring thus generated.

[47] 'Dritte Fortsetzung,' &c., 1766, s. 85.   Berkeley on the same subject, in
[48] 'Die Bastardbefruchtung,' &c.,   'Journal of Royal Hort. Soc.,' 1866,
1865, s. 92 ; see also the Rev. M. J.   p. 80.

The period of life at which the causes that induce variability act, is
another obscure subject, which has been discussed by various authors.[49]   In
some of the cases, to be given in the following chapter, of modifications from
the direct action of changed conditions, which are inherited, there can be no
doubt that the causes have acted on the mature or nearly mature animal.
On the other hand, monstrosities, which cannot be distinctly separated
from lesser variations, are often caused by the embryo being injured whilst
in the mother's womb or in the egg. Thus I. Geoffroy St. Hilaire[50] asserts
that poor women who work hard during their pregnancy, and the mothers
of illegitimate children troubled in their minds and forced to conceal their
state, are far more liable to give birth to monsters than women in easy
circumstances.   The eggs of the fowl when placed upright or otherwise
treated unnaturally frequently produce monstrous chickens.   It would,
however, appear that complex monstrosities are induced more frequently
during a rather late than during a very early period of embryonic life;
but this may partly result from some one part, which has been injured
during an early period, affecting by its abnormal growth other parts subsequently developed; and this would be less likely to occur with parts
injured at a later period.[51]   When any part or organ becomes monstrous
through abortion, a rudiment is generally left, and this likewise indicates
that its development had already commenced.

Insects sometimes have their antennæ or legs in a monstrous condition,
and yet the larvæ from which they are metamorphosed do not possess
either antennæ or legs; and in these cases, as Quatrefages[52] believes, we
are enabled to see the precise period at which the normal progress of development has been troubled. But the nature of the food given to a caterpillar
sometimes affects the colours of the moth, without the caterpillar itself
being affected; therefore it seems possible that other characters in the
mature insect might be indirectly modified through the larvæ.   There is
no reason to suppose that organs which have been rendered monstrous
have always been acted on during their development; the cause may have
acted on the organisation at a much earlier stage. It is even probable that
either the male or female sexual elements, or both, before their union, may
be affected in such a manner as to lead to modifications in organs developed at a late period of life; in nearly the same manner as a child may
inherit from his father a disease which does not appear until old age.

In accordance with the facts above given, which prove that in many
cases a close relation exists between variability and the sterility following
from changed conditions, we may conclude that the exciting cause often
acts at the earliest possible period, namely, on the sexual elements, before
impregnation has taken place.   That an affection of the female sexual
element may induce variability we may likewise infer as probable from the
occurrence of bud-variations; for a bud seems to be the analogue of an ovule.
But the male element is apparently much oftener affected by changed con-

---

[49] Dr. P. Lucas has given a history
of opinion on this subject : ' Héréd. Nat.,'
1847, tom. i. p. 175.

[50] 'Hist. des Anomalies,' tom. iii. p. 499.

[51] Idem, tom. iii. pp. 392, 502.

[52] *See* his interesting work, ' Méta-
morphoses de l'Homme,' &c., 1862, p.
129.

ditions, at least in a visible manner, than the female element or ovule; and we know from Gärtner's and Wichura's statements that a hybrid used as the father and crossed with a pure species gives a greater degree of variability to the offspring, than does the same hybrid when used as the mother. Lastly, it is certain that variability may be transmitted through either sexual element, whether or not originally excited in them, for Kölreuter and Gärtner[53] found that when two species were crossed, if either one was variable, the offspring were rendered variable.

Summary. — From the facts given in this chapter, we may conclude that the variability of organic beings under domestication, although so general, is not an inevitable contingent on growth and reproduction, but results from the conditions to which the parents have been exposed. Changes of any kind in the conditions of life, even extremely slight changes, often suffice to cause variability. Excess of nutriment is perhaps the most efficient single exciting cause. Animals and plants continue to be variable for an immense period after their first domestication; but the conditions to which they are exposed never long remain quite constant. In the course of time they can be habituated to certain changes, so as to become less variable; and it is possible that when first domesticated they may have been even more variable than at present. There is good evidence that the power of changed conditions accumulates; so that two, three, or more generations must be exposed to new conditions before any effect is visible. The crossing of distinct forms, which have already become variable, increases in the offspring the tendency to further variability, by the unequal commingling of the characters of the two parents, by the reappearance of long-lost characters, and by the appearance of absolutely new characters. Some variations are induced by the direct action of the surrounding conditions on the whole organisation, or on certain parts alone, and other variations are induced indirectly through the reproductive system being affected in the same manner as is so common with organic beings when removed from their natural conditions. The causes which induce variability act on the mature organism, on the embryo, and, as we have good reason to believe, on both sexual elements before impregnation has been effected.

<hr>

[53] 'Dritte Fortsetzung,' &c., s. 123; 'Bastarderzeugung,' s. 249.

CHAPTER XXIII.

DIRECT AND DEFINITE ACTION OF THE EXTERNAL CONDITIONS
OF LIFE.

SLIGHT MODIFICATIONS IN PLANTS FROM THE DEFINITE ACTION OF CHANGED CONDI-
TIONS, IN SIZE, COLOUR, CHEMICAL PROPERTIES, AND IN THE STATE OF THE TISSUES
— LOCAL DISEASES — CONSPICUOUS MODIFICATIONS FROM CHANGED CLIMATE OR
FOOD, ETC. — PLUMAGE OF BIRDS AFFECTED BY PECULIAR NUTRIMENT, AND BY
THE INOCULATION OF POISON — LAND-SHELLS — MODIFICATIONS OF ORGANIC
BEINGS IN A STATE OF NATURE THROUGH THE DEFINITE ACTION OF EXTERNAL CON-
DITIONS — COMPARISON OF AMERICAN AND EUROPEAN TREES — GALLS — EFFECTS
OF PARASITIC FUNGI — CONSIDERATIONS OPPOSED TO THE BELIEF IN THE POTENT
INFLUENCE OF CHANGED EXTERNAL CONDITIONS — PARALLEL SERIES OF VARIETIES
— AMOUNT OF VARIATION DOES NOT CORRESPOND WITH THE DEGREE OF CHANGE IN
THE CONDITIONS — BUD-VARIATION — MONSTROSITIES PRODUCED BY UNNATURAL
TREATMENT — SUMMARY.

IF we ask ourselves why this or that character has been modified under domestication, we are, in most cases lost in utter
darkness. Many naturalists, especially of the French school,
attribute every modification to the "monde ambiant," that is, to
changed climate, with all its diversities of heat and cold, dampness
and dryness, light and electricity, to the nature of the soil,
and to varied kinds and amount of food. By the term definite
action, as used in this chapter, I mean an action of such a nature
that, when many individuals of the same variety are exposed
during several generations to any change in their physical
conditions of life, all, or nearly all the individuals, are modified
in the same manner. A new sub-variety would thus be produced without the aid of selection.

I do not include under the term of definite action the effects
of habit or of the increased use and disuse of various organs.
Modifications of this nature, no doubt, are definitely caused by
the conditions to which the beings are subjected; but they depend much less on the nature of the conditions than on the laws
of growth; hence they are included under a distinct head in the

following chapter. We know, however, far too little of the causes and laws of variation to make a sound classification. The direct action of the conditions of life, whether leading to definite or indefinite results, is a totally distinct consideration from the effects of natural selection ; for natural selection depends on the survival under various and complex circumstances of the best-fitted individuals, but has no relation whatever to the primary cause of any modification of structure.

I will first give in detail all the facts which I have been able to collect, rendering it probable that climate, food, &c., have acted so definitely and powerfully on the organisation of our domesticated productions, that they have sufficed to form new subvarieties or races, without the aid of selection by man or of natural selection. I will then give the facts and considerations opposed to this conclusion, and finally we will weigh, as fairly as we can, the evidence on both sides.

When we reflect that distinct races of almost all our domesticated animals exist in each kingdom of Europe, and formerly even in each district of England, we are at first strongly inclined to attribute their origin to the definite action of the physical conditions of each country ; and this has been the conclusion of many authors. But we should bear in mind that man annually has to choose which animals shall be preserved for breeding, and which shall be slaughtered. We have also seen that both methodical and unconscious selection were formerly practised, and are now occasionally practised by the most barbarous races, to a much greater extent than might have been anticipated. Hence it is very difficult to judge how far the difference in conditions between, for instance, the several districts in England, could have sufficed without the aid of selection to modify the breeds which have been reared in each. It may be argued that, as numerous wild animals and plants have ranged during many ages throughout Great Britain, and still retain the same character, the difference in conditions between the several districts could not have modified in so marked a manner the various native races of cattle, sheep, pigs, and horses. The same difficulty of distinguishing between selection and the definite effects of the conditions of life, is encountered in a still higher degree when we compare closely allied natural

forms, inhabiting two countries, such as North America and
Europe, which do not differ greatly in climate, nature of soil, &c.,
for in this case natural selection will inevitably and rigorously
have acted during a long succession of ages.

From the importance of the difficulty just alluded to, it will be advisable
to give as large a body of facts as possible, showing that extremely slight
differences in treatment, either in different parts of the same country, or
during different seasons, certainly cause an appreciable effect, at least on
varieties which are already in an unstable condition. Ornamental flowers
are good for this purpose, as they are highly variable, and are carefully
observed. All floriculturists are unanimous that certain varieties are
affected by very slight differences in the nature of the artificial compost in
which they are grown, and by the natural soil of the district, and by the
season. Thus, a skilful judge, in writing on Carnations and Picotees,[1]
asks "where can Admiral Curzon be seen possessing the colour, size, and
"strength which it has in Derbyshire? Where can Flora's Garland be
"found equal to those at Slough? Where do high-coloured flowers revel
"better than at Woolwich and Birmingham? Yet in no two of these
"districts do the same varieties attain an equal degree of excellence,
"although each may be receiving the attention of the most skilful culti-
"vators." The same writer then recommends every cultivator to keep
five different kinds of soil and manure, "and to endeavour to suit the
"respective appetites of the plants you are dealing with, for without such
"attention all hope of general success will be vain." So it is with the
Dahlia:[2] the Lady Cooper rarely succeeds near London, but does admirably
in other districts; the reverse holds good with other varieties; and again,
there are others which succeed equally well in various situations. A skilful
gardener[3] states that he procured cuttings of an old and well-known
variety (pulchella) of Verbena, which from having been propagated in a
different situation presented a slightly different shade of colour; the two
varieties were afterwards multiplied by cuttings, being carefully kept
distinct; but in the second year they could hardly be distinguished, and
in the third year no one could distinguish them.

The nature of the season has an especial influence on certain varieties
of the Dahlia: in 1841 two varieties were pre-eminently good, and the
next year these same two were pre-eminently bad. A famous amateur[4]
asserts that in 1861 many varieties of the Rose came so untrue in character, "that it was hardly possible to recognise them, and the thought
"was not seldom entertained that the grower had lost his tally." The
same amateur[5] states that in 1862 two-thirds of his Auriculas produced
central trusses of flowers, and these are remarkable from not keeping true;

[1] 'Gardeners Chronicle,' 1853, p. 183.
[2] Mr. Wildman, 'Floricultural Soc.,' Feb. 7, 1843, reported in 'Gard. Chron.,' 1843, p. 86.
[3] Mr. Robson, in 'Journal of Horticulture, Feb. 13th, 1866, p. 122.
[4] 'Journal of Horticulture,' 1861, p. 24.
[5] Ibid., 1862, p. 83.

and he adds that in some seasons certain varieties of this plant all prove good, and the next season all prove bad; whilst exactly the reverse happens with other varieties. In 1845 the editor of the 'Gardener's Chronicle'[6] remarked how singular it was that this year many Calceolarias tended to assume a tubular form. With Heartsease[7] the blotched sorts do not acquire their proper character until hot weather sets in; whilst other varieties lose their beautiful marks as soon as this occurs.

Analogous facts have been observed with leaves: Mr. Beaton asserts[8] that he raised at Shrubland, during six years, twenty thousand seedlings from the Punch Pelargonium, and not one had variegated leaves; but at Surbiton, in Surrey, one-third, or even a greater proportion, of the seedlings from this same variety were more or less variegated. The soil of another district in Surrey has a strong tendency to cause variegation, as appears from information given me by Sir F. Pollock. Verlot[9] states that the variegated strawberry retains its character as long as grown in a dryish soil, but soon loses it when planted in fresh and humid soil. Mr. Salter, who is well known for his success in cultivating variegated plants, informs me that rows of strawberries were planted in his garden in 1859, in the usual way; and at various distances in one row, several plants simultaneously became variegated, and what made the case more extraordinary, all were variegated in precisely the same manner. These plants were removed, but during the three succeeding years other plants in the same row became variegated, and in no instance were the plants in any adjoining row affected.

The chemical qualities, odours, and tissues of plants are often modified by a change which seems to us slight. The Hemlock is said not to yield conicine in Scotland. The root of the *Aconitum napellus* becomes innocuous in frigid climates. The medicinal properties of the Digitalis are easily affected by culture. The Rhubarb flourishes in England, but does not produce the medicinal substance which makes the plant so valuable in Chinese Tartary. As the *Pistacia lentiscus* grows abundantly in the South of France, the climate must suit it, but it yields no mastic. The *Laurus sassafras* in Europe loses the odour proper to it in North America.[10] Many similar facts could be given, and they are remarkable because it might have been thought that definite chemical compounds would have been little liable to change either in quality or quantity.

The wood of the American Locust-tree (*Robinia*) when grown in England is nearly worthless, as is that of the Oak-tree when grown at the Cape of Good Hope.[11] Hemp and flax, as I hear from Dr. Falconer, flourish and yield plenty of seed on the plains of India, but their fibres are brittle

---

[6] 'Gard. Chron.,' 1845, p. 660.

[7] Ibid., 1863, p. 628.

[8] 'Journal of Hort.,' 1861, pp. 64, 309.

[9] 'Des Variétés,' &c., p. 76.

[10] Engel, ' Sur les Prop. Médicales des Plantes,' 1860, pp. 10, 25. On changes in the odours of plants, see

Dalibert's Experiments, quoted by Beckman, 'Inventions,' vol. ii. p. 344; and Nees, in Ferussac, ' Bull. des Sc. Nat.,' 1824, tom. i. p. 60. With respect to the rhubarb, &c., see also 'Gardener's Chronicle,' 1849, p. 355 ; 1862, p. 1123.

[11] Hooker, 'Flora Indica,' p. 32.

and useless. Hemp, on the other hand, fails to produce in England that resinous matter which is so largely used in India as an intoxicating drug.

The fruit of the Melon is greatly influenced by slight differences in culture and climate. Hence it is generally a better plan, according to Naudin, to improve an old kind than to introduce a new one into any locality. The seed of the Persian Melon produces near Paris fruit inferior to the poorest market kinds, but at Bordeaux yields delicious fruit.[12] Seed is annually brought from Thibet to Kashmir,[13] and produces fruit weighing from four to ten pounds, but plants raised from seed saved in Kashmir next year give fruit weighing only from two to three pounds. It is well known that American varieties of the Apple produce in their native land magnificent and brightly-coloured fruit, but in England of poor quality and a dull colour. In Hungary there are many varieties of the Kidney-bean, remarkable for the beauty of their seeds, but the Rev. M. J. Berkeley[14] found that their beauty could hardly ever be preserved in England, and in some cases the colour was greatly changed. We have seen in the ninth chapter, with respect to wheat, what a remarkable effect transportal from the North to the South of France, and reversely, produced on the weight of the grain.

When man can perceive no change in plants or animals which have been exposed to a new climate or to different treatment, insects can sometimes perceive a marked change. The same species of cactus has been carried to India from Canton, Manilla, Mauritius, and from the hot-houses of Kew, and there is likewise a so-called native kind, formerly introduced from South America; all these plants are alike in appearance, but the cochineal insect flourishes only on the native kind, on which it thrives prodigiously.[15] Humboldt remarks[16] that white men "born in the torrid zone walk barefoot with impunity in the same apartment where a European, recently landed, is exposed to the attacks of the *Pulex penetrans.*" This insect, the too well-known chigoe, must therefore be able to distinguish what the most delicate chemical analysis fails to distinguish, namely, a difference between the blood or tissues of a European and those of a white man born in the country. But the discernment of the chigoe is not so surprising as it at first appears; for

[12] Naudin, 'Annales des Sc. Nat., 4th series, Bot., tom. xi., 1859, p. 81. 'Gardener's Chronicle,' 1859, p. 464.

[13] Moorcroft's 'Travels,' &c., vol. ii. p. 143.

[14] 'Gardener's Chronicle,' 1861, p. 1113.

[15] Royle, 'Productive Resources of India,' p. 59.

[16] 'Personal Narrative,' Eng. translat., vol. v. p. 101. This statement has been confirmed by Karsten ('Beitrag zur Kenntniss der Rhynchoprion:' Moscow, 1864, s. 39), and by others.

according to Liebig [17] the blood of men with different complexions, though inhabiting the same country, emits a different odour.

Diseases peculiar to certain localities, heights, or climates, may be here briefly noticed, as showing the influence of external circumstances on the human body. Diseases confined to certain races of man do not concern us, for the constitution of the race may play the more important part, and this may have been determined by unknown causes. The Plica Polonica stands, in this respect, in a nearly intermediate position; for it rarely affects Germans, who inhabit the neighbourhood of the Vistula, where so many Poles are grievously affected; and on the other hand, it does not affect Russians, who are said to belong to the same original stock with the Poles.[18] The elevation of a district often governs the appearance of diseases; in Mexico the yellow fever does not extend above 924 mètres; and in Peru, people are affected with the *verugas* only between 600 and 1600 mètres above the sea; many other such cases could be given. A peculiar cutaneous complaint, called the *Bouton d'Alep*, affects in Aleppo and some neighbouring districts almost every native infant, and some few strangers; and it seems fairly well established that this singular complaint depends on drinking certain waters. In the healthy little island of St. Helena the scarlet-fever is dreaded like the Plague; analogous facts have been observed in Chili and Mexico.[19] Even in the different departments of France it is found that the various infirmities which render the conscript unfit for serving in the army, prevail with remarkable inequality, revealing, as Boudin observes, that many of them are endemic, which otherwise would never have been suspected.[20] Any one who will study the distribution of disease will be struck with surprise at what slight differences in the surrounding circumstances govern the nature and severity of the complaints by which man is at least temporarily affected.

The modifications as yet referred to have been extremely slight, and in most cases have been caused, as far as we can judge, by equally slight changes in the conditions. But can it be safely maintained that such changed conditions, if acting during a long series of generations, would not produce a marked effect? It is commonly believed that the people of the United States differ in appearance from the parent Anglo-Saxon race; and selection cannot have come into action within so short a period. A good observer [21] states that a general absence of fat,

[17] 'Organic Chemistry,' Eng. translat., 1st edit., p. 369.

[18] Prichard, 'Phys. Hist. of Mankind,' 1851, vol. i. p. 155.

[19] Darwin, 'Journal of Researches,' 1845, p. 434.

[20] These statements on disease are taken from Dr. Boudin's 'Géographie et de Statistique Médicales,' 1857, tom. i. p. xliv. and lii.; tom. ii. p. 315.

[21] E. Desor, quoted in the 'Anthrop. Rev.,' 1863, p. 180. For much confirmatory evidence, see Quatrefages, 'Unité de l'Espèce Humaine,' 1861, p. 131.

a thin and elongated neck, stiff and lank hair, are the chief characteristics. The change in the nature of the hair is supposed to be caused by the dryness of the atmosphere. If immigration into the United States were now stopped, who can say that the character of the whole people would not be greatly modified in the course of two or three thousand years?

The direct and definite action of changed conditions, in contradistinction to the accumulation of indefinite variations, seems to me so important that I will give a large additional body of miscellaneous facts. With plants, a considerable change of climate sometimes produces a conspicuous result. I have given in detail in the ninth chapter the most remarkable case known to me, namely, that in Germany several varieties of maize brought from the hotter parts of America were transformed in the course of only two or three generations. Dr. Falconer informs me that he has seen the English Ribston-pippin apple, a Himalayan oak, Prunus and Pyrus, all assume in the hotter parts of India a fastigate or pyramidal habit; and this fact is the more interesting, as a Chinese tropical species of Pyrus naturally has this habit of growth. Although in these cases the changed manner of growth seems to have been directly caused by the great heat, we know that many fastigate trees have originated in their temperate homes. In the Botanic Gardens of Ceylon the apple-tree [22] "sends out numerous runners under ground, which continually rise into small stems, and form a growth around the parent-tree." The varieties of the cabbage which produce heads in Europe fail to do so in certain tropical countries.[23] The *Rhododendron ciliatum* produced at Kew flowers so much larger and paler-coloured than those which it bears on its native Himalayan mountain, that Dr. Hooker[24] would hardly have recognised the species by the flowers alone. Many similar facts with respect to the colour and size of flowers could be given.

The experiments of Vilmorin and Buckman on carrots and parsnips prove that abundant nutriment produces a definite and inheritable effect on the so-called roots, with scarcely any change in other parts of the plant. Alum directly influences the colour of the flowers of the Hydrangea.[25] Dryness seems generally to favour the hairyness or villosity of plants. Gärtner found that hybrid Verbascums became extremely woolly when grown in pots. Mr. Masters, on the other hand, states that the *Opuntia leucotricha* "is well clothed with beautiful white hairs when grown in a "damp heat; but in a dry heat exhibits none of this peculiarity."[26] Slight variations of many kinds, not worth specifying in detail, are retained only as

[22] 'Ceylon,' by Sir J. E. Tennent, vol. i., 1859, p. 89.

[23] Godron, 'De l'Espèce,' tom. ii. p. 52.

[24] 'Journal of Horticultural Soc.,' vol. vii., 1852, p. 117.

[25] 'Journal of Hort. Soc.,' vol. i. p. 160.

[26] See Lecoq on the Villosity of Plants, 'Geograph. Bot.,' tom. iii. pp. 287, 291; Gärtner, 'Bastarderz.,' s. 261; Mr. Masters, on the Opuntia, in 'Gard. Chronicle,' 1846, p. 444.

long as plants are grown in certain soils, of which Sageret[27] gives from his own experience some instances. Odart, who insists strongly on the permanence of the varieties of the grape, admits[28] that some varieties, when grown under a different climate or treated differently, vary in an extremely slight degree, as in the tint of the fruit and in the period of ripening. Some authors have denied that grafting causes even the slightest difference in the scion; but there is sufficient evidence that the fruit is sometimes slightly affected in size and flavour, the leaves in duration, and the flowers in appearance.[29]

With animals there can be no doubt, from the facts given in the first chapter, that European dogs deteriorate in India, not only in their instincts but in structure; but the changes which they undergo are of such a nature, that they may be partly due to reversion to a primitive form, as in the case of feral animals. In parts of India the turkey becomes reduced in size, "with the pendulous appendage over the beak enormously developed."[30] We have seen how soon the wild duck, when domesticated, loses its true character, from the effects of abundant or changed food, or from taking little exercise. From the direct action of a humid climate and poor pasture the horse rapidly decreases in size in the Falkland Islands. From information which I have received, this seems likewise to be the case to a certain extent with sheep in Australia.

Climate definitely influences the hairy covering of animals; in the West Indies a great change is produced in the fleece of sheep, in about three generations. Dr. Falconer states[31] that the Thibet mastiff and goat, when brought down from the Himalaya to Kashmir, lose their fine wool. At Angora not only goats, but shepherd-dogs and cats, have fine fleecy hair, and Mr. Ainsworth[32] attributes the thickness of the fleece to the severe winters, and its silky lustre to the hot summers. Burnes states positively[33] that the Karakool sheep lose their peculiar black curled fleeces when removed into any other country. Even within the limits of England, I have been assured that with two breeds of sheep the wool was slightly changed by the flocks being pastured in different localities.[34] It has been asserted on good authority[35] that horses kept during several years in the deep coal-mines of Belgium become covered with velvety hair, almost like that on the mole. These cases probably stand in close relation to the natural change of coat in winter and summer. Naked varieties of several domestic animals have occasionally appeared; but there is no reason to

[27] 'Pom. Phys.,' p. 136.
[28] 'Ampelographie,' 1849, p. 19.
[29] Gärtner, 'Bastarderz.,' s. 606, has collected nearly all recorded facts. Andrew Knight (in 'Transact. Hort. Soc.,' vol. ii. p. 160) goes so far as to maintain that few varieties are absolutely permanent in character when propagated by buds or grafts.
[30] Mr. Blyth, in 'Annals and Mag. of Nat. Hist.,' vol. xx, 1847, p. 391.
[31] 'Natural History Review,' 1862, p. 113.
[32] 'Journal of Roy. Geographical Soc.,' vol. ix., 1839, p. 275.
[33] 'Travels in Bokhara,' vol. iii. p. 151.
[34] See also, on the influence of marshy pastures on the wool, Godron, 'L'Espèce,' tom. ii. p. 22.
[35] Isidore Geoffroy St. Hilaire, 'Hist. Nat. Gén.,' tom. iii. p. 438.

believe that this is in any way related to the nature of the climate to which they have been exposed.[36]

It appears at first sight probable that the increased size, the tendency to fatten, the early maturity and altered forms of our improved cattle, sheep, and pigs, have directly resulted from their abundant supply of food. This is the opinion of many competent judges, and probably is to a great extent true. But as far as form is concerned, we must not overlook the equal or more potent influence of lessened use on the limbs and lungs. We see, moreover, as far as size is concerned, that selection is apparently a more powerful agent than a large supply of food, for we can thus only account for the existence, as remarked to me by Mr. Blyth, of the largest and smallest breeds of sheep in the same country, of Cochin-China fowls and Bantams, of small Tumbler and large Runt pigeons, all kept together and supplied with abundant nourishment. Nevertheless there can be little doubt that our domesticated animals have been modified, independently of the increased or lessened use of parts, by the conditions to which they have been subjected, without the aid of selection. For instance, Prof. Rütimeyer[37] shows that the bones of all domesticated quadrupeds can be distinguished from those of wild animals by the state of their surface and general appearance. It is scarcely possible to read Nathusius's excellent 'Vorstudien,'[38] and doubt that, with the highly improved races of the pig, abundant food has produced a conspicuous effect on the general form of the body, on the breadth of the head and face, and even on the teeth. Nathusius rests much on the case of a purely bred Berkshire pig, which when two months old became diseased in its digestive organs, and was preserved for observation until nineteen months old; at this age it had lost several characteristic features of the breed, and had acquired a long, narrow head, of large size relatively to its small body, and elongated legs. But in this case and in some others we ought not to assume that, because certain characters are lost, perhaps through reversion, under one course of treatment, therefore that they had been at first directly produced by an opposite course.

In the case of the rabbit, which has become feral on the island of Porto Santo, we are at first strongly tempted to attribute the whole change— the greatly reduced size, the altered tints of the fur, and the loss of certain characteristic marks—to the definite action of the new conditions to which it has been exposed. But in all such cases we have to consider in addition the tendency to reversion to progenitors more or less remote, and the natural selection of the finest shades of difference.

The nature of the food sometimes either definitely induces certain peculiarities, or stands in some close relation with them. Pallas long ago asserted that the fat-tailed sheep of Siberia degenerated and lost their enormous tails when removed from certain saline pastures; and recently

---

[36] Azara has made some good remarks on this subject, 'Quadrupèdes du Paraguay,' tom. ii. p. 337. See an account of a family of naked mice produced in England, 'Proc. Zoolog. Soc.,' 1856, p. 38.

[37] 'Die Fauna der Pfahlbauten,' 1861, s. 15.

[38] 'Schweinschädel,' 1864, s. 99.

Erman[39] states that this occurs with the Kirgisian sheep when brought to Orenburgh.

It is well known that hemp-seed causes bullfinches and certain other birds to become black. Mr. Wallace has communicated to me some much more remarkable facts of the same nature. The natives of the Amazonian region feed the common green parrot (*Chrysotis festiva*, Linn.) with the fat of large Siluroid fishes, and the birds thus treated become beautifully variegated with red and yellow feathers. In the Malayan archipelago, the natives of Gilolo alter in an analogous manner the colours of another parrot, namely, the *Lorius garrulus*, Linn., and thus produce the *Lori rajah* or King-Lory. These parrots in the Malay Islands and South America, when fed by the natives on natural vegetable food, such as rice and plantains, retain their proper colours. Mr. Wallace has, also, recorded[40] a still more singular fact. "The Indians (of S. America) have " a curious art by which they change the colours of the feathers of many " birds. They pluck out those from the part they wish to paint, and " inoculate the fresh wound with the milky secretion from the skin of a " small toad. The feathers grow of a brilliant yellow colour, and on being " plucked out, it is said, grow again of the same colour without any fresh " operation."

Bechstein[41] does not entertain any doubt that seclusion from light affects, at least temporarily, the colours of cage-birds.

It is well known that the shells of land-mollusca are affected by the abundance of lime in different districts. Isidore Geoffroy St. Hilaire[42] gives the case of *Helix lactea*, which has recently been carried from Spain to the South of France and to the Rio Plata, and in both these countries now presents a distinct appearance, but whether this has resulted from food or climate is not known. With respect to the common oyster, Mr. F. Buckland informs me that he can generally distinguish the shells from different districts; young oysters brought from Wales and laid down in beds where "*natives*" are indigenous, in the short space of two months begin to assume the "native" character. M. Costa[43] has recorded a much more remarkable case of the same nature, namely, that young shells taken from the shores of England and placed in the Mediterranean, at once altered their manner of growth and formed prominent diverging rays, like those on the shells of the proper Mediterranean oyster. The same individual shell, showing both forms of growth, was exhibited before a society in Paris. Lastly, it is well known that caterpillars fed on different food sometimes either themselves acquire a different colour or produce moths different in colour.[44]

---

[39] 'Travels in Siberia,' Eng. translat., vol. i. p. 228.

[40] A. R. Wallace, 'Travels on the Amazon and Rio Negro,' p. 294.

[41] 'Naturgeschichte der Stubenvögel,' 1840, s. 262, 308.

[42] 'Hist. Nat. Gén.,' tom. iii. p. 402.

[43] 'Bull. de la Soc. Imp. d'Acclimat.,' tom. viii. p. 351.

[44] *See* an account of Mr. Gregson's experiments on the *Abraxus grossulariata*, 'Proc. Entomolog. Soc.,' Jan. 6th, 1862: these experiments have been confirmed by Mr. Greening, in 'Proc. of the Northern Entomolog. Soc.,' July 28th, 1862. For the effects of food on caterpillars, *see* a curious account by M. Michely, in 'Bull. de la

It would be travelling beyond my proper limits here to discuss how far organic beings in a state of nature are definitely modified by changed conditions. In my 'Origin of Species' I have given a brief abstract of the facts bearing on this point, and have shown the influence of light on the colours of birds, and of residence near the sea on the lurid tints of insects, and on the succulency of plants. Mr. Herbert Spencer[45] has recently discussed with much ability this whole subject on broad and general grounds. He argues, for instance, that with all animals the external and internal tissues are differently acted on by the surrounding conditions, and they invariably differ in intimate structure. So again the upper and lower surfaces of true leaves, as well as of stems and petioles, when these assume the function and occupy the position of leaves, are differently circumstanced with respect to light, &c., and apparently in consequence differ in structure. But, as Mr. Herbert Spencer admits, it is most difficult in all such cases to distinguish between the effects of the definite action of physical conditions and the accumulation through natural selection of inherited variations which are serviceable to the organism, and which have arisen independently of the definite action of these conditions.

Although we are not here concerned with organic beings in a state of nature, yet I may call attention to one case. Mr. Meehan,[46] in a remarkable paper, compares twenty-nine kinds of American trees, belonging to various orders, with their nearest European allies, all grown in close proximity in the same garden and under as nearly as possible the same conditions. In the American species Mr. Meehan finds, with the rarest exceptions, that the leaves fall earlier in the season, and assume before falling a brighter tint; that they are less deeply toothed or serrated; that the buds are smaller; that the trees are more diffuse in growth and have fewer branchlets; and, lastly, that the seeds are smaller—all in comparison with the corresponding European species. Now, considering that these trees belong to distinct orders, it is out of the question that the peculiarities just specified should have been inherited in the one continent from one progenitor, and in the other from another progenitor; and considering that the trees inhabit widely different stations, these peculiarities can hardly be supposed to be of any special

Soc. Imp. d'Acclimat.,' tom. viii. p. 563. For analogous facts from Dahlbom on Hymenoptera, see Westwood's 'Modern Class. of Insects,' vol. ii. p. 98. See also Dr. L. Müller, 'Die Abhängigkeit der Insecten,' 1867, s. 70.

[45] 'The Principles of Biology,' vol. ii. 1866. The present chapters were written before I had read Mr. Herbert Spencer's work, so that I have not been able to make so much use of it as I should otherwise probably have done.

[46] 'Proc. Acad. Nat. Soc. of Philadelphia,' Jan. 28th, 1862.

service to the two series in the Old and New Worlds; therefore these peculiarities cannot have been naturally selected. Hence we are led to infer that they have been definitely caused by the long-continued action of the different climate of the two continents on the trees.

Galls.—Another class of facts, not relating to cultivated plants, deserves attention. I allude to the production of galls. Every one knows the curious, bright-red, hairy productions on the wild rose-tree, and the various different galls produced by the oak. Some of the latter resemble fruit, with one face as rosy as the rosiest apple. These bright colours can be of no service either to the gall-forming insect or to the tree, and probably are the direct result of the action of the light, in the same manner as the apples of Nova Scotia or Canada are brighter coloured than English apples. The strongest upholder of the doctrine that organic beings are created beautiful to please mankind would not, I presume, extend this view to galls. According to Osten Sacken's latest revision, no less than fifty-eight kinds of galls are produced on the several species of oak, by Cynips with its sub-genera; and Mr. B. D. Walsh [47] states that he can add many others to the list. One American species of willow, the *Salix-humilis*, bears ten distinct kinds of galls. The leaves which spring from the galls of various English willows differ completely in shape from the natural leaves. The young shoots of junipers and firs, when punctured by certain insects, yield monstrous growths like flowers and cones; and the flowers of some plants become from the same cause wholly changed in appearance. Galls are produced in every quarter of the world; of several sent to me by Mr. Thwaites from Ceylon, some were as symmetrical as a composite flower when in bud, others smooth and spherical like a berry; some protected by long spines, others clothed with yellow wool formed of long cellular hairs, others with regularly tufted hairs. In some galls the internal structure is simple, but in others it is highly complex; thus M. Lucaze-Duthiers [48] has figured in the common ink-gall no less than seven concentric layers, composed of distinct tissue,

[47] *See* Mr. B. D. Walsh's excellent papers in 'Proc. Entomolog. Soc. Philadelphia,' Dec. 1866, p. 284. With respect to the willow, *see* Idem, 1864, p 546.

[48] *See* his admirable Histoire des Galles, in 'Annal. des Sc. Nat. Bot.,' 3rd series, tom. xix., 1853, p. 273.

namely, the epidermic, sub-epidermic, spongy, intermediate, and the hard protective layer formed of curiously thickened woody cells, and, lastly, the central mass abounding with starch-granules on which the larvæ feed.

Galls are produced by insects of various orders, but the greater number by species of Cynips. It is impossible to read M. Lucaze-Duthier's discussion and doubt that the poisonous secretion of the insect causes the growth of the gall, and every one knows how virulent is the poison secreted by wasps and bees, which belong to the same order with Cynips. Galls grow with extraordinary rapidity, and it is said that they attain their full size in a few days ;[49] it is certain that they are almost completely developed before the larvæ are hatched. Considering that many gall-insects are extremely small, the drop of secreted poison must be excessively minute ; it probably acts on one or two cells alone, which, being abnormally stimulated, rapidly increase by a process of self-division. Galls, as Mr. Walsh[50] remarks, afford good, constant, and definite characters, each kind keeping as true to form as does any independent organic being. This fact becomes still more remarkable when we hear that, for instance, seven out of the ten different kinds of galls produced on *Salix humilis* are formed by gall-gnats (*Cecidomyidæ*) which, " though essentially distinct species, yet resemble one another " so closely that in almost all cases it is difficult, and in some " cases impossible, to distinguish the full-grown insects one from " the other."[51] For in accordance with a wide-spread analogy we may safely infer that the poison secreted by insects so closely allied would not differ much in nature ; yet this slight difference is sufficient to induce widely different results. In some few cases the same species of gall-gnat produces on distinct species of willows galls which cannot be distinguished ; the *Cynips fecundatrix*, also, has been known to produce on the Turkish oak, to which it is not properly attached, exactly the same kind of gall as on the European oak.[52] These latter facts apparently prove that the nature of the poison is a much more powerful

[49] Kirby and Spence's 'Entomology,' 1818, vol. i. p. 450 ; Lucaze-Duthiers, idem, p. 284.

[50] 'Proc. Entomolog. Soc. Philadelphia,' 1864, p. 558.

[51] Mr. B. D. Walsh, idem, p. 633 ; and Dec. 1866, p. 275.

[52] Mr. B. D. Walsh, idem, 1864, p. 545, 411, 495; and Dec. 1866, p. 278. *See* also Lucaze-Duthiers.

agent in determining the form of the gall than the specific character of the tree which is acted on.

As the poisonous secretion of insects belonging to various orders has the special power of affecting the growth of various plants ;—as a slight difference in the nature of the poison suffices to produce widely different results ;—and lastly, as we know that the chemical compounds secreted by plants are eminently liable to be modified by changed conditions of life, we may believe it possible that various parts of a plant might be modified through the agency of its own altered secretions. Compare, for instance, the mossy and viscid calyx of a moss-rose, which suddenly appears through bud-variation on a Provence-rose, with the gall of red moss growing from the inoculated leaf of a wild rose, with each filament symmetrically branched like a microscopical spruce-fir, bearing a glandular tip and secreting odoriferous gummy matter.[53] Or compare, on the one hand, the fruit of the peach, with its hairy skin, fleshy covering, hard shell and kernel, and on the other hand one of the more complex galls with its epidermic, spongy, and woody layers, surrounding tissue loaded with starch granules. These normal and abnormal structures manifestly present a certain degree of resemblance. Or, again, reflect on the cases above given of parrots which have had their plumage brightly decorated through some change in their blood, caused by having been fed on certain fishes, or locally inoculated with the poison of a toad. I am far from wishing to maintain that the moss-rose or the hard shell of the peach-stone or the bright colours of birds are actually due to any chemical change in the sap or blood ; but these cases of galls and of parrots are excellently adapted to show us how powerfully and singularly external agencies may affect structure. With such facts before us, we need feel no surprise at the appearance of any modification in any organic being.

I may, also, here allude to the remarkable effects which parasitic fungi sometimes produce on plants. Reissek[54] has described a Thesium, affected by an Œcidium, which was greatly modified, and assumed some of the

---

[53] Lucaze-Duthiers, idem, pp. 325, 328.

[54] 'Linnæa,' vol. xvii., 1843 ; quoted by Dr. M. T. Masters, Royal Institution, March 16th, 1860.

characteristic features of certain allied species, or even genera. Suppose, says Reissek, "the condition originally caused by the fungus to become " constant in the course of time, the plant would, if found growing wild, " be considered as a distinct species or even as belonging to a new genus." I quote this remark to show how profoundly, yet in how natural a manner, this plant must have been modified by the parasitic fungus.

Facts and Considerations opposed to the belief that the Conditions of Life act in a potent manner in causing definite Modifications of Structure.

I have alluded to the slight differences in species when naturally living in distinct countries under different conditions ; and such differences we feel at first inclined, probably to a limited extent with justice, to attribute to the definite action of the surrounding conditions. But it must be borne in mind that there are a far greater number of animals and plants which range widely and have been exposed to great diversities of conditions, yet remain nearly uniform in character. Some authors, as previously remarked, account for the varieties of our culinary and agricultural plants by the definite action of the conditions to which they have been exposed in the different parts of Great Britain; but there are about 200 plants[55] which are found in every single English county ; these plants must have been exposed for an immense period to considerable differences of climate and soil, yet do not differ. So, again, some birds, insects, other animals, and plants range over large portions of the world, yet retain the same character.

Notwithstanding the facts previously given on the occurrence of highly peculiar local diseases and on the strange modifications of structure in plants caused by the inoculated poison of insects, and other analogous cases; still there are a multitude of variations—such as the modified skull of the niata ox and bulldog, the long horns of Caffre cattle, the conjoined toes of the solid-hoofed swine, the immense crest and protuberant skull of Polish fowls, the crop of the pouter-pigeon, and a host of other such cases —which we can hardly attribute to the definite action, in the sense before specified, of the external conditions of life. No doubt in every case there must have been some exciting cause; but as we see innumerable individuals exposed to nearly the same conditions, and one alone is affected, we may conclude that the constitution of the individual is of far higher

---

[55] Hewett C. Watson, 'Cybele Britannica,' vol. i., 1847, p. 11.

importance than the conditions to which it has been exposed. It seems, indeed, to be a general rule that conspicuous variations occur rarely, and in one individual alone out of many thousands, though all may have been exposed, as far as we can judge, to nearly the same conditions. As the most strongly marked variations graduate insensibly into the most trifling, we are led by the same train of thought to attribute each slight variation much more to innate differences of constitution, however caused, than to the definite action of the surrounding conditions.

We are led to the same conclusion by considering the cases, formerly alluded to, of fowls and pigeons, which have varied and will no doubt go on varying in directly opposite ways, though kept during many generations under nearly the same conditions. Some, for instance, are born with their beaks, wings, tails, legs, &c., a little longer, and others with these same parts a little shorter. By the long-continued selection of such slight individual differences, which occur in birds kept in the same aviary, widely different races could certainly be formed; and long-continued selection, important as is the result, does nothing but preserve the variations which appear to us to arise spontaneously.

In these cases we see that domesticated animals vary in an indefinite number of particulars, though treated as uniformly as is possible. On the other hand, there are instances of animals and plants, which, though exposed to very different conditions, both under nature and domestication, have varied in nearly the same manner. Mr. Layard informs me that he has observed amongst the Caffres of South Africa a dog singularly like an arctic Esquimaux dog. Pigeons in India present nearly the same wide diversities of colour as in Europe; and I have seen chequered and simply barred pigeons, and pigeons with blue and white loins, from Sierra Leone, Madeira, England, and India. New varieties of flowers are continually raised in different parts of Great Britain, but many of these are found by the judges at our exhibitions to be almost identical with old varieties. A vast number of new fruit-trees and culinary vegetables have been produced in North America: these differ from European varieties in the same general manner as the several varieties raised in Europe differ from each other; and no one has ever pretended that the climate of America has given to the many American varieties any general character by which they can be recognised. Nevertheless, from the facts previously advanced on the authority of Mr. Meehan with respect to American and European forest-trees, it would be rash to affirm that varieties raised in the two countries would not in the course of ages assume a distinctive character. Mr. Masters has recorded a striking fact[56] bearing on this subject: he raised numerous plants of *Hybiscus Syriacus* from seed collected in South Carolina and the Holy Land, where the parent-plants must have been exposed to considerably different conditions; yet the seedlings from both localities broke into two similar strains, one with obtuse leaves and purple or crimson flowers, and the other with elongated leaves and more or less pink flowers.

---

[56] 'Gardener's Chronicle,' 1857, p. 629.

We may, also, infer the prepotent influence of the constitution of the organism over the definite action of the conditions of life, from the several cases given in the earlier chapters of parallel series of varieties,—an important subject, hereafter to be more fully discussed. Sub-varieties of the several kinds of wheat, gourds, peaches, and other plants, and to a certain limited extent sub-varieties of the fowl, pigeon, and dog, have been shown either to resemble or to differ from each other in a closely corresponding and parallel manner. In other cases, a variety of one species resembles a distinct species; or the varieties of two distinct species resemble each other. Although these parallel resemblances no doubt often result from reversion to the former characters of a common progenitor; yet in other cases, when new characters first appear, the resemblance must be attributed to the inheritance of a similar constitution, and consequently to a tendency to vary in the same manner. We see something of a similar kind in the same monstrosity appearing and reappearing many times in the same animal, and, as Dr. Maxwell Masters has remarked to me, in the same plant.

We may at least conclude thus far, that the amount of modification which animals and plants have undergone under domestication, does not correspond with the degree to which they have been subjected to changed circumstances. As we know the parentage of domesticated birds far better than of most quadrupeds, we will glance through the list. The pigeon has varied in Europe more than almost any other bird; yet it is a native species, and has not been exposed to any extraordinary change of conditions. The fowl has varied equally, or almost equally, with the pigeon, and is a native of the hot jungles of India. Neither the peacock, a native of the same country, nor the guinea-fowl, an inhabitant of the dry deserts of Africa, has varied at all, or only in colour. The turkey, from Mexico, has varied but little. The duck, on the other hand, a native of Europe, has yielded some well-marked races; and as this is an aquatic bird, it must have been subjected to a far more serious change in its habits than the pigeon or even the fowl, which nevertheless have varied in a much higher degree. The goose, a native of Europe and aquatic like the duck, has varied less than any other domesticated bird, except the peacock.

Bud-variation is, also, important under our present point of view. In some few cases, as when all the eyes or buds on the same tuber of the potato, or all the fruit on the same plum-tree, or all the flowers on the same plant, have suddenly varied in the same manner, it might be argued that the varia-

tion had been definitely caused by some change in the conditions
to which the plants had been exposed; yet, in other cases, such
an admission is extremely difficult. As new characters sometimes appear by bud-variation, which do not occur in the parent-
species or in any allied species, we may reject, at least in these
cases, the idea that they are due to reversion. Now it is well
worth while to reflect maturely on some striking case of bud-
variation, for instance that of the peach. This tree has been
cultivated by the million in various parts of the world, has been
treated differently, grown on its own roots and grafted on various
stocks, planted as a standard, against a wall, and under glass;
yet each bud of each sub-variety keeps true to its kind. But
occasionally, at long intervals of time, a tree in England, or
under the widely-different climate of Virginia, produces a single
bud, and this yields a branch which ever afterwards bears nectarines. Nectarines differ, as every one knows, from peaches
in their smoothness, size, and flavour; and the difference is so
great, that some botanists have maintained that they are specifically distinct. So permanent are the characters thus suddenly
acquired, that a nectarine produced by bud-variation has propagated itself by seed. To guard against the supposition that
there is some fundamental distinction between bud and seminal
variation, it is well to bear in mind that nectarines have likewise been produced from the stone of the peach; and, reversely,
peaches from the stone of the nectarine. Now is it possible to
conceive external conditions more closely alike than those to
which the buds on the same tree are exposed? Yet one bud
alone, out of the many thousands borne by the same tree, has
suddenly without any apparent cause produced a nectarine. But
the case is even stronger than this, for the same flower-bud
has yielded a fruit, one-half or one-quarter a nectarine, and
the other half or three-quarters a peach. Again, seven or eight
varieties of the peach have yielded by bud-variation nectarines:
the nectarines thus produced, no doubt, differ a little from
each other; but still they are nectarines. Of course there
must be some cause, internal or external, to excite the peach-
bud to change its nature; but I cannot imagine a class of facts
better adapted to force on our minds the conviction that what
we call the external conditions of life are quite insignificant in

relation to any particular variation, in comparison with the organisation or constitution of the being which varies.

It is known from the labours of Geoffroy St. Hilaire, and recently from those of Dareste and others, that eggs of the fowl, if shaken, placed upright, perforated, covered in part with varnish, &c., produce monstrous chickens. Now these monstrosities may be said to be directly caused by such unnatural conditions, but the modifications thus induced are not of a definite nature. An excellent observer, M. Camille Dareste,[57] remarks " that the various species of monstrosities are not determined " by specific causes; the external agencies which modify the " development of the embryo act solely in causing a pertur- " bation—a perversion in the normal course of development." He compares the result to what we see in illness : a sudden chill, for instance, affects one individual alone out of many, causing either a cold, or sore-throat, rheumatism, or inflammation of the lungs or pleura. Contagious matter acts in an analogous manner.[58] We may take a still more specific instance : seven pigeons were struck by rattle-snakes ;[59] some suffered from convulsions; some had their blood coagulated, in others it was perfectly fluid ; some showed ecchymosed spots on the heart, others on the intestines, &c.; others again showed no visible lesion in any organ. It is well known that excess in drinking causes different diseases in different men; but men living under a cold and tropical climate are differently affected :[60] and in this case we see the definite influence of opposite conditions. The foregoing facts apparently give us as good an idea as we are likely for a long time to obtain, how in many cases external conditions act directly, though not definitely, in causing modifications of structure.

Summary.—There can be no doubt, from the facts given in the early part of this chapter, that extremely slight changes in

[57] 'Mémoire sur la Production Arti- ficielle des Monstrosités,' 1862, pp. 8-12 ; ' Recherches sur les Conditions, &c., chez les Monstres,' 1863, p. 6. An abstract is given of Geoffroy's Experiments by his son, in his ' Vie, Travaux, &c.,' 1847, p. 290.

[58] Paget, 'Lectures on Surgical Pathology,' 1853, vol. i. p. 483.

[59] 'Researches upon the Venom of the Rattle-snake,' Jan. 1861, by Dr. Mitchell, p. 67.

[60] Mr. Sedgwick, in ' British and Foreign Medico-Chirurg. Review, July 1863, p. 175.

the conditions of life sometimes act in a definite manner on our already variable domesticated productions; and, as the action of changed conditions in causing general or indefinite variability is accumulative, so it may be with their definite action. Hence it is possible that great and definite modifications of structure may result from altered conditions acting during a long series of generations. In some few instances a marked effect has been produced quickly on all, or nearly all, the individuals which have been exposed to some considerable change of climate, food, or other circumstance. This has occurred, and is now occurring, with European men in the United States, with European dogs in India, with horses in the Falkland Islands, apparently with various animals at Angora, with foreign oysters in the Mediterranean, and with maize grown in Europe from tropical seed. We have seen that the chemical compounds secreted by plants and the state of their tissues are readily affected by changed conditions. In some cases a relation apparently exists between certain characters and certain conditions, so that if the latter be changed the character is lost—as with cultivated flowers, with some few culinary plants, with the fruit of the melon, with fat-tailed sheep, and other sheep having peculiar fleeces.

The production of galls, and the change of plumage in parrots when fed on peculiar food or when inoculated by the poison of a toad, prove to us what great and mysterious changes in structure and colour may be the definite result of chemical changes in the nutrient fluids or tissues.

We have also reason to believe that organic beings in a state of nature may be modified in various definite ways by the conditions to which they have been long exposed, as in the case of American trees in comparison with their representatives in Europe. But in all such cases it is most difficult to distinguish between the definite results of changed conditions, and the accumulation through natural selection of serviceable variations which have arisen independently of the nature of the conditions. If, for instance, a plant had to be modified so as to become fitted to inhabit a humid instead of an arid station, we have no reason to believe that variations of the right kind would occur more frequently if the parent-plant inhabited a station a little more humid than usual.   Whether the station was unusually dry or humid, variations adapting the plant in a slight degree for directly opposite habits of life would occasionally arise, as we have reason to believe from what we know in other cases.

In most, perhaps in all cases, the organisation or constitution of the being which is acted on, is a much more important element than the nature of the changed conditions, in determining the nature of the variation.   We have evidence of this in the appearance of nearly similar modifications under different conditions, and of different modifications under apparently nearly the same conditions.   We have still better evidence of this in closely parallel varieties being frequently produced from distinct races, or even distinct species, and in the frequent recurrence of the same monstrosity in the same species.   We have also seen that the degree to which domesticated birds have varied, does not stand in any close relation with the amount of change to which they have been subjected.

To recur once again to bud-variations.   When we reflect on the millions of buds which many trees have produced, before some one bud has varied, we are lost in wonder what the precise cause of each variation can be.   Let us recall the case given by Andrew Knight of the forty-year-old tree of the yellow magnum bonum plum, an old variety which has been propagated by grafts on various stocks for a very long period throughout Europe and North America, and on which a single bud suddenly produced the red magnum bonum.   We should also bear in mind that distinct varieties, and even distinct species,—as in the case of peaches, nectarines, and apricots,—of certain roses and camellias,—although separated by a vast number of generations from any progenitor in common, and although cultivated under diversified conditions, have yielded by bud-variation closely analogous varieties.   When we reflect on these facts we become deeply impressed with the conviction that in such cases the nature of the variation depends but little on the conditions to which the plant has been exposed, and not in any especial manner on its individual character, but much more on the general nature or constitution, inherited from some remote progenitor, of the whole group of allied beings to which the plant belongs.   We are thus driven to conclude that in most cases the conditions of life play a subordinate part in causing any particular modification; like that which a spark plays, when a mass of combustibles bursts into flame—the nature of the flame depending on the combustible matter, and not on the spark.

No doubt each slight variation must have its efficient cause; but it is as hopeless an attempt to discover the cause of each as to say why a chill or a poison affects one man differently from another. Even with modifications resulting from the definite action of the conditions of life, when all or nearly all the individuals, which have been similarly exposed, are similarly affected, we can rarely see the precise relation between cause and effect. In the next chapter it will be shown that the increased use or disuse of various organs, produces an inherited effect. It will further be seen that certain variations are bound together by correlation and other laws. Beyond this we cannot at present explain either the causes or manner of action of Variation.

Finally, as indefinite and almost illimitable variability is the usual result of domestication and cultivation, with the same part or organ varying in different individuals in different or even in directly opposite ways; and as the same variation, if strongly pronounced, usually recurs only after long intervals of time, any particular variation would generally be lost by crossing, reversion, and the accidental destruction of the varying individuals, unless carefully preserved by man. Hence, although it must be admitted that new conditions of life do sometimes definitely affect organic beings, it may be doubted whether well-marked races have often been produced by the direct action of changed conditions without the aid of selection either by man or nature.

CHAPTER XXIV.

LAWS OF VARIATION — USE AND DISUSE, ETC.

NISUS FORMATIVUS, OR THE CO-ORDINATING POWER OF THE ORGANISATION — ON THE
EFFECTS OF THE INCREASED USE AND DISUSE OF ORGANS — CHANGED HABITS OF
LIFE — ACCLIMATISATION WITH ANIMALS AND PLANTS — VARIOUS METHODS BY
WHICH THIS CAN BE EFFECTED — ARRESTS OF DEVELOPMENT — RUDIMENTARY
ORGANS.

In this and the two following chapters I shall discuss, as well
as the difficulty of the subject permits, the several laws which
govern Variability.   These may be grouped under the effects of
use and disuse, including changed habits and acclimatisation—
arrests of development—correlated variation—the cohesion of
homologous parts—the variability of multiple parts—compensation of growth—the position of buds with respect to the axis
of the plant—and lastly, analogous variation. These several
subjects so graduate into each other that their distinction is
often arbitrary.

It may be convenient first briefly to discuss that co-ordinating
and reparative power which is common, in a higher or lower
degree, to all organic beings, and which was formerly designated by physiologists as the *nisus formativus*.

Blumenbach and others[1] have insisted that the principle which permits
a Hydra, when cut into fragments, to develop itself into two or more
perfect animals, is the same with that which causes a wound in the higher
animals to heal by a cicatrice.   Such cases as that of the Hydra are
evidently analogous with the spontaneous division or fissiparous generation
of the lowest animals, and likewise with the budding of plants.   Between
these extreme cases and that of a mere cicatrice we have every gradation.
Spallanzani,[2] by cutting off the legs and tail of a Salamander, got in the
course of three months six crops of these members; so that 687 perfect
bones were reproduced by one animal during one season.   At whatever

---

[1] 'An Essay on Generation,' Eng.
translat., p. 18; Paget, 'Lectures on
Surgical Pathology,' 1853, vol. i. p. 209.

[2] 'An Essay on Animal Reproduction,' Eng. translat., 1769, p. 79.

point the limb was cut off, the deficient part, and no more, was exactly reproduced. Even with man, as we have seen in the twelfth chapter, when treating of polydactylism, the entire limb whilst in an embryonic state, and supernumerary digits, are occasionally, though imperfectly, reproduced after amputation. When a diseased bone has been removed, a new one sometimes "gradually assumes the regular form, and all the " attachments of muscles, ligaments, &c., become as complete as before."[3]

This power of regrowth does not, however, always act perfectly: the reproduced tail of a lizard differs in the forms of the scales from the normal tail: with certain Orthopterous insects the large hind legs are reproduced of smaller size:[4] the white cicatrice which in the higher animals unites the edges of a deep wound is not formed of perfect skin, for elastic tissue is not produced till long afterwards.[5] "The activity of the *nisus* "*formativus*," says Blumenbach, "is in an inverse ratio to the age of the " organised body." To this may be added that its power is greater in animals the lower they are in the scale of organisation; and animals low in the scale correspond with the embryos of higher animals belonging to the same class. Newport's observations[6] afford a good illustration of this fact, for he found that " myriapods, whose highest development scarcely " carries them beyond the larvæ of perfect insects, can regenerate limbs and " antennæ up to the time of their last moult;" and so can the larvæ of true insects, but not the mature insect. Salamanders correspond in development with the tadpoles or larvæ of the tailless Batrachians, and both possess to a large extent the power of regrowth; but not so the mature tailless Batrachians.

Absorption often plays an important part in the repairs of injuries. When a bone is broken, and does not unite, the ends are absorbed and rounded, so that a false joint is formed; or if the ends unite, but overlap, the projecting parts are removed.[7] But absorption comes into action, as Virchow remarks, during the normal growth of bones; parts which are solid during youth become hollowed out for the medullary tissue as the bone increases in size. In trying to understand the many well-adapted cases of regrowth when aided by absorption, we should remember that most parts of the organisation, even whilst retaining the same form, undergo constant renewal; so that a part which was not renewed would naturally be liable to complete absorption.

Some cases, usually classed under the so-called *nisus formativus*, at first appear to come under a distinct head; for not only are old structures reproduced, but structures which appear new are formed. Thus, after inflammation " false membranes," furnished with blood-vessels, lymphatics, and nerves, are developed; or a fœtus escapes from the Fallopian tubes, and falls into the abdomen, "nature pours out a quantity of plastic " lymph, which forms itself into organised membrane, richly supplied with " blood-vessels," and the fœtus is nourished for a time. In certain cases of

[3] Carpenter's 'Principles of Comp. Physiology,' 1854, p. 479.

[4] Charlesworth's ' Mag. of Nat. Hist.,' vol. i., 1837, p. 145.

[5] Paget, ' Lectures on Surgical Pathology,' vol. i. p. 239.

[6] Quoted by Carpenter, 'Comp. Phys., p. 479.   [7] Paget, ' Lectures,' &c., p. 257.

hydrocephalus the open and dangerous spaces in the skull are filled up with new bones, which interlock by perfect serrated sutures.[8] But most physiologists, especially on the Continent, have now given up the belief in plastic lymph or blastema, and Virchow[9] maintains that every structure, new or old, is formed by the proliferation of pre-existing cells. On this view false membranes, like cancerous or other tumours, are merely abnormal developments of normal growths; and we can thus understand how it is that they resemble adjoining structures; for instance, that " false membrane in the " serous cavities acquires a covering of epithelium exactly like that which " covers the original serous membrane; adhesions of the iris may become " black apparently from the production of pigment-cells like those of the " uvea."[10]

No doubt the power of reparation, though not always quite perfect, is an admirable provision, ready for various emergencies, even for those which occur only at long intervals of time.[11] Yet this power is not more wonderful than the growth and development of every single creature, more especially of those which are propagated by fissiparous generation. This subject has been here noticed, because we may infer that, when any part or organ is either greatly increased in size or wholly suppressed through variation and continued selection, the co-ordinating power of the organisation will continually tend to bring all the parts again into harmony with each other.

On the Effects of the Increased Use and Disuse of Organs.

It is notorious, and we shall immediately adduce proofs, that increased use or action strengthens muscles, glands, sense-organs, &c.; and that disuse, on the other hand, weakens them. I have not met with any clear explanation of this fact in works on Physiology. Mr. Herbert Spencer[12] maintains that when muscles are much used, or when intermittent pressure is applied to the epidermis, an excess of nutritive matter exudes from the vessels, and that this gives additional development to the adjoining parts. That an increased flow of blood towards an organ leads to its greater development is probable, if not certain. Mr. Paget[13] thus accounts for the long, thick, and dark-coloured hair which occasionally grows, even in young children, near oldstanding inflamed surfaces or fractured bones. When Hunter

[8] These cases are given by Blumenbach in his 'Essay on Generation,' pp. 52, 54.

[9] 'Cellular Pathology,' trans. by Dr. Chance, 1860, pp. 27, 441.

[10] Paget, 'Lectures on Pathology,'

vol. i., 1853, p. 357.

[11] Paget, idem, p. 150.

[12] 'The Principles of Biology,' vol. ii., 1866, chap. 3-5.

[13] 'Lectures on Pathology,' 1853, vol. i. p. 71.

inserted the spur of a cock into the comb, which is well supplied
with blood-vessels, it grew in one case in a spiral direction to a
length of six inches, and in another case forward, like a horn, so
that the bird could not touch the ground with its beak.   But
whether Mr. Herbert Spencer's view of the exudation of nutritive
matter due to increased movement and pressure, will fully
account for the augmented size of bones, ligaments, and especially of internal glands and nerves, seems doubtful.   According to the interesting observations of M. Sedillot,[14] when
a portion of one bone of the leg or fore-arm of an animal is
removed and is not replaced by growth, the associated bone
enlarges till it attains a bulk equal to that of the two bones, of
which it has to perform the functions.   This is best exhibited
in dogs in which the tibia has been removed; the companion
bone, which is naturally almost filiform and not one-fifth the size
of the other, soon acquires a size equal to or greater than the
tibia.   Now, it is at first difficult to believe that increased weight
acting on a straight bone could, by alternately increased and
diminished pressure, cause nutritive matter to exude from the
vessels which permeate the periosteum.   Nevertheless, the observations adduced by Mr. Spencer,[15] on the strengthening of the
bowed bones of rickety children, along their concave sides, leads
to the belief that this is possible.

Mr. H. Spencer has also shown that the ascent of the sap in
trees is aided by the rocking movement caused by the wind;
and the sap strengthens the trunk " in proportion to the stress
" to be borne; since the more severe and the more repeated the
" strains, the greater must be the exudation from the vessels
" into the surrounding tissue, and the greater the thickening of
" this tissue by secondary deposits." [16]   But woody trunks may
be formed of hard tissue without their having been subjected
to any movement, as we see with ivy closely attached to old
walls.   In all these cases, it is very difficult to disentangle the
effects of long-continued selection from those consequent on
the increased action or movement of the part.   Mr. H. Spencer [17]
acknowledges this difficulty, and gives as an instance the spines

[14] ' Comptes Rendus,' Sept. 26th, 1864, p. 539.
[15] ' The Principles of Biology,' vol. ii. p. 243.
[16] Idem, vol. ii. p. 269.   [17] Idem, vol. ii. p. 273.

or thorns of trees, and the shells of nuts. Here we have extremely hard woody tissue without the possibility of any movement to cause exudation, and without, as far as we can see, any
other directly exciting cause; and as the hardness of these
parts is of manifest service to the plant, we may look at the
result as probably due to the selection of so-called spontaneous
variations. Every one knows that hard work thickens the
epidermis on the hands; and when we hear that with infants
long before their birth the epidermis is thicker on the palms and
soles of the feet than on any other part of the body, as was observed with admiration by Albinus,[18] we are naturally inclined
to attribute this to the inherited effects of long-continued use
or pressure. We are tempted to extend the same view even to
the hoofs of quadrupeds; but who will pretend to determine
how far natural selection may have aided in the formation of
structures of such obvious importance to the animal?

That use strengthens the muscles may be seen in the limbs of artisans
who follow different trades; and when a muscle is strengthened, the tendons,
and the crests of bone to which they are attached, become enlarged; and
this must likewise be the case with the blood-vessels and nerves. On
the other hand, when a limb is not used, as by Eastern fanatics, or when
the nerve supplying it with nervous power is effectually destroyed, the
muscles wither. So again, when the eye is destroyed the optic nerve becomes atrophied, sometimes even in the course of a few months.[19] The
Proteus is furnished with branchiæ as well as with lungs: and Schreibers[20]
found than when the animal was compelled to live in deep water the
branchiæ were developed to thrice their ordinary size, and the lungs were
partially atrophied. When, on the other hand, the animal was compelled to
live in shallow water, the lungs became larger and more vascular, whilst
the branchiæ disappeared in a more or less complete degree. Such modifications as these are, however, of comparatively little value for us, as we do
not actually know that they tend to be inherited.

In many cases there is reason to believe that the lessened use of various
organs has affected the corresponding parts in the offspring. But there
is no good evidence that this ever follows in the course of a single generation. It appears, as in the case of general or indefinite variability, that
several generations must be subjected to changed habits for any appreciable result. Our domestic fowls, ducks, and geese have almost lost, not

---

[18] Paget, 'Lectures on Pathology,'
vol. ii. p. 209.

[19] Müller's 'Phys.,' Eng. translat.,
pp. 54, 791. Prof. Reed has given
('Physiological and Anat. Researches,'

p. 10) a curious account of the atrophy
of the limbs of rabbits after the destruction of the nerve.

[20] Quoted by Lecoq, in 'Geograph.
Bot.,' tom. i., 1854, p. 182.

only in the individual but in the race, their power of flight; for we do not see a chicken, when frightened, take flight like a young pheasant. Hence I was led carefully to compare the limb-bones of fowls, ducks, pigeons, and rabbits, with the same bones in the wild parent-species. As the measurements and weights were fully given in the earlier chapters, I need here only recapitulate the results. With domestic pigeons, the length of the sternum, the prominence of its crest, the length of the scapulæ and furcula, the length of the wings as measured from tip to tip of the radius, are all reduced relatively to the same parts in the wild pigeon. The wing and tail feathers, however, are increased in length, but this may have as little connection with the use of the wings or tail, as the lengthened hair on a dog with the amount of exercise which the breed has habitually taken. The feet of pigeons, except in the long-beaked races, are reduced in size. With fowls the crest of the sternum is less prominent, and is often distorted or monstrous; the wing-bones have become lighter relatively to the leg-bones, and are apparently a little shorter in comparison with those of the parent-form, the *Gallus bankiva*. With ducks, the crest of the sternum is affected in the same manner as in the foregoing cases: the furcula, coracoids, and scapulæ are all reduced in weight relatively to the whole skeleton: the bones of the wings are shorter and lighter, and the bones of the legs longer and heavier, relatively to each other, and relatively to the whole skeleton, in comparison with the same bones in the wild-duck. The decreased weight and size of the bones, in the foregoing cases, is probably the indirect result of the reaction of the weakened muscles on the bones. I failed to compare the feathers of the wings of the tame and wild duck; but Gloger[21] asserts that in the wild duck the tips of the wing-feathers reach almost to the end of the tail, whilst in the domestic duck they often hardly reach to its base. He remarks, also, on the greater thickness of the legs, and says that the swimming membrane between the toes is reduced; but I was not able to detect this latter difference.

With the domesticated rabbit the body, together with the whole skeleton, is generally larger and heavier than in the wild animal, and the leg-bones are heavier in due proportion; but whatever standard of comparison be taken, neither the leg-bones nor the scapulæ have increased in length proportionally with the increased dimensions of the rest of the skeleton. The skull has become in a marked manner narrower, and, from the measurements of its capacity formerly given, we may conclude, that this narrowness results from the decreased size of the brain, consequent on the mentally inactive life led by these closely-confined animals.

We have seen in the eighth chapter that silk-moths, which have been kept during many centuries closely confined, emerge from their cocoons with their wings distorted, incapable of flight, often greatly reduced in size, or even, according to Quatrefages, quite rudimentary. This condition of the wings may be largely owing to the same kind of monstrosity which often affects wild Lepidoptera when artificially reared from the cocoon; or it may

---

[21] 'Das Abändern der Vögel,' 1833, s. 74.

be in part due to an inherent tendency, which is common to the females of many Bombycidæ, to have their wings in a more or less rudimentary state; but part of the effect may probably be attributed to long-continued disuse.

From the foregoing facts there can be no doubt that certain parts of the skeleton in our anciently domesticated animals, have been modified in length and weight by the effects of decreased or increased use; but they have not been modified, as shown in the earlier chapters, in shape or structure. We must, however, be cautious in extending this latter conclusion to animals living a free life; for these will occasionally be exposed during successive generations to the severest competition. With wild animals it would be an advantage in the struggle for life that every superfluous and useless detail of structure should be removed or absorbed; and thus the reduced bones might ultimately become changed in structure. With highly-fed domesticated animals, on the other hand, there is no economy of growth, nor any tendency to the elimination of trifling and superfluous details of structure.

Turning now to more general observations, Nathusius has shown that, with the improved races of the pig, the shortened legs and snout, the form of the articular condyles of the occiput, and the position of the jaws with the upper canine teeth projecting in a most anomalous manner in front of the lower canines, may be attributed to these parts not having been fully exercised. For the highly-cultivated races do not travel in search of food, nor root up the ground with their ringed muzzles. These modifications of structure, which are all strictly inherited, characterise several improved breeds, so that they cannot have been derived from any single domestic or wild stock.[22] With respect to cattle, Professor Tanner has remarked that the lungs and liver in the improved breeds " are found to be considerably " reduced in size when compared with those possessed by animals " having perfect liberty;"[23] and the reduction of these organs affects the general shape of the body. The cause of the reduced lungs in highly-bred animals which take little exercise is ob-

[22] Nathusius, 'Die Racen des Schweines,' 1860, s. 53, 57; 'Vorstudien .... Schweineschædel,' 1864, s. 103, 130, 133.

[23] 'Journal of Agriculture of Highland Soc.,' July, 1860, p. 321.

vious; and perhaps the liver may be affected by the nutritious and artificial food on which they largely subsist.

It is well known that, when an artery is tied, the anastomising branches, from being forced to transmit more blood, increase in diameter; and this increase cannot be accounted for by mere extension, as their coats gain in strength. Mr. Herbert Spencer[24] has argued that with plants the flow of sap from the point of supply to the growing part first elongates the cells in this line; and that the cells then become confluent, thus forming the ducts; so that, on this view, the vessels in plants are formed by the mutual reaction of the flowing sap and cellular tissue. Dr. W. Turner has remarked,[25] with respect to the branches of arteries, and likewise to a certain extent with nerves, that the great principle of compensation frequently comes into play; for "when two nerves pass to adjacent cutaneous areas, " an inverse relation as regards size may subsist between them; a deficiency " in one may be supplied by an increase in the other, and thus the area of " the former may be trespassed on by the latter nerve." But how far in these cases the difference in size in the nerves and arteries is due to original variation, and how far to increased use or action, is not clear.

In reference to glands, Mr. Paget observes that "when one kidney is " destroyed the other often becomes much larger, and does double work."[26] If we compare the size of the udders and their power of secretion in cows which have been long domesticated, and in certain goats in which the udders nearly touch the ground, with the size and power of secretion of these organs in wild or half-domesticated animals, the difference is great. A good cow with us daily yields more than five gallons, or forty pints of milk, whilst a first-rate animal, kept, for instance, by the Damaras of South Africa,[27] " rarely gives more than two or three pints of milk daily, and, should her " calf be taken from her, she absolutely refuses to give any." We may attribute the excellence of our cows, and of certain goats, partly to the continued selection of the best milking animals, and partly to the inherited effects of the increased action, through man's art, of the secreting glands.

It is notorious, as was remarked in the twelfth chapter, that short-sight is inherited; and if we compare watchmakers or engravers with, for instance, sailors, we can hardly doubt that vision continually directed towards a near object permanently affects the structure of the eye.

Veterinarians are unanimous that horses become affected with spavins, splints, ringbones, &c., from being shod, and from travelling on hard roads, and they are almost equally unanimous that these injuries are transmitted. Formerly horses were not shod in North Carolina, and it has been asserted that they did not then suffer from these diseases of the legs and feet.[28]

---

[24] 'Principles of Biology,' vol. ii. p. 263.

[25] 'Natural History Review,' vol. iv., Oct. 1864, p. 617.

[26] 'Lectures on Surgical Pathology,' 1853, vol. i. p. 27.

[27] Andersson, 'Travels in South Africa,' p. 318. For analogous cases in South America, see Aug. St. Hilaire, 'Voyage dans le Province de Goyaz,' tom. i. p. 71.

[28] Birckell's 'Nat. Hist. of North Carolina,' 1739, p. 53.

Our domesticated quadrupeds are all descended, as far as is known, from species having erect ears; yet few kinds can be named, of which at least one race has not drooping ears. Cats in China, horses in parts of Russia, sheep in Italy and elsewhere, the guinea-pig in Germany, goats and cattle in India, rabbits, pigs, and dogs in all long-civilised countries, have dependent ears. With wild animals, which constantly use their ears like funnels to catch every passing sound, and especially to ascertain the direction whence it comes, there is not, as Mr. Blyth has remarked, any species with drooping ears except the elephant. Hence the incapacity to erect the ears is certainly in some manner the result of domestication; and this incapacity has been attributed by various authors[29] to disuse, for animals protected by man are not compelled habitually to use their ears. Col. Hamilton Smith[30] states that in ancient effigies of the dog, " with the exception of one Egyptian instance, " no sculpture of the earlier Grecian era produces representations " of hounds with completely drooping ears; those with them half " pendulous are missing in the most ancient; and this character " increases, by degrees, in the works of the Roman period." Godron also has remarked that " the pigs of the ancient Egyptians " had not their ears enlarged and pendent."[31] But it is remarkable that the drooping of the ears, though probably the effect of disuse, is not accompanied by any decrease in size; on the contrary, when we remember that animals so different as fancy rabbits, certain Indian breeds of the goat, our petted spaniels, bloodhounds, and other dogs, have enormously elongated ears, it would appear as if disuse actually caused an increase in length. With rabbits, the drooping of the much elongated ears has affected even the structure of the skull.

The tail of no wild animal, as remarked to me by Mr. Blyth, is curled; whereas pigs and some races of dogs have their tails much curled. This deformity, therefore, appears to be the result of domestication, but whether in any way connected with the lessened use of the tail is doubtful.

[29] Livingstone, quoted by Youatt on Sheep, p. 142. Hodgson, in ' Journal of Asiatic Soc. of Bengal,' vol. xvi., 1847, p. 1006. &c. &c.

[30] ' Naturalist Library,' Dogs, vol. ii. 1840, p. 104.

[31] ' De l'Espèce,' tom. i., 1859, p. 367.

The epidermis on our hands is easily thickened, as every one knows, by hard work. In a district of Ceylon the sheep have " horny callosities that defend their knees, and which arise from " their habit of kneeling down to crop the short herbage, and " this distinguishes the Jaffna flocks from those of other portions " of the island;" but it is not stated whether this peculiarity is inherited.[32]

The mucous membrane which lines the stomach is continuous with the external skin of the body; therefore it is not surprising that its texture should be affected by the nature of the food consumed, but other and more interesting changes likewise follow. Hunter long ago observed that the muscular coat of the stomach of a gull (*Larus tridactylus*) which had been fed for a year chiefly on grain was thickened; and, according to Dr. Edmondston, a similar change periodically occurs in the Shetland Islands in the stomach of the *Larus argentatus*, which in the spring frequents the corn-fields and feeds on the seed. The same careful observer has noticed a great change in the stomach of a raven which had been long fed on vegetable food. In the case of an owl (*Strix grallaria*) similarly treated, Menetries states that the form of the stomach was changed, the inner coat became leathery, and the liver increased in size. Whether these modifications in the digestive organs would in the course of generations become inherited is not known.[33]

The increased or diminished length of the intestines, which apparently results from changed diet, is a more remarkable case, because it is characteristic of certain animals in their domesticated condition, and therefore must be inherited. The complex absorbent system, the blood-vessels, nerves, and muscles, are necessarily all modified together with the intestines. According to Daubenton, the intestines of the domestic cat are one-third longer than those of the wild cat of Europe; and although this species is not the parent-stock of the domestic animal, yet, as Isidore Geoffroy has remarked, the several species

[32] 'Ceylon,' by Sir J. E. Tennent, 1859, vol. ii. p. 531.
[33] For the foregoing statements, *see* Hunter's 'Essays and Observations,' 1861, vol. ii. p. 329; Dr. Edmondston, as quoted in Macgillivray's 'British Birds,' vol. v. p. 550; Menetries, as quoted in Bronn's ' Geschichte der Natur,' B. ii. s. 110.

of cats are so closely allied that the comparison is probably a fair one. The increased length appears to be due to the domestic cat being less strictly carnivorous in its diet than any wild feline species; I have seen a French kitten eating vegetables as readily as meat. According to Cuvier, the intestines of the domesticated pig exceed greatly in proportionate length those of the wild boar. In the tame and wild rabbit the change is of an opposite nature, and probably results from the nutritious food given to the tame rabbit.[34]

Changed Habits of Life, independently of the Use or Disuse of particular Organs.—This subject, as far as the mental powers of animals are concerned, so blends into instinct, on which I shall treat in a future work, that I will here only remind the reader of the many cases which occur under domestication, and which are familiar to every one—for instance the tameness of our animals—the pointing or retrieving of dogs—their not attacking the smaller animals kept by man—and so forth. How much of these changes ought to be attributed to inherited habit, and how much to the selection of individuals which have varied in the desired manner, irrespectively of the special circumstances under which they have been kept, can seldom be told. We have already seen that animals may be habituated to a changed diet; but a few additional instances may here be given.

In the Polynesian Islands and in China the dog is fed exclusively on vegetable matter, and the taste for this kind of food is to a certain extent inherited.[35] Our sporting dogs will not touch the bones of game birds, whilst other dogs devour them with greediness. In some parts of the world sheep have been largely fed on fish. The domestic hog is fond of barley, the wild boar is said to disdain it; and the disdain is partially inherited, for some young wild pigs bred in captivity showed an aversion for this grain, whilst others of the same brood relished it.[36] One of my relations bred some young pigs from

[34] These statements on the intestines are taken from Isidore Geoffroy St. Hilaire, 'Hist. Nat. Gén.,' tom. iii. pp. 427, 441.

[35] Gilbert White, 'Nat. Hist. Sel-

bourne,' 1825, vol. ii. p. 121.

[36] Burdach, 'Traité de Phys.,' tom. ii. p. 267, as quoted by Dr. P. Lucas, 'L'Héréd. Nat.,' tom. i. p. 388.

a Chinese sow by a wild Alpine boar; they lived free in the
park, and were so tame that they came to the house to be
fed; but they would not touch swill, which was devoured by
the other pigs.   An animal when once accustomed to an unnatural diet, which can generally be effected only during youth,
dislikes its proper food, as Spallanzani found to be the case
with a pigeon which had been long fed on meat.   Individuals
of the same species take to new food with different degrees of
readiness; one horse, it is stated, soon learned to eat meat,
whilst another would have perished from hunger rather than
have partaken of it.[37]

The caterpillars of the *Bombyx hesperus* feed in a state of
nature on the leaves of the *Café diable*, but, after having been
reared on the Ailanthus, they would not touch the *Café diable*,
and actually died of hunger.[38]

It has been found possible to accustom marine fish to live
in fresh water; but as such changes in fish, and other marine
animals, have been chiefly observed in a state of nature, they
do not properly belong to our present subject.   The period
of gestation and of maturity, as shown in the earlier chapters,—
the season and the frequency of the act of breeding,—have all
been greatly modified under domestication.   With the Egyptian
goose the rate of change in the season has been recorded.[39]
The wild drake pairs with one female, the domestic drake is
polygamous.   Certain breeds of fowls have lost the habit of
incubation.   The paces of the horse, and the manner of flight
in certain breeds of the pigeon, have been modified, and are
inherited.   The voice differs much in certain fowls and pigeons.
Some breeds are clamorous and others silent, as in the Call and
common duck, or in the Spitz and pointer dog.   Every one
knows how dogs differ from each other in their manner of
hunting, and in their ardour after different kinds of game or
vermin.

With plants the period of vegetation is easily changed and
is inherited, as in the case of summer and winter wheat, barley,

[37] This and several other cases are
given by Colin, 'Physiologie Comp. des
Animaux Dom.,' 1854, tom. i. p. 426.

[38] M. Michely de Cayenne, in 'Bull.

Soc. d'Acclimat,' tom. viii., 1861, p.
563.

[39] Quatrefages, 'Unité de l'Espèce
Humaine,' 1861, p. 79.

and vetches; but to this subject we shall immediately return under acclimatisation. Annual plants sometimes become perennial under a new climate, as I hear from Dr. Hooker is the case with the stock and mignonette in Tasmania. On the other hand, perennials sometimes become annuals, as with the Ricinus in England, and as, according to Captain Mangles, with many varieties of the heartsease. Von Berg[40] raised from seed of *Verbascum phœnicium*, which is usually a biennial, both annual and perennial varieties. Some deciduous bushes become evergreen in hot countries.[41] Rice requires much water, but there is one variety in India which can be grown without irrigation.[42] Certain varieties of the oat and of our other cereals are best fitted for certain soils.[43] Endless similar facts could be given in the animal and vegetable kingdoms. They are noticed here because they illustrate analogous differences in closely allied natural species, and because such changed habits of life, whether due to use and disuse, or to the direct action of external conditions, or to so-called spontaneous variation, would be apt to lead to modifications of structure.

Acclimatisation.—From the previous remarks we are naturally led to the much disputed subject of acclimatisation. There are two distinct questions: Do varieties descended from the same species differ in their power of living under different climates? And secondly, if they so differ, how have they become thus adapted? We have seen that European dogs do not succeed well in India, and it is asserted,[44] that no one has succeeded in there keeping the Newfoundland long alive; but then it may be argued, probably with truth, that these northern breeds are specifically distinct from the native dogs which flourish in India. The same remark may be made with respect to different breeds of sheep, of which, according to Youatt,[45] not one brought "from " a torrid climate lasts out the second year," in the Zoological Gardens. But sheep are capable of some degree of acclimatisation, for Merino sheep bred at the Cape of Good Hope have been found

[40] 'Flora,' 1835, B. ii. p. 504.
[41] Alph. De Candolle, 'Géograph. Bot.,' tom. ii. p. 1078.
[42] Royle, 'Illustrations of the Botany of the Himalaya,' p. 19.
[43] 'Gardener's Chronicle,' 1850, pp. 204, 219.
[44] Rev. R. Everest, 'Journal As. Soc. of Bengal,' vol. iii. p. 19.
[45] Youatt on Sheep, 1838, p. 491.

far better adapted for India than those imported from England.[46]
It is almost certain that the breeds of the fowl are descended
from the same species; but the Spanish breed, which there is
good reason to believe originated near the Mediterranean,[47]
though so fine and vigorous in England, suffers more from frost
than any other breed. The Arrindy silk-moth introduced from
Bengal, and the Ailanthus moth from the temperate province of
Shan Tung, in China, belong to the same species, as we may
infer from their identity in the caterpillar, cocoon, and mature
states;[48] yet they differ much in constitution: the Indian form
" will flourish only in warm latitudes," the other is quite hardy
and withstands cold and rain.

Plants are more strictly adapted to climate than are animals. The latter
when domesticated withstand such great diversities of climate, that we
find nearly the same species in tropical and temperate countries; whilst
the cultivated plants are widely dissimilar. Hence a larger field is open
for inquiry in regard to the acclimatisation of plants than of animals. It
is no exaggeration to say that with almost every plant which has long
been cultivated varieties exist, which are endowed with constitutions fitted
for very different climates; I will select only a few of the more striking
cases, as it would be tedious to give all. In North America numerous
fruit-trees have been raised, and in horticultural publications,—for instance,
in Downing,—lists are given of the varieties which are best able to withstand the severe climate of the northern States and Canada. Many American
varieties of the pear, plum, and peach are excellent in their own country,
but until recently hardly one was known that succeeded in England; and
with apples,[49] not one succeeds. Though the American varieties can withstand a severer winter than ours, the summer here is not hot enough.
Fruit-trees have originated in Europe as in America with different constitutions, but they are not here much noticed, as the same nurserymen do
not supply a wide area. The Forelle pear flowers early, and when the
flowers have just set, and this is the critical period, they have been
observed, both in France and England, to withstand with complete impunity a frost of 18° and even 14° Fahr. ,which killed the flowers, whether fully
expanded or in bud, of all other kinds of pears.[50] This power in the flower
of resisting cold and afterwards producing fruit does not invariably depend, as we know on good authority,[51] on general constitutional vigour.

---

[46] Royle, ' Prod. Resources of India,'
p. 153.

[47] Tegetmeier, ' Poultry Book,' 1866,
p. 102.

[48] Dr. R. Paterson, in a paper communicated to Bot. Soc. of Canada,
quoted in the ' Reader,' 1863. Nov. 13th.

[49] See remarks by Editor in ' Gard.
Chronicle,' 1848, p. 5.

[50] ' Gard. Chronicle,' 1860, p. 938.
Remarks by Editor and quotation from
Decaisne.

[51] J. de Jonghe, of Brussels, in
' Gard. Chronicle,' 1857, p. 612.

In proceeding northward, the number of varieties which are enabled to resist the climate rapidly decreases, as may be seen in the list of the varieties of the cherry, apple, and pear, which can be cultivated in the neighbourhood of Stockholm.[52] Near Moscow, Prince Troubetzkoy planted for experiment in the open ground several varieties of the pear, but one alone, the *Poire sans Pepins*, withstood the cold of winter.[53] We thus see that our fruit-trees, like distinct species of the same genus, certainly differ from each other in their constitutional adaptation to different climates.

With the varieties of many plants, the adaptation to climate is often very close. Thus it has been proved by repeated trials "that few if any of the " English varieties of wheat are adapted for cultivation in Scotland;"[54] but the failure in this case is at first only in the quantity, though ultimately in the quality, of the grain produced. The Rev. J. M. Berkeley sowed wheat-seed from India, and got " the most meagre ears," on land which would certainly have yielded a good crop from English wheat.[55] In these cases varieties have been carried from a warmer to a cooler climate; in the reverse case, as "when wheat was imported directly from France into the " West Indian Islands, it produced either wholly barren spikes or fur-" nished with only two or three miserable seeds, while West Indian seed " by its side yielded an enormous harvest."[56] Here is another case of close adaptation to a slightly cooler climate; a kind of wheat which in England may be used indifferently either as a winter or summer variety, when sown under the warmer climate of Grignan, in France, behaved exactly as if it had been a true winter wheat.[57]

Botanists believe that all the varieties of maize belong to the same species; and we have seen that in North America, in proceeding north-ward, the varieties cultivated in each zone produce their flowers and ripen their seed within shorter and shorter periods. So that the tall, slowly maturing southern varieties do not succeed in New England, and the New English varieties do not succeed in Canada. I have not met with any statement that the southern varieties are actually injured or killed by a degree of cold which the northern varieties withstand with impunity, though this is probable; but the production of early flowering and early seeding varieties deserves to be considered as one form of acclimatisation. Hence it has been found possible, according to Kalm, to cultivate maize further and further northwards in America. In Europe, also, as we learn from the evidence given by Alph. De Candolle, the culture of maize has extended since the end of the last century thirty leagues north of its former boundary.[58] On the authority of the great Linnæus,[59] I may quote an

---

[52] Ch. Martius, ' Voyage Bot. Côtes Sept. de la Norvège,' p. 26.

[53] 'Journal de l'Acad. Hort. de Gand,' quoted in ' Gard. Chron.,' 1859, p. 7.

[54] 'Gard. Chronicle,' 1851, p. 396.

[55] Idem., 1862, p. 235.

[56] On the authority of Labat, quoted in ' Gard. Chron.,' 1862, p. 235.

[57] MM. Edwards and Colin, ' Annal. des Sc. Nat.,' 2nd series, Bot., tom. v. p. 22.

[58] ' Géograph. Bot.,' p. 337.

[59] 'Swedish Acts,' Eng. translat., 1739-40, vol. i. Kalm, in his ' Travels,' vol. ii. p. 166, gives an analogous case with cotton-plants raised in New Jersey from Carolina seed.

analogous case, namely, that in Sweden tobacco raised from home-grown seed ripens its seed a month sooner and is less liable to miscarry than plants raised from foreign seed.

With the Vine, differently from the maize, the line of practical culture has retreated a little southward since the middle ages; [60] but this seems due to commerce, including that of wine, being now freer or more easy. Nevertheless the fact of the vine not having spread northward shows that acclimatisation has made no progress during several centuries. There is, however, a marked difference in the constitution of the several varieties,—some being hardy, whilst others, like the muscat of Alexandria, require a very high temperature to come to perfection. According to Labat,[61] vines taken from France to the West Indies succeed with extreme difficulty, whilst those imported from Madeira, or the Canary Islands, thrive admirably.

Gallesio gives a curious account of the naturalisation of the Orange in Italy. During many centuries the sweet orange was propagated exclusively by grafts, and so often suffered from frosts that it required protection. After the severe frost of 1709, and more especially after that of 1763, so many trees were destroyed that seedlings from the sweet orange were raised, and, to the surprise of the inhabitants, their fruit was found to be sweet. The trees thus raised were larger, more productive, and hardier than the former kinds; and seedlings are now continually raised. Hence Gallesio concludes that much more was effected for the naturalisation of the orange in Italy by the accidental production of new kinds during a period of about sixty years, than had been effected by grafting old varieties during many ages.[62] I may add that Risso[63] describes some Portuguese varieties of the orange as extremely sensitive to cold, and as much tenderer than certain other varieties.

The peach was known to Theophrastus, 322 B.C.[64] According to the authorities quoted by Dr. F. Rolle,[65] it was tender when first introduced into Greece, and even in the island of Rhodes only occasionally bore fruit. If this be correct, the peach, in spreading during the last two thousand years over the middle parts of Europe, must have become much hardier. At the present day different varieties differ much in hardiness: some French varieties will not succeed in England; and near Paris, the *Pavie de Bonneuil* does not ripen its fruit till very late, even when grown on a wall; " it is, therefore, only fit for a very hot southern climate." [66]

I will briefly give a few other cases. A variety of *Magnolia grandiflora*, raised by M. Roy, withstands cold several degrees lower than that which any other variety can resist. With camellias there is much difference in hardiness. One particular variety of Noisette rose withstood the severe frost of 1860 " untouched and hale amidst a universal destruction of other

[60] De Candolle, 'Géograph. Bot.,' p. 339.

[61] 'Gard. Chronicle,' 1862, p. 235.

[62] Gallesio, 'Teoria della Riproduzione Veg.,' 1816, p. 125; and 'Traité du Citrus,' 1811, p. 359.

[63] 'Essai sur l'Hist. des Orangers,' 1813, p. 20, &c.

[64] Alph. De Candolle, 'Géograph. Bot.,' p. 882.

[65] 'Ch. Darwin's Lehre von der Entstehung,' &c., 1862, s. 87.

[66] Decaisne, quoted in 'Gard. Chronicle,' 1865, p. 271.

"Noisettes." In New York the "Irish yew is quite hardy, but the com-
"mon yew is liable to be cut down." I may add that there are varieties
of the sweet potato (*Convolvulus batatas*) which are suited for warmer,
as well as for colder, climates.[67]

The plants as yet mentioned have been found capable of
resisting an unusual degree of cold or heat, when fully grown.
The following cases refer to plants whilst young. In a large
bed of young Araucarias of the same age, growing close together and equally exposed, it was observed,[68] after the unusually
severe winter of 1860-61, that, "in the midst of the dying,
"numerous individuals remained on which the frost had abso-
"lutely made no kind of impression." Dr. Lindley, after
alluding to this and other similar cases, remarks, "Among
"the lessons which the late formidable winter has taught us,
"is that, even in their power of resisting cold, individuals of
"the same species of plants are remarkably different." Near
Salisbury, there was a sharp frost on the night of May 24th,
1836, and all the French beans (*Phaseolus vulgaris*) in a bed
were killed except about one in thirty, which completely
escaped.[69] On the same day of the month, but in the year
1864, there was a severe frost in Kent, and two rows of scarlet-
runners (*P. multiflorus*) in my garden, containing 390 plants of
the same age and equally exposed, were all blackened and killed
except about a dozen plants. In an adjoining row of "Fulmer's
dwarf bean" (*P. vulgaris*), one single plant escaped. A still
more severe frost occurred four days afterwards, and of the dozen
plants which had previously escaped only three survived; these
were not taller or more vigorous than the other young plants,
but they escaped completely, with not even the tips of their
leaves browned. It was impossible to behold these three plants,
with their blackened, withered, and dead brethren all round
them, and not see at a glance that they differed widely in constitutional power of resisting frost.

This work is not the proper place to show that wild plants

[67] For the magnolia, *see* Loudon's
'Gard. Mag.,' vol. xiii., 1837, p. 21.
For camellias and roses, *see* 'Gard.
Chron.,' 1860, p. 384. For the yew,
'Journal of Hort.,' March 3rd, 1863,
p. 174. For sweet potatoes, *see* Col.

von Siebold, in 'Gard. Chron.,' 1855,
p. 822.

[68] The Editor, 'Gard. Chron.,' 1861,
p. 239.

[69] Loudon's 'Gard. Mag.,' vol. xii.,
1836, p. 378.

of the same species, naturally growing at different altitudes
or under different latitudes, become to a certain extent acclimatised, as is proved by the different behaviour of their seedlings when raised in England. In my 'Origin of Species'
I have alluded to some cases, and I could add others. One
instance must suffice: Mr. Grigor, of Forres,[70] states that
seedlings of the Scotch fir (*Pinus sylvestris*), raised from seed
from the Continent and from the forests of Scotland, differ
much. "The difference is perceptible in one-year-old, and more
" so in two-year-old seedlings; but the effects of the winter on
" the second year's growth almost uniformly makes those from
" the Continent quite brown, and so damaged, that by the month
" of March they are quite unsaleable, while the plants from the
" native Scotch pine, under the same treatment, and standing
" alongside, although considerably shorter, are rather stouter and
" quite green, so that the beds of the one can be known from the
" other when seen from the distance of a mile." Closely similar
facts have been observed with seedling larches.

Hardy varieties would alone be valued or noticed in Europe; whilst
tender varieties, requiring more warmth, would generally be neglected;
but such occasionally arise. Thus Loudon [71] describes a Cornish variety
of the elm which is almost an evergreen, and of which the shoots are
often killed by the autumnal frosts, so that its timber is of little value.
Horticulturists know that some varieties are much more tender than
others: thus all the varieties of the broccoli are more tender than cabbages; but there is much difference in this respect in the sub-varieties of
the broccoli; the pink and purple kinds are a little hardier than the white
Cape broccoli, " but they are not to be depended on after the thermometer
" falls below 24° Fahr.:" the Walcheren broccoli is less tender than the
Cape, and there are several varieties which will stand much severer cold
than the Walcheren.[72] Cauliflowers seed more freely in India than cabbages.[73] To give one instance with flowers: eleven plants raised from a
hollyhock, called the *Queen of the Whites*,[74] were found to be much more
tender than various other seedlings. It may be presumed that all tender
varieties would succeed better under a climate warmer than ours. With
fruit-trees, it is well known that certain varieties, for instance of the
peach, stand forcing in a hot-house better than others; and this shows

---

[70] 'Gardener's Chron.,' 1865, p. 699.

[71] 'Arboretum et Fruticetum,' vol.
iii. p. 1376.

[72] Mr. Robson, in ' Journal of Horticulture,' 1861, p. 23.

[73] Dr. Bonavia, 'Report of the Agri.-
Hort. Soc. of Oudh,' 1866.

[74] 'Cottage Gardener,' 1860, April,
24th, p. 57.

either pliability of organisation or some constitutional difference. The same individual cherry-tree, when forced, has been observed during successive years gradually to change its period of vegetation.[75] Few pelargoniums can resist the heat of a stove, but *Alba multiflora* will, as a most skilful gardener asserts, " stand pine-apple top and bottom heat the whole " winter, without looking any more drawn than if it had stood in a com- " mon greenhouse; and *Blanche Fleur* seems as if it had been made on " purpose for growing in winter, like many bulbs, and to rest all summer." [76] There can hardly be a doubt that the *Alba multiflora* pelargonium must have a widely different constitution from that of most other varieties of this plant; it would probably withstand even an equatorial climate.

We have seen that according to Labat the vine and wheat require acclimatisation in order to succeed in the West Indies. Similar facts have been observed at Madras: " two parcels of mignonette-seed, one direct " from Europe, the other saved at Bangalore (of which the mean tempera- " ture is much below that of Madras) were sown at the same time : they both " vegetated equally favourably, but the former all died off a few days after " they appeared above ground; the latter still survive, and are vigorous " healthy plants." So again, "turnip and carrot seed saved at Hyderabad " are found to answer better at Madras than seed from Europe or from the " Cape of Good Hope." [77] Mr. J. Scott, of the Calcutta Botanic Gardens, informs me that seeds of the sweet-pea (*Lathyrus odoratus*) imported from England produce plants, with thick, rigid stems and small leaves, which rarely blossom and never yield seed; plants raised from French seed blossom sparingly, but all the flowers are sterile; on the other hand, plants raised from sweet-peas grown near Darjeeling in Upper India, but originally derived from England, can be successfully cultivated on the plains of India; for they flower and seed profusely, and their stems are lax and scandent. In some of the foregoing cases, as Dr. Hooker has remarked to me, the greater success may perhaps be attributed to the seeds having been more fully ripened under a more favourable climate; but this view can hardly be extended to so many cases, including plants, which, from being cultivated under a climate hotter than their native one, become fitted for a still hotter climate. We may therefore safely conclude that plants can to a certain extent become accustomed to a climate either hotter or colder than their own; although these latter cases have been more frequently observed.

We will now consider the means by which acclimatisation may be effected, namely, through the spontaneous appearance of varieties having a different constitution, and through the effects of use or habit. In regard to the first process, there is no evidence that a change in the constitution of the off-

[75] 'Gardener's Chronicle,' 1841, p. 291.

[76] Mr. Beaton, in 'Cottage Gardener,' March 20th, 1860, p. 377. Queen Mab

will also stand stove heat, see ' Gard. Chronicle,' 1845, p. 226.

[77] 'Gardener's Chronicle,' 1841, p 439.

spring necessarily stands in any direct relation with the nature
of the climate inhabited by the parents. On the contrary,
it is certain that hardy and tender varieties of the same species
appear in the same country. New varieties thus spontaneously arising become fitted to slightly different climates in
two different ways; firstly, they may have the power, either as
seedlings or when full-grown, of resisting intense cold, as with
the Moscow pear, or of resisting intense heat, as with some kinds
of Pelargonium, or the flowers may withstand severe frost, as
with the Forelle pear. Secondly, plants may become adapted
to climates widely different from their own, from flowering and
fruiting either earlier or later in the season. In both these
cases the power of acclimatisation by man consists simply in the
selection and preservation of new varieties. But without any
direct intention on his part of securing a hardier variety, acclimatisation may be unconsciously effected by merely raising
tender plants from seed, and by occasionally attempting their
cultivation further and further northwards, as in the case of
maize, the orange, and the peach.

How much influence ought to be attributed to inherited habit
or custom in the acclimatisation of animals and plants is a
much more difficult question. In many cases natural selection
can hardly have failed to have come into play and complicated
the result. It is notorious that mountain sheep resist severe
weather and storms of snow which would destroy lowland
breeds; but then mountain sheep have been thus exposed from
time immemorial, and all delicate individuals will have been
destroyed, and the hardiest preserved. So with the Arrindy
silk-moths of China and India; who can tell how far natural
selection may have taken a share in the formation of the two
races, which are now fitted for such widely different climates?
It seems at first probable that the many fruit-trees, which are
so well fitted for the hot summers and cold winters of North
America, in contrast with their poor success under our climate,
have become adapted through habit; but when we reflect on
the multitude of seedlings annually raised in that country,
and that none would succeed unless born with a fitting constitution, it is possible that mere habit may have done nothing
towards their acclimatisation. On the other hand, when we

hear that Merino sheep, bred during no great number of generations at the Cape of Good Hope—that some European plants
raised during only a few generations in the cooler parts of
India, withstand the hotter parts of that country much better
than the sheep or seeds imported directly from England, we
must attribute some influence to habit. We are led to the
same conclusion when we hear from Naudin[78] that the races of
melons, squashes, and gourds, which have long been cultivated
in Northern Europe, are comparatively more precocious, and
need much less heat for maturing their fruit, than the varieties
of the same species recently brought from tropical regions. In
the reciprocal conversion of summer and winter wheat, barley,
and vetches into each other, habit produces a marked effect
in the course of a very few generations. The same thing apparently occurs with the varieties of maize, which, when carried
from the Southern to the Northern States of America, or into
Germany, soon become accustomed to their new homes. With
vine-plants taken to the West Indies from Madeira, which are
said to succeed better than plants brought directly from France,
we have some degree of acclimatisation in the individual, independently of the production of new varieties by seed.

The common experience of agriculturists is of some value,
and they often advise persons to be cautious in trying in one
country the productions of another. The ancient agricultural
writers of China recommend the preservation and cultivation
of the varieties peculiar to each country. During the classical
period, Columella wrote, " Vernaculum pecus peregrino longe
" præstantius est." [79]

I am aware that the attempt to acclimatise either animals or
plants has been called a vain chimæra. No doubt the attempt
in most cases deserves to be thus called, if made independently of the production of new varieties endowed with a different constitution. Habit, however much prolonged, rarely
produces any effect on a plant propagated by buds; it apparently acts only through successive seminal generations.

---

[78] Quoted by Asa Gray, in 'Am.
Journ. of Sci.,' 2nd series, Jan. 1865,
p. 106.

[79] For China, *see* ' Mémoire sur les

Chinois,' tom. xi., 1786, p. 60. Columella
is quoted by Carlier, in 'Journal de
Physique,' tom. xxiv. 1784.

The laurel, bay, laurestinus, &c., and the Jerusalem artichoke, which are propagated by cuttings or tubers, are probably now as tender in England as when first introduced; and this appears to be the case with the potato, which until recently was seldom multiplied by seed. With plants propagated by seed, and with animals, there will be little or no acclimatisation unless the hardier individuals are either intentionally or unconsciously preserved. The kidney-bean has often been advanced as an instance of a plant which has not become hardier since its first introduction into Britain. We hear, however, on excellent authority,[80] that some very fine seed, imported from abroad, produced plants "which blossomed most profusely, but were " nearly all but abortive, whilst plants grown alongside from " English seed podded abundantly;" and this apparently shows some degree of acclimatisation in our English plants. We have also seen that seedlings of the kidney-bean occasionally appear with a marked power of resisting frost; but no one, as far as I can hear, has ever separated such hardy seedlings, so as to prevent accidental crossing, and then gathered their seed, and repeated the process year after year. It may, however, be objected with truth that natural selection ought to have had a decided effect on the hardiness of our kidney-beans; for the tenderest individuals must have been killed during every severe spring, and the hardier preserved. But it should be borne in mind that the result of increased hardiness would simply be that gardeners, who are always anxious for as early a crop as possible, would sow their seed a few days earlier than formerly. Now, as the period of sowing depends much on the soil and elevation of each district, and varies with the season; and as new varieties have often been imported from abroad, can we feel sure that our kidney-beans are not somewhat hardier? I have not been able, by searching old horticultural works, to answer this question satisfactorily.

On the whole the facts now given show that, though habit does something towards acclimatisation, yet that the spontaneous appearance of constitutionally different individuals is a far more effective agent. As no single instance has been recorded, either with animals or plants, of hardier individuals having been long and steadily selected, though such selection is admitted to be indispensable for the improvement of any other charaćter, it is not surprising that man has done little in the acclimatisation of domesticated animals and cultivated plants. We need not, however, doubt that under nature new races and new species would become adapted to widely different climates, by spontaneous variation, aided by habit, and regulated by natural selection.

Arrests of Development: Rudimentary and Aborted Organs.

These subjects are here introduced because there is reason to believe that rudimentary organs are in many cases the result of disuse. Modifications of structure from arrested development, so great or so serious as to deserve to be called monstrosities, are of common occurrence, but, as they differ much from any normal structure, they require here only a passing notice. When a part or organ is arrested during its embryonic growth, a rudiment is generally left. Thus the whole head may be represented by a soft nipple-like projection, and the limbs by mere papillæ. These rudiments of limbs are sometimes inherited, as has been observed in a dog.[81]

Many lesser anomalies in our domesticated animals appear to be due to arrested development. What the cause of the arrest may be, we seldom know, except in the case of direct injury to the embryo within the egg or womb. That the cause does not generally act at a very early embryonic period we may infer from the affected organ seldom being wholly aborted, —a rudiment being generally preserved. The external ears are represented by mere vestiges in a Chinese breed of sheep; and in another breed, the tail is reduced " to a little button, suffocated, in a manner, by fat."[82] In tailless dogs and cats a stump is left; but I do not know whether it includes at an early embryonic age rudiments of all the caudal vertebræ. In certain breeds of fowls the comb and wattles are reduced to rudiments; in the Cochin-China breed scarcely more than rudiments of spurs exist. With polled Suffolk cattle, "rudiments of horns can often be felt " at an early age;"[83] and with species in a state of nature, the relatively greater development of rudimentary organs at an early period of life is highly characteristic of such organs. With hornless breeds of cattle and sheep, another and singular kind of rudiment has been observed, namely, minute dangling horns attached to the skin alone, and which are often shed and grow again. With hornless goats, according to Desmarest,[84]

---

81 Isid. Geoffroy St. Hilaire, ' Hist. Nat. des Anomalies,' 1836, tom. ii. pp. 210, 223, 224, 395; ' Philosoph. Transact.,' 1775, p. 313.

82 Pallas, quoted by Youatt on Sheep, p. 25.

83 Youatt on Cattle, 1834, p. 174.

84 ' Encyclop. Méthod.,' 1820, p. 483 : see p. 500, on the Indian zebu casting its horns. Similar cases in European cattle were given in the third chapter.

the bony protuberances which properly support the horns exist as mere rudiments.

With cultivated plants it is far from rare to find the petals, stamens, and pistils represented by rudiments, like those observed in natural species. So it is with the whole seed in many fruits; thus near Astrakhan there is a grape with mere traces of seeds, "so small and lying so near the "stalk that they are not perceived in eating the grape." [85] In certain varieties of the gourd, the tendrils, according to Naudin, are represented by rudiments or by various monstrous growths. In the broccoli and cauliflower the greater number of the flowers are incapable of expansion, and include rudimentary organs. In the Feather hyacinth (*Muscari comosum*) the upper and central flowers are brightly coloured but rudimentary; under cultivation the tendency to abortion travels downwards and outwards, and all the flowers become rudimentary; but the abortive stamens and pistils are not so small in the lower as in the upper flowers. In the *Viburnum opulus*, on the other hand, the outer flowers naturally have their organs of fructification in a rudimentary state, and the corolla is of large size; under cultivation, the change spreads to the centre, and all the flowers become affected; thus the well-known Snow-ball bush is produced. In the Compositæ, the so-called doubling of the flowers consists in the greater development of the corolla of the central florets, generally accompanied with some degree of sterility; and it has been observed [86] that the progressive doubling invariably spreads from the circumference to the centre,—that is, from the ray florets, which so often include rudimentary organs, to those of the disc. I may add, as bearing on this subject, that, with Asters, seeds taken from the florets of the circumference have been found to yield the greatest number of double flowers. [87] In these several cases we have a natural tendency in certain parts to become rudimentary, and this under culture spreads either to, or from, the axis of the plant. It deserves notice, as showing how the same laws govern the changes which natural species and artificial varieties undergo, that in a series of species in the genus Carthamus, one of the Compositæ, a tendency in the seeds to the abortion of the pappus may be traced extending from the circumference to the centre of the disc: thus, according to A. de Jussieu, [88] the abortion is only partial in *Carthamus creticus*, but more extended in *C. lanatus*; for in this species two or three alone of the central seeds are furnished with a pappus, the surrounding seeds being either quite naked or furnished with a few hairs; and lastly, in *C. tinctorius*, even the central seeds are destitute of pappus, and the abortion is complete.

With animals and plants under domestication, when an organ disappears, leaving only a rudiment, the loss has generally been sudden, as with hornless and tailless breeds; and such cases may be ranked as inherited monstrosities. But in some few cases the loss has been gradual, and

---

[85] Pallas, 'Travels,' Eng. translat., vol. i. p. 243.

[86] Mr. Beaton, in 'Journal of Horticulture,' May 21, 1861, p. 133.

[87] Lecoq, 'De la Fécondation,' 1862, p. 233.

[88] 'Annales du Muséum,' tom. vi. p. 319.

has been partly effected by selection, as with the rudimentary combs and wattles of certain fowls. We have also seen that the wings of some domesticated birds have been slightly reduced by disuse, and the great reduction of the wings in certain silk-moths, with mere rudiments left, has probably been aided by disuse.

With species in a state of nature, rudimentary organs are so extremely common that scarcely one can be named which is wholly free from a blemish of this nature. Such organs are generally variable, as several naturalists have observed; for, being useless, they are not regulated by natural selection, and they are more or less liable to reversion. The same rule certainly holds good with parts which have become rudimentary under domestication. We do not know through what steps under nature rudimentary organs have passed in being reduced to their present condition; but we so incessantly see in species of the same group the finest gradations between an organ in a rudimentary and perfect state, that we are led to believe that the passage must have been extremely gradual. It may be doubted whether a change of structure so abrupt as the sudden loss of an organ would ever be of service to a species in a state of nature; for the conditions to which all organisms are closely adapted usually change very slowly. Even if an organ did suddenly disappear in some one individual by an arrest of development, intercrossing with the other individuals of the same species would cause it to reappear in a more or less perfect manner, so that its final reduction could only be effected by the slow process of continued disuse or natural selection. It is much more probable that, from changed habits of life, organs first become of less and less use, and ultimately superfluous; or their place may be supplied by some other organ; and then disuse, acting on the offspring through inheritance at corresponding periods of life, would go on reducing the organ; but as most organs could be of no use at an early embryonic period, they would not be affected by disuse; consequently they would be preserved at this stage of growth, and would remain as rudiments. In addition to the effects of disuse, the principle of economy of growth, already alluded to in this chapter, would lead to the still further reduction of all superfluous parts. With respect to the final and total suppression or abortion of any organ, another and distinct principle, which will be discussed in the chapter on pangenesis, probably takes a share in the work.

With animals and plants reared by man there is no severe or recurrent struggle for existence, and the principle of economy will not come into action. So far, indeed, is this from being the case, that in some instances organs, which are naturally rudimentary in the parent-species, become partially redeveloped in the domesticated descendants. Thus cows, like most other ruminants, properly have four active and two rudimentary mammæ; but in our domesticated animals, the latter occasionally become considerably developed and yield milk. The atrophied mammæ, which, in male domesticated animals, including man, have in some rare cases grown to full size and secreted milk, perhaps offer an analogous case. The hind feet of dogs include rudiments of a fifth toe, and in certain large breeds these toes, though still rudimentary, become considerably developed

and are furnished with claws. In the common Hen, the spurs and comb
are rudimentary, but in certain breeds these become, independently of
age or disease of the ovaria, well developed. The stallion has canine teeth,
but the mare has only traces of the alveoli, which, as I am informed by
the eminent veterinary Mr. G. T. Brown, frequently contain minute irregular nodules of bone. These nodules, however, sometimes become developed into imperfect teeth, protruding through the gums and coated
with enamel; and occasionally they grow to a third or even a fourth of
the length of the canines in the stallion. With plants I do not know
whether the redevelopment of rudimentary organs occurs more frequently
under culture than under nature. Perhaps the pear-tree may be a case in
point, for when wild it bears thorns, which though useful as a protection
are formed of branches in a rudimentry condition, but, when the tree is
cultivated, the thorns are reconverted into branches.

Finally, though organs which must be classed as rudimentary frequently occur in our domesticated animals and
cultivated plants, these have generally been formed suddenly,
through an arrest of development. They usually differ in
appearance from the rudiments which so frequently characterise
natural species. In the latter, rudimentary organs have been
slowly formed through continued disuse, acting by inheritance
at a corresponding age, aided by the principle of the economy
of growth, all under the control of natural selection. With
domesticated animals, on the other hand, the principle of economy is far from coming into action, and their organs, although
often slightly reduced by disuse, are not thus almost obliterated
with mere rudiments left.

CHAPTER XXV.

LAWS OF VARIATION, *continued* — CORRELATED VARIABILITY.

EXPLANATION OF TERM — CORRELATION AS CONNECTED WITH DEVELOPMENT — MODIFICATIONS CORRELATED WITH THE INCREASED OR DECREASED SIZE OF PARTS — CORRELATED VARIATION OF HOMOLOGOUS PARTS — FEATHERED FEET IN BIRDS ASSUMING THE STRUCTURE OF THE WINGS — CORRELATION BETWEEN THE HEAD AND THE EXTREMITIES — BETWEEN THE SKIN AND DERMAL APPENDAGES — BETWEEN THE ORGANS OF SIGHT AND HEARING — CORRELATED MODIFICATIONS IN THE ORGANS OF PLANTS — CORRELATED MONSTROSITIES — CORRELATION BETWEEN THE SKULL AND EARS — SKULL AND CREST OF FEATHERS — SKULL AND HORNS — CORRELATION OF GROWTH COMPLICATED BY THE ACCUMULATED EFFECTS OF NATURAL SELECTION — COLOUR AS CORRELATED WITH CONSTITUTIONAL PECULIARITIES.

ALL the parts of the organisation are to a certain extent connected or correlated together; but the connexion may be so slight that it hardly exists, as with compound animals or the buds on the same tree. Even in the higher animals various parts are not at all closely related; for one part may be wholly suppressed or rendered monstrous without any other part of the body being affected. But in some cases, when one part varies, certain other parts always, or nearly always, simultaneously vary; they are then subject to the law of correlated variation. Formerly I used the somewhat vague expression of correlation of growth, which may be applied to many large classes of facts. Thus, all the parts of the body are admirably co-ordinated for the peculiar habits of life of each organic being, and they may be said, as the Duke of Argyll insists in his 'Reign of Law,' to be correlated for this purpose. Again, in large groups of animals certain structures always co-exist; for instance, a peculiar form of stomach with teeth of peculiar form, and such structures may in one sense be said to be correlated. But these cases have no necessary connexion with the law to be discussed in the present chapter; for we do not know that the initial or primary variations of the several parts were in any way related; slight modifications or individual differences may have been preserved, first in one and then in another part, until the final and perfectly co-adapted structure was acquired; but to this subject I shall presently recur. Again, in many groups of animals the males alone are furnished with weapons, or are ornamented with gay colours; and these characters manifestly stand in some sort of correlation with the male reproductive organs, for when the latter are destroyed these characters disappear. But it was shown in the twelfth chapter that the very same peculiarity may become attached at any age to either sex, and afterwards be exclusively transmitted by the same sex at a corresponding age. In these cases we have inheritance limited by, or correlated with, both sex and age; but we have no reason for supposing that the original cause of the variation was necessarily connected with the reproductive organs, or with the age of the affected being.

In cases of true correlated variation, we are sometimes able to see the nature of the connexion; but in most cases the bond is hidden from us, and certainly differs in different cases. We can seldom say which of two correlated parts first varies, and induces a change in the other; or whether the two are simultaneously produced by some distinct cause. Correlated variation is an important subject for us; for when one part is modified through continued selection, either by man or under nature, other parts of the organisation will be unavoidably modified. From this correlation it apparently follows that, with our domesticated animals and plants, varieties rarely or never differ from each other by some single character alone.

One of the simplest cases of correlation is that a modification which arises during an early stage of growth tends to influence the subsequent development of the same part, as well as of other and intimately connected parts. Isidore Geoffroy St. Hilaire states[1] that this may constantly be observed with monstrosities

---

[1] 'Hist. des Anomalies,' tom. iii. p. 392. Prof. Huxley applies the same principle in accounting for the remarkable, though normal, differences in the arrangement of the nervous system in the Mollusca, in his great paper on the Morphology of the Cephalous Mollusca, in 'Phil. Transact.,' 1853, p. 56.

in the animal kingdom; and Moquin-Tandon[2] remarks, that, as with plants the axis cannot become monstrous without in some way affecting the organs subsequently produced from it, so axial anomalies are almost always accompanied by deviations of structure in the appended parts. We shall presently see that with short-muzzled races of the dog certain histological changes in the basal elements of the bones arrest their development and shorten them, and this affects the position of the subsequently developed molar teeth. It is probable that certain modifications in the larvæ of insects would affect the structure of the mature insects. But we must be very careful not to extend this view too far, for, during the normal course of development, certain members in the same group of animals are known to pass through an extraordinary course of change, whilst other and closely allied members arrive at maturity with little change of structure.

Another simple case of correlation is that with the increased or decreased dimensions of the whole body, or of any particular part, certain organs are increased or diminished in number, or are otherwise modified. Thus pigeon-fanciers have gone on selecting pouters for length of body, and we have seen that their vertebræ are generally increased in number, and their ribs in breadth. Tumblers have been selected for their small bodies, and their ribs and primary wing-feathers are generally lessened in number. Fantails have been selected for their large, widely-expanded tails, with numerous tail-feathers, and the caudal vertebræ are increased in size and number. Carriers have been selected for length of beak, and their tongues have become longer, but not in strict accordance with the length of beak. In this latter breed and in others having large feet, the number of the scutellæ on the toes is greater than in the breeds with small feet. Many similar cases could be given. In Germany it has been observed that the period of gestation is longer in large-sized than in small-sized breeds of cattle. With our highly-improved animals of all kinds the period of maturity has advanced, both with respect to the full growth of the body and the period of reproduction; and, in correspondence with this, the teeth are now developed earlier than formerly, so that,

[2] ' Eléments de Tératologie Veg.,' 1841, p. 113.

to the surprise of agriculturists, the ancient rules for judging the age of an animal by the state of its teeth are no longer trustworthy.[3]

Correlated Variation of Homologous Parts.—Parts which are homologous tend to vary in the same manner; and this is what might have been expected, for such parts are identical in form and structure during an early period of embryonic development, and are exposed in the egg or womb to similar conditions. The symmetry, in most kinds of animals, of the corresponding or homologous organs on the right and left sides of the body, is the simplest case in point; but this symmetry sometimes fails, as with rabbits having only one ear, or stags with one horn, or with many-horned sheep which sometimes carry an additional horn on one side of their heads. With flowers which have regular corollas, the petals generally vary in the same manner, as we see in the same complicated and elegant pattern, on the flowers of the Chinese pink; but with irregular flowers, though the petals are of course homologous, this symmetry often fails, as with the varieties of the *Antirrhinum* or snapdragon, or that variety of the kidney-bean (*Phaseolus multiflorus*) which has a white standard-petal.

In the vertebrata the front and hind limbs are homologous, and they tend to vary in the same manner, as we see in long and short-legged, or in thick and thin-legged races of the horse and dog. Isidore Geoffroy[4] has remarked on the tendency of supernumerary digits in man to appear, not only on the right and left sides, but on the upper and lower extremities. Meckel has insisted[5] that, when the muscles of the arm depart in number or arrangement from their proper type, they almost always imitate those of the leg; and so conversely the varying muscles of the leg imitate the normal muscles of the arm.

In several distinct breeds of the pigeon and fowl, the legs and the two outer toes are heavily feathered, so that in the trumpeter pigeon they appear like little wings. In the feather-legged bantam the " boots" or feathers, which grow from the outside of the leg and generally from the two outer toes, have,

---

[3] Prof. J. B. Simonds, on the Age of the Ox, Sheep, &c., quoted in ' Gard. Chronicle,' 1854, p. 588.

[4] ' Hist. des Anomalies,' tom. i. p. 674.
[5] Quoted by Isid. Geoffroy, idem, tom. i. p. 635.

according to the excellent authority of Mr. Hewitt,[6] been seen
to exceed the wing-feathers in length, and in one case were
actually nine and a half inches in length! As Mr. Blyth has
remarked to me, these leg-feathers resemble the primary wing-
feathers, and are totally unlike the fine down which naturally
grows on the legs of some birds, such as grouse and owls. Hence
it may be suspected that excess of food has first given redundancy to the plumage, and then that the law of homologous
variation has led to the development of feathers on the legs, in
a position corresponding with those on the wing, namely, on the
outside of the tarsi and toes. I am strengthened in this belief
by the following curious case of correlation, which for a long
time seemed to me utterly inexplicable, namely, that in pigeons
of any breed, if the legs are feathered, the two outer toes are
partially connected by skin. These two outer toes correspond
with our third and fourth toes. Now, in the wing of the pigeon
or any other bird, the first and fifth digits are wholly aborted;
the second is rudimentary and carries the so-called " bastard-
wing;" whilst the third and fourth digits are completely united
and enclosed by skin, together forming the extremity of the
wing. So that in feather-footed pigeons, not only does the exterior surface support a row of long feathers, like wing-feathers,
but the very same digits which in the wing are completely united
by skin become partially united by skin in the feet; and thus
by the law of the correlated variation of homologous parts we
can understand the curious connection of feathered legs and
membrane between the two outer toes.

Andrew Knight[7] has remarked that the face or head and the
limbs vary together in general proportions. Compare, for instance,
the head and limbs of a dray and race-horse, or of a greyhound
and mastiff. What a monster a greyhound would appear with
the head of a mastiff! The *modern* bulldog, however, has fine
limbs, but this is a recently-selected character. From the
measurements given in the sixth chapter, we clearly see that
in all the breeds of the pigeon the length of the beak and the
size of the feet are correlated. The view which, as before explained, seems the most probable is, that disuse in all cases tends

---

[6] 'The Poultry Book,' by W. B. Tegetmeier, 1866, p. 250.

[7] A. Walker on Intermarriage, 1838, p. 160.

to diminish the feet, the beak becoming at the same time through correlation shorter; but that in those few breeds in which length of beak has been a selected point, the feet, notwithstanding disuse, have through correlation increased in size.

With the increased length of the beak in pigeons, not only the tongue increases in length, but likewise the orifice of the nostrils. But the increased length of the orifice of the nostrils perhaps stands in closer correlation with the development of the corrugated skin or wattle at the base of the beak; for when there is much wattle round the eyes, the eyelids are greatly increased or even doubled in length.

There is apparently some correlation even in colour between the head and the extremities. Thus with horses a large white star or blaze on the forehead is generally accompanied by white feet.[8] With white rabbits and cattle, dark marks often co-exist on the tips of the ears and on the feet. In black and tan dogs of different breeds, tan-coloured spots over the eyes and tan-coloured feet almost invariably go together. These latter cases of connected colouring may be due either to reversion or to analogous variation,—subjects to which we shall hereafter return,—but this does not necessarily determine the question of their original correlation. If those naturalists are correct who maintain that the jaw-bones are homologous with the limb-bones, then we can understand why the head and limbs tend to vary together in shape and even in colour; but several highly competent judges dispute the correctness of this view.

The lopping forwards and downwards of the immense ears of fancy rabbits is in part due to the disuse of the muscles, and in part to the weight and length of the ears, which have been increased by selection during many generations. Now, with the increased size and changed direction of the ears, not only has the bony auditory meatus become changed in outline, direction, and greatly in size, but the whole skull has been slightly modified. This could be clearly seen in "half-lops"— that is, in rabbits with one ear alone lopping forward—for the opposite sides of their skulls were not strictly symmetrical. This seems to me a curious instance of correlation, between hard

<hr>

[8] 'The Farrier and Naturalist,' vol. i., 1828, p. 456.

bones and organs so soft and flexible, as well as so unimportant
under a physiological point of view, as the external ears. The
result no doubt is largely due to mere mechanical action, that
is, to the weight of the ears, on the same principle that the
skull of a human infant is easily modified by pressure.

The skin and the appendages of hair, feathers, hoofs, horns, and
teeth, are homologous over the whole body. Every one knows
that the colour of the skin and that of the hair usually vary
together; so that Virgil advises the shepherd to look whether the
mouth and tongue of the ram are black, lest the lambs should
not be purely white. With poultry and certain ducks we have
seen that the colour of the plumage stands in some connexion
with the colour of the shell of the egg,—that is, with the
mucous membrane which secretes the shell. The colour of
the skin and hair, and the odour emitted by the glands of the
skin, are said[9] to be connected, even in the same race of men.
Generally the hair varies in the same way all over the body in
length, fineness, and curliness. The same rule holds good with
feathers, as we see with the laced and frizzled breeds both of
fowls and pigeons. In the common cock the feathers on the neck
and loins are always of a particular shape, called hackles: now
in the Polish breed, both sexes are characterised by a tuft of
feathers on the head; but through correlation these feathers
in the male always assume the form of hackles. The wing and
tail-feathers, though arising from parts not homologous, vary
in length together; so that long or short winged pigeons generally
have long or short tails. The case of the Jacobin-pigeon is
more curious, for the wing and tail feathers are remarkably
long; and this apparently has arisen in correlation with the
elongated and reversed feathers on the back of the neck, which
form the hood.

The hoofs and hair are homologous appendages; and a careful
observer, namely Azara,[10] states that in Paraguay horses of various
colours are often born with their hair curled and twisted like that
on the head of a negro. This peculiarity is strongly inherited.
But what is remarkable is that the hoofs of these horses " are
" absolutely like those of a mule." The hair also of the mane and
tail is invariably much shorter than usual, being only from four

[9] Godron, 'Sur l'Espèce,' tom. ii. p. 217.
[10] 'Quadrupèdes du Paraguay,' tom. ii. p. 333.

to twelve inches in length; so that curliness and shortness of
the hair are here, as with the negro, apparently correlated.

With respect to the horns of sheep, Youatt [11] remarks that
" multiplicity of horns is not found in any breed of much value:
" it is generally accompanied by great length and coarseness of
" the fleece." Several tropical breeds of sheep, which are clothed
with hair instead of wool, have horns almost like those of a goat.
Sturm [12] expressly declares that in different races the more the
wool is curled the more the horns are spirally twisted. We have
seen in the third chapter, where other analogous facts have been
given, that the parent of the Mauchamp breed, so famous for
its fleece, had peculiarly shaped horns. The inhabitants of
Angora assert [13] that " only the white goats which have horns
" wear the fleece in the long curly locks that are so much
" admired; those which are not horned having a comparatively
" close coat." From these cases we may conclude that the
hair or wool and the horns vary in a correlated manner. Those
who have tried hydropathy are aware that the frequent application of cold water stimulates the skin; and whatever stimulates the skin tends to increase the growth of the hair, as is well
shown in the abnormal growth of hair near old inflamed surfaces.
Now, Professor Low [14] is convinced that with the different races
of British cattle thick skin and long hair depend on the hu-
midity of the climate which they inhabit. We can thus see
how a humid climate might act on the horns—in the first place
directly on the skin and hair, and secondly by correlation on
the horns. The presence or absence of horns, moreover, both in
the case of sheep and cattle, acts, as will presently be shown, by
some sort of correlation on the skull.

With respect to hair and teeth, Mr. Yarrell [15] found many of
the teeth deficient in three hairless "*Ægyptian*" dogs, and in a
hairless terrier. The incisors, canines, and premolars suffered
most, but in one case all the teeth, except the large tubercular molar on each side, were deficient. With man several
striking cases have been recorded [16] of inherited baldness with in-

[11] On Sheep, p. 142.

[12] 'Ueber Racen, Kreuzungen, &c.,'
1825, s. 24.

[13] Quoted from Conolly, in 'The
Indian Field,' Feb. 1859, vol. ii. p. 266.

[14] 'Domesticated Animals of the
British Islands,' pp. 307, 368.

[15] 'Proceedings Zoolog. Soc.,' 1833,
p. 113.

[16] Sedgwick, ' Brit. and Foreign
Medico-Chirurg. Review,' April 1863,
p. 453.

herited deficiency, either complete or partial, of the teeth. We see the same connexion in those rare cases in which the hair has been renewed in old age, for this has "usually been accompanied by a renewal of the teeth." I have remarked in a former part of this volume that the great reduction in the size of the tusks in domestic boars probably stands in close relation with their diminished bristles, due to a certain amount of protection; and that the reappearance of the tusks in boars, which have become feral and are fully exposed to the weather, probably depends on the reappearance of the bristles. I may add, though not strictly connected with our present point, that an agriculturist[17] asserts that "pigs with little hair on their bodies are most liable to " lose their tails, showing a weakness of the tegumental structure. " It may be prevented by crossing with a more hairy breed."

In the previous cases deficient hair, and teeth deficient in number or size, are apparently connected. In the following cases abnormally redundant hair, and teeth either deficient or redundant, are likewise connected. Mr. Crawfurd[18] saw at the Burmese Court a man, thirty years old, with his whole body, except the hands and feet, covered with straight silky hair, which on the shoulders and spine was five inches in length. At birth the ears alone were covered. He did not arrive at puberty, or shed his milk teeth, until twenty years old; and at this period he acquired five teeth in the upper jaw, namely four incisors and one canine, and four incisor teeth in the lower jaw; all the teeth were small. This man had a daughter, who was born with hair within her ears; and the hair soon extended over her body. When Captain Yule[19] visited the Court, he found this girl grown up; and she presented a strange appearance with even her nose densely covered with soft hair. Like her father, she was furnished with incisor teeth alone. The King had with difficulty bribed a man to marry her, and of her two children, one, a boy fourteen months old, had hair growing out of his ears, with a beard and moustache. This strange peculiarity had, therefore, been inherited for three generations, with the molar teeth deficient in the grandfather and mother; whether

[17] 'Gard. Chronicle,' 1849, p. 205.
[18] 'Embassy to the Court of Ava,' vol. i. p. 320.
[19] 'Narrative of a Mission to the Court of Ava in 1855,' p. 94.

these teeth would likewise fail in the infant could not be told. Here is another case communicated to me by Mr. Wallace on the authority of Dr. Purland, a dentist: Julia Pastrana, a Spanish dancer, was a remarkably fine woman, but she had a thick masculine beard and a hairy forehead; she was photographed, and her stuffed skin was exhibited as a show; but what concerns us is, that she had in both the upper and lower jaw an irregular double set of teeth, one row being placed within the other, of which Dr. Purland took a cast. From the redundancy of the teeth her mouth projected, and her face had a gorilla-like appearance. These cases and those of the hairless dogs forcibly call to mind the fact, that the two orders of mammals—namely, the Edentata and Cetacea—which are the most abnormal in their dermal covering, are likewise the most abnormal either by deficiency or redundancy of teeth.

The organs of sight and hearing are generally admitted to be homologous, both with each other and with the various dermal appendages; hence these parts are liable to be abnormally affected in conjunction. Mr. White Cowper says "that in all cases of " double microphthalmia brought under his notice he has at the " same time met with defective development of the dental sys- " tem." Certain forms of blindness seem to be associated with the colour of the hair; a man with black hair and a woman with light-coloured hair, both of sound constitution, married and had nine children, all of whom were born blind; of these children, five "with dark hair and brown iris were afflicted " with amaurosis; the four others, with light-coloured hair and " blue iris, had amaurosis and cataract conjoined." Several cases could be given, showing that some relation exists between various affections of the eyes and ears; thus Liebreich states that out of 241 deaf-mutes in Berlin, no less than fourteen suffered from the rare disease called pigmentary retinitis. Mr. White Cowper and Dr. Earle have remarked that inability to distinguish different colours, or colour-blindness, "is often " associated with a corresponding inability to distinguish " musical sounds."[20]

[20] These statements are taken from Mr. Sedgwick, in the 'Medico-Chirurg. Review,' July 1861, p. 198; April 1863, pp. 455 and 458. Liebreich is quoted by Professor Devay, in his 'Mariages Consanguins,' 1862, p. 116.

Here is a more curious case: white cats, if they have blue
eyes, are almost always deaf. I formerly thought that the rule
was invariable, but I have heard of a few authentic exceptions.
The first two notices were published in 1829, and relate to
English and Persian cats: of the latter, the Rev. W. T. Bree
possessed a female, and he states " that of the offspring pro-
" duced at one and the same birth, such as, like the mother,
" were entirely white (with blue eyes) were, like her, invariably
" deaf; while those that had the least speck of colour on their
" fur, as invariably possessed the usual faculty of hearing." [21]
The Rev. W. Darwin Fox informs me that he has seen more
than a dozen instances of this correlation in English, Persian,
and Danish cats; but he adds " that, if one eye, as I have
" several times observed, be not blue, the cat hears. On the
" other hand, I have never seen a white cat with eyes of the com-
" mon colour that was deaf." In France Dr. Sichel [22] has observed during twenty years similar facts; he adds the remarkable case of the iris beginning, at the end of four months, to
grow dark-coloured, and then the cat first began to hear.

This case of correlation in cats has struck many persons as
marvellous. There is nothing unusual in the relation between
blue eyes and white fur; and we have already seen that the
organs of sight and hearing are often simultaneously affected.
In the present instance the cause probably lies in a slight arrest
of development in the nervous system in connection with the
sense-organs. Kittens during the first nine days, whilst their
eyes are closed, appear to be completely deaf; I have made a
great clanging noise with a poker and shovel close to their heads,
both when they were asleep and awake, without producing any
effect. The trial must not be made by shouting close to their ears,
for they are, even when asleep, extremely sensitive to a breath
of air. Now, as long as the eyes continue closed, the iris is no
doubt blue, for in all the kittens which I have seen this colour
remains for some time after the eyelids open. Hence, if we suppose the development of the organs of sight and hearing to be
arrested at the stage of the closed eyelids, the eyes would re-

[21] Loudon's ' Mag. of Nat. Hist.,' vol.
i., 1829, pp. 66, 178. See also Dr. P.
Lucas, 'L'Héréd. Nat.,' tom. i. p. 428,
on the inheritance of deafness in cats.
[22] 'Annales des Sc. Nat.' Zoolog., 3rd
series, 1847, tom. viii. p. 239.

main permanently blue and the ears would be incapable of per-
ceiving sound ; and we should thus understand this curious case.
As, however, the colour of the fur is determined long before
birth, and as the blueness of the eyes and the whiteness of the
fur are obviously connected, we must believe that some primary
cause acts at an early period.

The instances of correlated variability hitherto given have
been chiefly drawn from the animal kingdom, and we will now
turn to plants.   Leaves, sepals, petals, stamens, and pistils are
all homologous.   In double flowers we see that the stamens and
pistils vary in the same manner, and assume the form and colour
of the petals.   In the double columbine (*Aquilegia vulgaris*), the
successive whorls of stamens are converted into cornucopias,
which are enclosed within each other and resemble the petals.
In hose-and-hose flowers the sepals mock the petals.   In some
cases the flowers and leaves vary together in tint : in all the
varieties of the common pea, which have purple flowers, a
purple mark may be seen on the stipules.   In other cases the
leaves and fruit and seeds vary together in colour, as in a
curious pale-leaved variety of the sycamore, which has recently
been described in France,[23] and as in the purple-leaved hazel, in
which the leaves, the husk of the nut, and the pellicle round
the kernel are all coloured purple.[24]   Pomologists can predict
to a certain extent, from the size and appearance of the
leaves of their seedlings, the probable nature of the fruit; for,
as Van Mons remarks,[25] variations in the leaves are generally
accompanied by some modification in the flower, and consequently in the fruit.   In the Serpent melon, which has a
narrow tortuous fruit above a yard in length, the stem of the
plant, the peduncle of the female flower, and the middle lobe
of the leaf, are all elongated in a remarkable manner.   On the
other hand, several varieties of Cucurbita, which have dwarfed
stems, all produce, as Naudin remarks with surprise, leaves of
the same peculiar shape.   Mr. G. Maw informs me that all the
varieties of the scarlet Pelargoniums which have contracted or
imperfect leaves have contracted flowers : the difference between

[23] 'Gardener's Chron.,' 1864, p. 1202.
[24] Verlot gives several other instances, 'Des Variétés,' 1865, p. 72.
[25] 'Arbres Fruitiers,' 1836, tom. ii. pp. 204, 226.

"Brilliant" and its parent "Tom Thumb" is a good instance of this. It may be suspected that the curious case described by Risso,[26] of a variety of the Orange which produces on the young shoots rounded leaves with winged petioles, and afterwards elongated leaves on long but wingless petioles, is connected with the remarkable change in form and nature which the fruit undergoes during its development.

In the following instance we have the colour and form of the petals apparently correlated, and both dependent on the nature of the season. An observer, skilled in the subject, writes,[27] " I noticed, during the year 1842, that every Dahlia, of " which the colour had any tendency to scarlet, was deeply " notched—indeed to so great an extent as to give the petals the " appearance of a saw; the indentures were, in some instances, " more than a quarter of an inch deep." Again, Dahlias which have their petals tipped with a different colour from the rest are very inconstant, and during certain years some, or even all the flowers, become uniformly coloured; and it has been observed with several varieties,[28] that when this happens the petals grow much elongated and lose their proper shape. This, however, may be due to reversion, both in colour and form, to the aboriginal species.

In this discussion on correlation, we have hitherto treated of cases in which we can partly understand the bond of connexion ; but I will now give cases in which we cannot even conjecture, or can only very obscurely see, what is the nature of the bond. Isidore Geoffroy St. Hilaire, in his work on Monstrosities, insists,[29] "que certaines anomalies coexistent rarement entr'elles, " d'autres fréquemment, d'autres enfin presque constamment, " malgré la différence très-grande de leur nature, et quoiqu'elles " puissent paraître complètement indépendantes les unes des " autres." We see something analogous in certain diseases : thus I hear from Mr. Paget that in a rare affection of the

[26] 'Annales du Muséum,' tom. xx. p. 188.

[27] 'Gardener's Chron.,' 1843, p. 877.

[28] Ibid., 1845, p. 102.

[29] 'Hist. des Anomalies,' tom. iii. p. 402. See also M. Camille Dareste, 'Recherches sur les Conditions,' &c., 1863, pp. 16, 48.

renal capsules (of which the functions are unknown), the skin
becomes bronzed; and in hereditary syphilis, both the milk and
the second teeth assume a peculiar and characteristic form.
Professor Rolleston, also, informs me that the incisor teeth are
sometimes furnished with a vascular rim in correlation with
intra-pulmonary deposition of tubercles.   In other cases of
phthisis and of cyanosis the nails and finger-ends become
clubbed like acorns.   I believe that no explanation has been
offered of these and of many other cases of correlated disease.

What can be more curious and less intelligible than the fact
previously given, on the authority of Mr. Tegetmeier, that
young pigeons of all breeds, which when mature have white,
yellow, silver-blue, or dun-coloured plumage, come out of the
egg almost naked; whereas pigeons of other colours when first
born are clothed with plenty of down ?   White Pea-fowls, as has
been observed both in England and France,[30] and as I have
myself seen, are inferior in size to the common coloured kind;
and this cannot be accounted for by the belief that albinism is
always accompanied by constitutional weakness; for white or
albino moles are generally larger than the common kind.

To turn to more important characters: the niata cattle of the
Pampas are remarkable from their short foreheads, upturned
muzzles, and curved lower jaws.   In the skull the nasal and premaxillary bones are much shortened, the maxillaries are excluded
from any junction with the nasals, and all the bones are slightly
modified, even to the plane of the occiput.   From the analogical
case of the dog, hereafter to be given, it is probable that the
shortening of the nasal and adjoining bones is the proximate
cause of the other modifications in the skull, including the
upward curvature of the lower jaw, though we cannot follow
out the steps by which these changes have been effected.

Polish fowls have a large tuft of feathers on their heads; and
their skulls are perforated by numerous holes, so that a pin can
be driven into the brain without touching any bone.   That
this deficiency of bone is in some way connected with the tuft
of feathers is clear from tufted ducks and geese likewise having

[30] Rev. E. S. Dixon, 'Ornamental Poultry,' 1848, p. 111; Isidore Geoffroy,
'Hist. Anomalies,' tom. i. p. 211.

perforated skulls. The case would probably be considered by some authors as one of balancement or compensation. In the chapter on Fowls, I have shown that with Polish fowls the tuft of feathers was probably at first small; by continued selection it became larger, and then rested on a fleshy or fibrous mass; and finally, as it became still larger, the skull itself became more and more protuberant until it acquired its present extraordinary structure. Through correlation with the protuberance of the skull, the shape and even the relative connexion of the premaxillary and nasal bones, the shape of the orifice of the nostrils, the breadth of the frontal bone, the shape of the post-lateral processes of the frontal and squamosal bones, and the direction of the bony cavity of the ear, have all been modified. The internal configuration of the skull and the whole shape of the brain have likewise been altered in a truly marvellous manner.

After this case of the Polish fowl it would be superfluous to do more than refer to the details previously given on the manner in which the changed form of the comb, in various breeds of the fowl, has affected the skull, causing by correlation crests, protuberances, and depressions on its surface.

With our cattle and sheep the horns stand in close connexion with the size of the skull, and with the shape of the frontal bones; thus Cline[31] found that the skull of a horned ram weighed five times as much as that of a hornless ram of the same age. When cattle become hornless, the frontal bones are "materially diminished in breadth towards the poll;" and the cavities between the bony plates "are not so deep, nor do " they extend beyond the frontals." [32]

It may be well here to pause and observe how the effects of correlated variability, of the increased use of parts, and of the accumulation through natural selection of so-called spontaneous variations, are in many cases inextricably commingled. We may borrow an illustration from Mr. Herbert Spencer, who remarks that, when the Irish elk acquired its gigantic horns, weighing above one hundred pounds, numerous co-ordinated

[31] 'On the Breeding of Domestic Animals,' 1829, p. 6.
[32] Youatt on Cattle, 1834, p. 283.

changes of structure would have been indispensable,—namely, a thickened skull to carry the horns; strengthened cervical vertebræ, with strengthened ligaments; enlarged dorsal verte- bræ to support the neck, with powerful fore-legs and feet; all these parts being supplied with proper muscles, blood-vessels, and nerves.   How then could these admirably co-ordinated modifications of structure have been acquired?   According to the doctrine which I maintain, the horns of the male elk were slowly gained through sexual selection,—that is, by the best- armed males conquering the worse-armed, and leaving a greater number of descendants.   But it is not at all necessary that the several parts of the body should have simultaneously varied. Each stag presents individual differences, and in the same district those which had slightly heavier horns, or stronger necks, or stronger bodies, or were the most courageous, would secure the greater number of does, and consequently leave a greater number of offspring.   The offspring would inherit, in a greater or less degree, these same qualities, would occasionally intercross with each other, or with other individuals varying in some favourable manner; and of their offspring, those which were the best endowed in any respect would continue multi- plying; and so onwards, always progressing, sometimes in one direction, and sometimes in another, towards the present excel- lently co-ordinated structure of the male elk.   To make this clear, let us reflect on the probable steps, as shown in the twentieth chapter, by which our race and dray-horses have arrived at their present state of excellence; if we could view the whole series of intermediate forms between one of these animals and an early unimproved progenitor, we should behold a vast number of animals, not equally improved in each generation throughout their entire structure, but sometimes a little more in one point, and sometimes in another, yet on the whole gradu- ally approaching in character to our present race or dray- horses, which are so admirably fitted in the one case for fleetness and in the other for draught.

Although natural selection would thus [33] tend to give to the

[33] Mr. Herbert Spencer ('Principles of Biology,' 1864, vol. i. pp. 452, 468) takes a different view; and in one place

remarks: "We have seen reason to " think that, as fast as essential faculties " multiply, and as fast as the number of

male elk its present structure, yet it is probable that the inherited influence of use has played an equal or more important
part. As the horns gradually increased in weight, the muscles
of the neck, with the bones to which they are attached, would
increase in size and strength; and these parts would react on
the body and legs. Nor must we overlook the fact that certain
parts of the skull and the extremities would, judging by analogy,
tend from the first to vary in a correlated manner. The increased
weight of the horns would also act directly on the skull, in the
same manner as, when one bone is removed in the leg of a
dog, the other bone, which has to carry the whole weight of the
body, increases in thickness. But from the facts given with
respect to horned and hornless cattle, it is probable that the
horns and skull would immediately act on each other through
the principle of correlation. Lastly, the growth and subsequent wear and tear of the augmented muscles and bones
would require an increased supply of blood, and consequently
an increased supply of food; and this again would require
increased powers of mastication, digestion, respiration, and
excretion.

Colour as Correlated with Constitutional Peculiarities.

It is an old belief that with man there is a connexion between
complexion and constitution; and I find that some of the best
authorities believe in this to the present day.[34] Thus Dr.
Beddoe by his tables shows[35] that a relation exists between
liability to consumption and the colour of the hair, eyes, and
skin. It has been affirmed[36] that, in the French army which
invaded Russia, soldiers having a dark complexion, from the

"organs that co-operate in any given
"function increases, indirect equilibra-
"tion through natural selection becomes
"less and less capable of producing spe-
"cific adaptations; and remains fully
"capable only of maintaining the gene-
"ral fitness of constitution to conditions."
This view that natural selection can do
little in modifying the higher animals
surprises me, seeing that man's selection has undoubtedly effected much
with our domesticated quadrupeds and
birds.

[34] Dr. Prosper Lucas apparently dis-
believes in any such connexion,'L'Héréd.
Nat.,' tom. ii. pp. 88-94.

[35] 'British Medical Journal,' 1862, p.
433.

[36] Boudin, 'Géograph. Médicale,'
tom. i. p. 406.

southern parts of Europe, withstood the intense cold better than those with lighter complexions from the north; but no doubt such statements are liable to error.

In the second chapter on Selection I have given several cases proving that with animals and plants differences in colour are correlated with constitutional differences, as shown by greater or less immunity from certain diseases, from the attacks of parasitic plants and animals, from burning by the sun, and from the action of certain poisons. When all the individuals of any one variety possess an immunity of this nature, we cannot feel sure that it stands in any sort of correlation with their colour; but when several varieties of the same species, which are similarly coloured, are thus characterised, whilst other coloured varieties are not thus favoured, we must believe in the existence of a correlation of this kind. Thus in the United States purple-fruited plums of many kinds are far more affected by a certain disease than green or yellow-fruited varieties. On the other hand, yellow-fleshed peaches of various kinds suffer from another disease much more than the white-fleshed varieties. In the Mauritius red sugar-canes are much less affected by a particular disease than the white canes. White onions and verbenas are the most liable to mildew; and in Spain the green-fruited grapes suffered from the vine-disease more than other coloured varieties. Dark-coloured pelargoniums and verbenas are more scorched by the sun than varieties of other colours. Red wheats are believed to be hardier than white; whilst red-flowered hyacinths were more injured during one particular winter in Holland than other coloured varieties. With animals, white terriers suffer most from the distemper, white chickens from a parasitic worm in their tracheæ, white pigs from scorching by the sun, and white cattle from flies; but the caterpillars of the silk-moth which yield white cocoons suffered in France less from the deadly parasitic fungus than those producing yellow silk.

The cases of immunity from the action of certain vegetable poisons, in connexion with colour, are more interesting, and are at present wholly inexplicable. I have already given a remarkable instance, on the authority of Professor Wyman, of all the hogs, excepting those of a black colour, suffering severely in Virginia from eating the root of the *Lachnanthes tinctoria*.

According to Spinola and others,[37] buckwheat (*Polygonum fago-pyrum*), when in flower, is highly injurious to white or white-spotted pigs, if they are exposed to the heat of the sun, but is quite innocuous to black pigs. By two accounts, the *Hypericum crispum* in Sicily is poisonous to white sheep alone; their heads swell, their wool falls off, and they often die; but this plant, according to Lecce, is poisonous only when it grows in swamps; nor is this improbable, as we know how readily the poisonous principle in plants is influenced by the conditions under which they grow.

Three accounts have been published in Eastern Prussia, of white and white-spotted horses being greatly injured by eating mildewed and honeydewed vetches; every spot of skin bearing white hairs becoming inflamed and gangrenous. The Rev. J. Rodwell informs me that his father turned out about fifteen cart-horses into a field of tares which in parts swarmed with black aphides, and which no doubt were honeydewed, and probably mildewed; the horses, with two exceptions, were chestnuts and bays with white marks on their faces and pasterns, and the white parts alone swelled and became angry scabs. The two bay horses with no white marks entirely escaped all injury. In Guernsey, when horses eat fools' parsley (*Æthusa cynapium*) they are sometimes violently purged; and this plant "has a peculiar effect on the nose and lips, causing "deep cracks and ulcers, particularly on horses with white "muzzles."[38] With cattle, independently of the action of any poison, cases have been published by Youatt and Erdt of cutaneous diseases with much constitutional disturbance (in one instance after exposure to a hot sun) affecting every single point which bore a white hair, but completely passing over other parts of the body. Similar cases have been observed with horses.[39]

We thus see that not only do those parts of the skin which bear white hair differ in a remarkable manner from those bearing

---

[37] This fact and the following cases, when not stated to the contrary, are taken from a very curious paper by Prof. Heusinger, in 'Wochenschrift für Heilkunde,' May 1846, s. 277.

[38] Mr. Mogford, in the 'Veterinarian,' quoted in 'The Field,' Jan. 22, 1861, p. 545.

[39] 'Edinburgh Veterinary Journal,' Oct. 1860, p. 347.

hair of any other colour, but that in addition some great constitutional difference must stand in correlation with the colour
of the hair; for in the above-mentioned cases, vegetable poisons
caused fever, swelling of the head, as well as other symptoms,
and even death, to all the white or white-spotted animals.

CHAPTER XXVI.

LAWS OF VARIATION, *continued* — SUMMARY.

ON THE AFFINITY AND COHESION OF HOMOLOGOUS PARTS — ON THE VARIABILITY OF
MULTIPLE AND HOMOLOGOUS PARTS — COMPENSATION OF GROWTH — MECHANICAL
PRESSURE — RELATIVE POSITION OF FLOWERS WITH RESPECT TO THE AXIS OF THE
PLANT, AND OF SEEDS IN THE CAPSULE, AS INDUCING VARIATION — ANALOGOUS OR
PARALLEL VARIETIES — SUMMARY OF THE THREE LAST CHAPTERS.

On the Affinity of Homologous Parts.—THIS law was first
generalised by Geoffroy Saint Hilaire, under the expression of
La loi de l'affinité de soi pour soi. It has been fully discussed and
illustrated by his son, Isidore Geoffroy, with respect to monsters
in the animal kingdom,[1] and by Moquin-Tandon, with respect
to monstrous plants.   When similar or homologous parts,
whether belonging to the same embryo or to two distinct
embryos, are brought during an early stage of development into
contact, they often blend into a single part or organ ; and this
complete fusion indicates some mutual affinity between the
parts, otherwise they would simply cohere.   Whether any
power exists which tends to bring homologous parts into contact seems more doubtful.   The tendency to complete fusion is
not a rare or exceptional fact.   It is exhibited in the most
striking manner by double monsters.   Nothing can be more extraordinary than the manner, as shown in various published plates,
in which the corresponding parts of two embryos become intimately fused together.   This is perhaps best seen in monsters
with two heads, which are united, summit to summit, or face to
face, or, Janus-like, back to back, or obliquely side to side.   In
one instance of two heads united almost face to face, but a little
obliquely, four ears were developed, and on one side a perfect
face, which was manifestly formed by the union of two half-

---

[1] 'Hist des Anomalies,' 1832, tom. i. pp. 22. 537-556 ; tom. iii. p. 462.

faces. Whenever two bodies or two heads are united, each bone, muscle, vessel, and nerve on the line of junction seems to seek out its fellow, and becomes completely fused with it. Lereboullet,[2] who carefully studied the development of double monsters in fishes, observed in fifteen instances the steps by which two heads gradually became fused into one. In this and other such cases, no one, I presume, supposes that the two already formed heads actually blend together, but that the corresponding parts of each head grow into one during the further progress of development, accompanied as it always is with incessant absorption and renovation. Double monsters were formerly thought to be formed by the union of two originally distinct embryos developed upon distinct vitelli; but now it is admitted that "their production is due to the sponta- " neous divarication of the embryonic mass into two halves;"[3] this, however, is effected by different methods. But the belief that double monsters originate from the division of one germ, does not necessarily affect the question of subsequent fusion, or render less true the law of the affinity of homologous parts.

The cautious and sagacious J. Müller,[4] when speaking of Janus-like monsters, says, that "without the supposition that some " kind of affinity or attraction is exerted between corresponding " parts, unions of this kind are inexplicable." On the other hand, Vrolik, and he is followed by others, disputes this con- clusion, and argues from the existence of a whole series of mon- strosities, graduating from a perfectly double monster to a mere rudiment of an additional digit, that " an excess of formative " power" is the cause and origin of every monstrous duplicity. That there are two distinct classes of cases, and that parts may be doubled independently of the existence of two embryos, is certain; for a single embryo, or even a single adult animal, may produce doubled organs. Thus Valentin, as quoted by Vrolik, injured the caudal extremity of an embryo, and three days afterwards it produced rudiments of a double pelvis and of double hind limbs.

---

[2] 'Comptes Rendus,' 1855, pp. 855, 1029.

[3] Carpenter's 'Comp. Phys.,' 1854, p. 480; see also Camille Dareste, 'Comptes Rendus,' March 20th, 1865, p. 562.

[4] 'Elements of Physiology,' Eng. translat., vol. i., 1838, p. 412. With respect to Vrolik, see Todd's 'Cyclop. of Anat. and Phys.,' vol. iv., 1849-52, p. 973.

Hunter and others have observed lizards with their tails reproduced and doubled. When Bonnet divided longitudinally the foot of the salamander, several additional digits were occasionally formed. But neither these cases, nor the perfect series from a double monster to an additional digit, seem to me opposed to the belief that corresponding parts have a mutual affinity, and consequently tend to fuse together. A part may be doubled and remain in this state, or the two parts thus formed may afterwards through the law of affinity become blended; or two homologous parts in two separate embryos may, through the same principle, unite and form a single part.

The law of the affinity and fusion of similar parts applies to the homologous organs of the same individual animal, as well as to double monsters. Isidore Geoffroy gives a number of instances of two or more digits, of two whole legs, of two kidneys, and of several teeth becoming symmetrically fused together in a more or less perfect manner. Even the two eyes have been known to unite into a single eye, forming a cyclopean monster, as have the two ears, though naturally standing so far apart. As Geoffroy remarks, these facts illustrate in an admirable manner the normal fusion of various organs which during an early embryonic period are double, but which afterwards always unite into a single median organ. Organs of this nature are generally found in a permanently double condition in other members of the same class. These cases of normal fusion appear to me to afford the strongest support in favour of the present law. Adjoining parts which are not homologous sometimes cohere; but this cohesion appears to result from mere juxtaposition, and not from mutual affinity.

In the vegetable kingdom Moquin-Tandon[5] gives a long list of cases, showing how frequently homologous parts, such as leaves, petals, stamens, and pistils, as well as aggregates of homologous parts, such as buds, flowers, and fruit, become blended into each other with perfect symmetry. It is interesting to examine a compound flower of this nature, formed of exactly double the proper number of sepals, petals, stamens, and pistils, with each whorl of organs circular, and with no trace left of the

[5] 'Tératologie Vég.' 1841, livre iii.

process of fusion. The tendency in homologous parts to unite
during their early development, Moquin-Tandon considers as
one of the most striking laws governing the production of
monsters. It apparently explains a multitude of cases, both in
the animal and vegetable kingdoms; it throws a clear light on
many normal structures which have evidently been formed by
the union of originally distinct parts, and it possesses, as we
shall see in a future chapter, much theoretical interest.

On the Variability of Multiple and Homologous Parts.—
Isidore Geoffroy[6] insists that, when any part or organ is
repeated many times in the same animal, it is particularly
liable to vary both in number and structure. With respect to
number, the proposition may, I think, be considered as fully
established; but the evidence is chiefly derived from organic
beings living under their natural conditions, with which we are
not here concerned. When the vertebræ, or teeth, or rays in
the fins of fishes, or feathers in the tails of birds, or petals,
stamens, pistils, and seeds in plants, are very numerous, the
number is generally variable. The explanation of this simple
fact is by no means obvious. With respect to the variability in
structure of multiple parts, the evidence is not so decisive; but
the fact, as far as it may be trusted, probably depends on multiple parts being of less physiological importance than single
parts; consequently their perfect standard of structure has been
less rigorously enforced by natural selection.

Compensation of Growth, or Balancement.—This law, as applied
to natural species, was propounded by Goethe and Geoffroy St.
Hilaire at nearly the same time. It implies that, when much
organised matter is used in building up some one part, other
parts are starved and become reduced. Several authors,
especially botanists, believe in this law; others reject it. As
far as I can judge, it occasionally holds good; but its importance has probably been exaggerated. It is scarcely possible
to distinguish between the supposed effects of such compensation
of growth, and the effects of long-continued selection, which

6 ' Hist. des Anomalies,' tom. iii. pp. 4, 5, 6.

may at the same time lead to the augmentation of one part and the diminution of another. There can be no doubt that an organ may be greatly increased without any corresponding diminution in the adjoining parts. To recur to our former illustration of the Irish elk, it may be asked what part has suffered in consequence of the immense development of the horns?

It has already been observed that the struggle for existence does not bear hard on our domesticated productions; consequently the principle of economy of growth will seldom affect them, and we ought not to expect to find frequent evidence of compensation. We have, however, some such cases. Moquin-Tandon describes a monstrous bean,[7] in which the stipules were enormously developed, and the leaflets apparently in consequence completely aborted; this case is interesting, as it represents the natural condition of *Lathyrus aphaca*, with its stipules of great size, and its leaves reduced to mere threads, which act as tendrils. De Candolle[8] has remarked that the varieties of *Raphanus sativus* which have small roots yield numerous seed, valuable from containing oil, whilst those with large roots are not productive in this latter respect; and so it is with *Brassica asperifolia*. The varieties of the potato which produce tubers very early in the season rarely bear flowers; but Andrew Knight,[9] by checking the growth of the tubers, forced the plants to flower. The varieties of *Cucurbita pepo* which produce large fruit yield, according to Naudin, few in number; whilst those producing small fruit yield a vast number. Lastly, I have endeavoured to show in the eighteenth chapter that with many cultivated plants unnatural treatment checks the full and proper action of the reproductive organs, and they are thus rendered more or less sterile; consequently, in the way of compensation, the fruit becomes greatly enlarged, and, in double flowers, the petals are greatly increased in number.

With animals, it has been found difficult to produce cows which should first yield much milk, and afterwards be capable of

[7] 'Tératologie Vég.,' p. 156. *See* also my paper on climbing plants in 'Journal of Linn. Soc. Bot.,' vol. ix., 1865, p. 114.

[8] 'Mémoires du Muséum,' &c., tom. viii. p. 178.

[9] Loudon's 'Encyclop. of Gardening,' p. 820.

fattening well. With fowls which have large topknots and beards the comb and wattles are generally much reduced in size. Perhaps the entire absence of the oil-gland in fantail pigeons may be connected with the great development of their tails.

Mechanical Pressure as a Cause of Modifications.—In some few cases there is reason to believe that mere mechanical pressure has affected certain structures. Every one knows that savages alter the shape of their infants' skulls by pressure at an early age; but there is no reason to believe that the result is ever inherited. Nevertheless Vrolik and Weber[10] maintain that the shape of the human head is influenced by the shape of the mother's pelvis. The kidneys in different birds differ much in form, and St. Ange[11] believes that this is determined by the form of the pelvis, which again, no doubt, stands in close relation with their various habits of locomotion. In snakes, the viscera are curiously displaced, in comparison with their position in other vertebrates; and this has been attributed by some authors to the elongation of their bodies; but here, as in so many previous cases, it is impossible to disentangle any direct result of this kind from that consequent on natural selection. Godron has argued[12] that the normal abortion of the spur on the inner side of the flower in Corydalis, is caused by the buds being closely pressed at a very early period of growth, whilst under ground, against each other and against the stem. Some botanists believe that the singular difference in the shape both of the seed and corolla, in the interior and exterior florets in certain compositous and umbelliferous plants, is due to the pressure to which the inner florets are subjected; but this conclusion is doubtful.

The facts just given do not relate to domesticated productions, and therefore do not strictly concern us. But here is a more appropriate case: H. Müller[13] has shown that in short-

[10] Prichard, 'Phys. Hist. of Mankind,' 1851, vol. i. p. 324.

[11] 'Annales des Sc. Nat.,' 1st series, tom. xix. p. 327.

[12] 'Comptes Rendus,' Dec. 1864, p. 1039.

[13] Ueber Fötale Rachites, 'Würzburger Medicin. Zeitschrift,' 1860, B. i. s. 265.

faced races of the dog some of the molar teeth are placed in a slightly different position from that which they occupy in other dogs, especially in those having elongated muzzles; and as he remarks, any inherited change in the arrangement of the teeth deserves notice, considering their classificatory importance. This difference in position is due to the shortening of certain facial bones, and the consequent want of space; and the shortening results from a peculiar and abnormal state of the basal cartilages of the bones.

Relative Position of Flowers with respect to the Axis, and of Seeds in the Capsule, as inducing Variation.

In the thirteenth chapter various peloric flowers were described, and their production was shown to be due either to arrested development, or to reversion to a primordial condition. Moquin-Tandon has remarked that the flowers which stand on the summit of the main stem or of a lateral branch are more liable to become peloric than those on the sides;[14] and he adduces, amongst other instances, that of *Teucrium campanulatum*. In another Labiate plant grown by me, viz. the *Galeobdolon luteum*, the peloric flowers were always produced on the summit of the stem, where flowers are not usually borne. In Pelargonium, a *single* flower in the truss is frequently peloric, and when this occurs I have during several years invariably observed it to be the central flower. This is of such frequent occurrence that one observer[15] gives the names of ten varieties flowering at the same time, in every one of which the central flower was peloric. Occasionally more than one flower in the truss is peloric, and then of course the additional ones must be lateral. These flowers are interesting as showing how the whole structure is correlated. In the common Pelargonium the upper sepal is produced into a nectary which coheres with the flower-peduncle; the two upper petals differ a little in shape from the three lower ones, and are marked with dark shades of colour; the stamens are graduated in length and upturned. In the peloric flowers, the nectary aborts; all the petals become alike both in shape and colour; the stamens are generally reduced in number and become straight, so that the whole flower resembles that of the allied genus Erodium. The correlation between these changes is well shown when one of the two upper petals alone loses its dark mark, for in this case the nectary does not entirely abort, but is usually much reduced in length.[16]

---

[14] 'Tératologie Vég.,' p. 192. Dr. M. Masters informs me that he doubts the truth of this conclusion; but the facts to be given seem to be sufficient to establish it.

[15] 'Journal of Horticulture,' July 2nd, 1861, p. 253.

[16] It would be worth trial to fertilise with the same pollen the central and lateral flowers of the pelargonium, and of some other highly cultivated plants, protecting them of course from insects :

Morren has described [17] a marvellous flask-shaped flower of the Calceolaria, nearly four inches in length, which was almost completely peloric; it grew on the summit of the plant, with a normal flower on each side; Prof. Westwood also has described [18] three similar peloric flowers, which all occupied a central position on the flower-branches. In the Orchideous genus, Phalænopsis, the terminal flower has been seen to become peloric.

In a Laburnum-tree I observed that about a fourth part of the racemes produced terminal flowers which had lost their papilionaceous structure. These were produced after almost all the other flowers on the same racemes had withered. The most perfectly pelorised examples had six petals, each marked with black striæ like those on the standard-petal. The keel seemed to resist the change more than the other petals. Dutrochet has described [19] an exactly similar case in France, and I believe these are the only two instances of pelorism in the laburnum which have been recorded. Dutrochet remarks that the racemes on this tree do not properly produce a terminal flower, so that, as in the case of the Galeobdolon, their position as well as their structure are both anomalies, which no doubt are in some manner related. Dr. Masters has briefly described another leguminous plant,[20] namely, a species of clover, in which the uppermost and central flowers were regular or had lost their papilionaceous structure. In some of these plants the flower-heads were also proliferous.

Lastly, Linaria produces two kinds of peloric flowers, one having simple petals, and the other having them all spurred. The two forms, as Naudin remarks,[21] not rarely occur on the same plant, but in this case the spurred form almost invariably stands on the summit of the spike.

The tendency in the terminal or central flower to become peloric more frequently than other flowers, probably results from "the bud which stands "on the end of a shoot receiving the most sap; it grows out into a stronger "shoot than those situated lower down."[22] I have discussed the connection between pelorism and a central position, partly because some few plants are known normally to produce a terminal flower different in structure from the lateral ones; but chiefly on account of the following case, in which we see a tendency to variability or to reversion connected with the same position. A great judge of Auriculas [23] states that when an Auricula throws up a side bloom it is pretty sure to keep its character; but that if it grows from the centre or heart of the plant, whatever the colour of the edging ought to be, "it is just as likely to come in any other "class as in the one to which it properly belongs." This is so notorious a

then to sow the seed separately, and observe whether the one or the other lot of seedlings varied the most.

[17] Quoted in 'Journal of Horticulture,' Feb. 24, 1863, p. 152.

[18] 'Gardener's Chronicle,' 1866, p. 612. For the Phalænopsis, see idem, 1867, p. 211.

[19] Mémoires . . des Végétaux,' 1837, tom. ii. p. 170.

[20] 'Journal of Horticulture,' July 23, 1861, p. 311.

[21] 'Nouvelles Archives du Muséum,' tom. i. p. 137.

[22] Hugo von Mohl, 'The Vegetable Cell,' Eng. tr., 1852, p. 76.

[23] The Rev. H. H. Dombrain, in 'Journal of Horticulture,' 1861, June 4th, p. 174; and June 25th, p. 234; 186?, April 29th, p. 83.

fact, that some florists regularly pinch off the central trusses of flowers. Whether in the highly improved varieties the departure of the central trusses from their proper type is due to reversion, I do not know. Mr. Dombrain insists that, whatever may be the commonest kind of imperfection in each variety, this is generally exaggerated in the central truss. Thus one variety "sometimes has the fault of producing a little green " floret in the centre of the flower," and in central blooms these become excessive in size. In some central blooms, sent to me by Mr. Dombrain, all the organs of the flower were rudimentary in structure, of minute size, and of a green colour, so that by a little further change all would have been converted into small leaves. In this case we clearly see a tendency to prolification—a term which, I may explain to those who have never attended to botany, means the production of a branch or flower, or head of flowers, out of another flower. Now Dr. Masters[24] states that the central or uppermost flower on a plant is generally the most liable to prolification. Thus, in the varieties of the Auricula, the loss of their proper character and a tendency to prolification, and in other plants a tendency to prolification and pelorism, are all connected together, and are due either to arrested development, or to reversion to a former condition.

The following is a more interesting case; Metzger[25] cultivated in Germany several kinds of maize brought from the hotter parts of America, and he found, as has been previously described, that in two or three generations the grains became greatly changed in form, size, and colour; and with respect to two races he expressly states that in the first generation, whilst the lower grains on each head retained their proper character, the uppermost grains already began to assume that character which in the third generation all the grains acquired. As we do not know the aboriginal parent of the maize, we cannot tell whether these changes are in any way connected with reversion.

In the two following cases, reversion, as influenced by the position of the seed in the capsule, evidently acts. The Blue Imperial pea is the offspring of the Blue Prussian, and has larger seed and broader pods than its parent. Now Mr. Masters, of Canterbury, a careful observer and a raiser of new varieties of the pea, states[26] that the Blue Imperial always has a strong tendency to revert to its parent-stock, and the reversion " occurs in this manner: the last (or uppermost) pea in the pod is fre- " quently much smaller than the rest; and if these small peas are care- " fully collected and sown separately, very many more, in proportion, " will revert to their origin, than those taken from the other parts of the " pod." Again M. Chaté[27] says that in raising seedling stocks he succeeds in getting eighty per cent. to bear double flowers, by leaving only a few of the secondary branches to seed; but in addition to this, " at the time " of extracting the seeds, the upper portion of the pod is separated and

---

[24] 'Transact. Linn. Soc.,' vol. xxiii., 1861, p. 360.

[25] 'Die Getreidearten,' 1843, s. 208, 209.

[26] 'Gardener's Chronicle,' 1850, p. 198.

[27] Quoted in 'Gardener's Chron.,' 1866, p. 74.

" placed aside, because it has been ascertained that the plants coming from
" the seeds situated in this portion of the pod, give eighty per cent. of single
"flowers." Now the production of single-flowering plants from the seed of
double-flowering plants is clearly a case of reversion. These latter facts,
as well as the connection between a central position and pelorism and
prolification, show in an interesting manner how small a difference—namely
a little greater freedom in the flow of sap towards one part of the same
plant—determines important changes of structure.

Analogous or Parallel Variation.—By this term I wish to
express that similar characters occasionally make their appearance in the several varieties or races descended from the same
species, and more rarely in the offspring of widely distinct species.
We are here concerned, not as hitherto with the causes of variation, but with the results; but this discussion could not have
been more conveniently introduced elsewhere. The cases of
analogous variation, as far as their origin is concerned, may be
grouped, disregarding minor subdivisions, under two main heads;
firstly, those due to unknown causes having acted on organic
beings with nearly the same constitution, and which consequently
vary in an analogous manner; and secondly, those due to the
reappearance of characters which were possessed by a more
or less remote progenitor. But these two main divisions can
often be only conjecturally separated, and graduate, as we shall
presently see, into each other.

Under the first head of analogous variations, not due to reversion, we
have the many cases of trees belonging to quite different orders which
have produced pendulous and fastigate varieties. The beech, hazel, and
barberry have given rise to purple-leaved varieties; and as Bernhardi
has remarked,[28] a multitude of plants, as distinct as possible, have yielded
varieties with deeply-cut or laciniated leaves. Varieties descended from
three distinct species of Brassica have their stems, or so-called roots, enlarged into globular masses. The nectarine is the offspring of the peach;
and the varieties of both these trees offer a remarkable parallelism in the
fruit being white, red, or yellow fleshed—in being clingstones or free-
stones—in the flowers being large or small—in the leaves being serrated
or crenated, furnished with globose or reniform glands, or quite destitute
of glands. It should be remarked that each variety of the nectarine has
not derived its character from a corresponding variety of the peach. The
several varieties also of a closely allied genus, namely the apricot, differ
from each other in nearly the same parallel manner. There is no reason

---

[28] ' Ueber den Begriff der Pflanzenart,' 1834, s. 14.

to believe that in any of these cases long-lost characters have reappeared, and in most of them this certainly has not occurred.

Three species of Cucurbita have yielded a multitude of races, which correspond so closely in character that, as Naudin insists, they may be arranged in an almost strictly parallel series. Several varieties of the melon are interesting from resembling in important characters other species, either of the same genus or of allied genera; thus, one variety has fruit so like, both externally and internally, the fruit of a perfectly distinct species, namely, the cucumber, as hardly to be distinguished from it; another has long cylindrical fruit twisting about like a serpent; in another the seeds adhere to portions of the pulp; in another the fruit, when ripe, suddenly cracks and falls into pieces; and all these highly remarkable peculiarities are characteristic of species belonging to allied genera. We can hardly account for the appearance of so many unusual characters by reversion to a single ancient form; but we must believe that all the members of the family have inherited a nearly similar constitution from an early progenitor. Our cereal and many other plants offer similar cases.

With animals we have fewer cases of analogous variation, independently of direct reversion. We see something of the kind in the resemblance between the short-muzzled races of the dog, such as the pug and bulldog; in feather-footed races of the fowl, pigeon, and canary-bird; in horses of the most different races presenting the same range of colour; in all black-and-tan dogs having tan-coloured eye-spots and feet, but in this latter case reversion may possibly have played a part. Low has remarked [29] that several breeds of cattle are "sheeted,"—that is, have a broad band of white passing round their bodies like a sheet; this character is strongly inherited and sometimes originates from a cross; it may be the first step in reversion to an original or early type, for, as was shown in the third chapter, white cattle with dark ears, feet, and tip of tail formerly existed, and now exist in a feral or semi-feral condition in several quarters of the world.

Under our second main division, namely, of analogous variations due to reversion, the best cases are afforded by animals, and by none better than by pigeons. In all the most distinct breeds sub-varieties occasionally appear coloured exactly like the parent rock-pigeon, with black wing-bars, white loins, banded tail, &c.; and no one can doubt that these characters are simply due to reversion. So with minor details; turbits properly have white tails, but occasionally a bird is born with a dark-coloured and banded tail; pouters properly have white primary wing-feathers, but not rarely a "sword-flighted" bird, that is, one with the few first primaries dark-coloured, appears; and in these cases we have characters proper to the rock-pigeon, but new to the breed, evidently appearing from reversion. In some domestic varieties the wing-bars, instead of being simply black, as in the rock-pigeon, are beautifully edged with different zones of colour, and they then present a striking analogy with the wing-bars in certain natural species of the same family, such as *Phaps chalcoptera*; and this may probably be accounted for by

---

[29] 'Domesticated Animals,' 1845, p. 351.

all the forms descended from the same remote progenitor having a tendency to vary in the same manner. Thus also we can perhaps understand the fact of some Laugher-pigeons cooing almost like turtle-doves, and of several races having peculiarities in their flight, for certain natural species (viz. *C. torquatrix* and *palumbus*) display singular vagaries in this respect. In other cases a race, instead of imitating in character a distinct species, resembles some other race; thus certain runts tremble and slightly elevate their tails, like fantails; and turbits inflate the upper part of their œsophagus, like pouter-pigeons.

It is a common circumstance to find certain coloured marks persistently characterising all the species of a genus, but differing much in tint; and the same thing occurs with the varieties of the pigeon: thus, instead of the general plumage being blue with the wing-bars black, there are snow-white varieties with red bars, and black varieties with white bars; in other varieties the wing-bars, as we have seen, are elegantly zoned with different tints. The Spot pigeon is characterised by the whole plumage being white, excepting the tail and a spot on the forehead; but these parts may be red, yellow, or black. In the rock-pigeon and in many varieties the tail is blue, with the outer edges of the outer feathers white; but in one sub-variety of the monk-pigeon we have a reversed variation, for the tail is white, except the outer edges of the outer feathers, which are black.[30]

With some species of birds, for instance with gulls, certain coloured parts appear as if almost washed out, and I have observed exactly the same appearance in the terminal dark tail-bar in certain pigeons, and in the whole plumage of certain varieties of the duck. Analogous facts in the vegetable kingdom could be given.

Many sub-varieties of the pigeon have reversed and somewhat lengthened feathers on the back part of their heads, and this is certainly not due to reversion to the parent-species, which shows no trace of such structure; but when we remember that sub-varieties of the fowl, turkey, canary-bird, duck, and goose, all have topknots or reversed feathers on their heads; and when we remember that scarcely a single large natural group of birds can be named, in which some members have not a tuft of feathers on their heads, we may suspect that reversion to some extremely remote form has come into action.

Several breeds of the fowl have either spangled or pencilled feathers; and these cannot be derived from the parent-species, the *Gallus bankiva*; though of course it is possible that an early progenitor of this species may have been spangled, and a still earlier or a later progenitor may have been pencilled. But as many gallinaceous birds are spangled or pencilled, it is a more probable view that the several domestic breeds of the fowl have acquired this kind of plumage from all the members of the family inheriting a tendency to vary in a like manner. The same principle may account for the ewes in certain breeds of sheep being hornless, like the females of some other hollow-horned ruminants; it may account for certain domestic cats having slightly-tufted ears, like those of the lynx; and for the skulls of domestic rabbits often differing from each

---

[30] Bechstein, ' Naturgeschichte Deutschlands,' Band iv., 1795, s. 31.

other in the same characters by which the skulls of the various species of the genus Lepus differ.

I will only allude to one other case, already discussed. Now that we know that the wild parent of the ass has striped legs, we may feel confident that the occasional appearance of stripes on the legs of the domestic ass is due to direct reversion; but this will not account for the lower end of the shoulder-stripe being sometimes angularly bent or slightly forked. So, again, when we see dun and other coloured horses with stripes on the spine, shoulders, and legs, we are led, from reasons formerly given, to believe that they reappear from direct reversion to the wild parent-horse. But when horses have two or three shoulder-stripes with one of them occasionally forked at the lower end, or when they have stripes on their faces, or as foals are faintly striped over nearly their whole bodies, with the stripes angularly bent one under the other on the forehead, or irregularly branched in other parts, it would be rash to attribute such diversified characters to the reappearance of those proper to the aboriginal wild horse. As three African species of the genus are much striped, and as we have seen that the crossing of the unstriped species often leads to the hybrid offspring being conspicuously striped—bearing also in mind that the act of crossing certainly causes the reappearance of long-lost characters—it is a more probable view that the above-specified stripes are due to reversion, not to the immediate wild parent-horse, but to the striped progenitor of the whole genus.

I have discussed this subject of analogous variation at considerable length, because, in a future work on natural species, it will be shown that the varieties of one species frequently mock distinct species—a fact in perfect harmony with the foregoing cases, and explicable only on the theory of descent. Secondly, because these facts are important from showing, as remarked in a former chapter, that each trifling variation is governed by law, and is determined in a much higher degree by the nature of the organisation, than by the nature of the conditions to which the varying being has been exposed. Thirdly, because these facts are to a certain extent related to a more general law, namely, that which Mr. B. D. Walsh[31] has called the " Law of *Equable Variability*," or, as he explains it, " if any given character is very variable in one species of a " group, it will tend to be variable in allied species; and if any " given character is perfectly constant in one species of a group, " it will tend to be constant in allied species."

This leads me to recall a discussion in the chapter on Selection, in which it was shown that with domestic races, which are

[31] ' Proc. Entomolog. Soc. of Philadelphia,' Oct. 1863, p. 213.

now undergoing rapid improvement, those parts or characters
which are the most valued vary the most. This naturally follows from recently selected characters continually tending to
revert to their former less improved standard, and from their
being still acted on by the same agencies, whatever these may
be, which first caused the characters in question to vary. The
same principle is applicable to natural species, for, as stated
in my ' Origin of Species,' generic characters are less variable
than specific characters; and the latter are those which have
been modified by variation and natural selection, since the period
when all the species belonging to the same genus branched off
from a common progenitor, whilst generic characters are those
which have remained unaltered from a much more remote epoch,
and accordingly are now less variable. This statement makes
a near approach to Mr. Walsh's law of Equable Variability.
Secondary sexual characters, it may be added, rarely serve to
characterise distinct genera, for they usually differ much in the
species of the same genus, and are highly variable in the individuals of the same species; we have also seen in the earlier
chapters of this work how variable secondary sexual characters
become under domestication.

Summary of the three previous Chapters, on the Laws of Variation.

In the twenty-third chapter we have seen that changed conditions occasionally act in a definite manner on the organisation,
so that all, or nearly all, the individuals thus exposed become
modified in the same manner. But a far more frequent result
of changed conditions, whether acting directly on the organisation or indirectly through the reproductive system being affected,
is indefinite and fluctuating variability. In the three latter
chapters we have endeavoured to trace some of the laws by
which such variability is regulated.

Increased use adds to the size of a muscle, together with the
blood-vessels, nerves, ligaments, the crests of bone to which these
are attached, the whole bone and other connected bones. So it
is with various glands. Increased functional activity strengthens
the sense-organs. Increased and intermittent pressure thickens
the epidermis; and a change in the nature of the food sometimes modifies the coats of the stomach, and increases or

decreases the length of the intestines. Continued disuse, on the other hand, weakens and diminishes all parts of the organisation. Animals which during many generations have taken but little exercise, have their lungs reduced in size, and as a consequence the bony fabric of the chest, and the whole form of the body, become modified. With our anciently domesticated birds, the wings have been little used, and they are slightly reduced; with their decrease, the crest of the sternum, the scapulæ, coracoids, and furcula, have all been reduced.

With domesticated animals, the reduction of a part from disuse is never carried so far that a mere rudiment is left, but we have good reason to believe that this has often occurred under nature. The cause of this difference probably is that with domestic animals not only sufficient time has not been granted for so profound a change, but that, from not being exposed to a severe struggle for life, the principle of the economy of organisation does not come into action. On the contrary, we sometimes see that structures which are rudimentary in the parent-species become partially redeveloped in their domesticated progeny. When rudiments are formed or left under domestication, they are the result of a sudden arrest of development, and not of long-continued disuse with the absorption of all superfluous parts; nevertheless they are of interest, as showing that rudiments are the relics of organs once perfectly developed.

Corporeal, periodical, and mental habits, though the latter have been almost passed over in this work, become changed under domestication, and the changes are often inherited. Such changed habits in any organic being, especially when living a free life, would often lead to the augmented or diminished use of various organs, and consequently to their modification. From long-continued habit, and more especially from the occasional birth of individuals with a slightly different constitution, domestic animals and cultivated plants become to a certain extent acclimatised, or adapted to a climate different from that proper to the parent-species.

Through the principle of correlated variability, when one part varies other parts vary,—either simultaneously, or one after the other. Thus an organ modified during an early embryonic period affects other parts subsequently developed. When an organ, such as the beak, increases or decreases in length, adjoining or correlated parts, as the tongue and the orifice of the nostrils, tend to vary in the same manner. When the whole body increases or decreases in size, various parts become modified; thus with pigeons the ribs increase or decrease in number and breadth. Homologous parts, which are identical during their early development and are exposed to similar conditions, tend to vary in the same or in some connected manner,— as in the case of the right and left sides of the body, of the front and hind limbs, and even of the head and limbs. So it is with the organs of sight and hearing; for instance, white cats with blue eyes are almost always deaf. There is a manifest relation throughout the body between the skin and its various appendages of hair, feathers, hoofs, horns, and teeth. In Paraguay, horses with curly hair have hoofs like those of a mule; the wool and the horns of sheep vary together; hairless dogs are deficient in their teeth; men with redundant hair have abnormal teeth, either deficient or in excess. Birds with long wing-feathers usually have long tail-feathers. When long feathers grow from the outside of the legs and toes of pigeons, the two outer toes are connected by membrane; for the whole leg tends to assume the structure of the wing. There is a manifest relation between a crest of feathers on the head and a marvellous amount of change in the skull of various fowls; and in a lesser degree, between the greatly elongated, lopping ears of rabbits and the structure of their skulls. With plants, the leaves, various parts of the flower, and the fruit, often vary together in a correlated manner.

In some cases we find correlation without being able even to conjecture what is the nature of the connexion, as with various correlated monstrosities and diseases. This is likewise the case with the colour of the adult pigeon, in connexion with the presence of down on the young bird. Numerous curious instances have been given of peculiarities of constitution, in correlation with colour, as shown by the immunity of individuals of some one colour from certain diseases, from the attacks of parasites, and from the action of certain vegetable poisons.

Correlation is an important subject; for with species, and in a lesser degree with domestic races, we continually find that certain parts have been greatly modified to serve some useful purpose; but we almost invariably find that other parts have likewise been more or less modified, without our being able to discover any advantage in the change. No doubt great caution is necessary in coming to this conclusion, for it is difficult to overrate our ignorance on the use of various parts of the organisation; but from what we have now seen, we may believe that many modifications are of no direct service, having arisen in correlation with other and useful changes.

Homologous parts during their early development evince an affinity for each other,—that is, they tend to cohere and fuse together much more readily than other parts. This tendency to fusion explains a multitude of normal structures. Multiple and homologous organs are especially liable to vary in number and probably in form. As the supply of organised matter is not unlimited, the principle of compensation sometimes comes into action; so that, when one part is greatly developed, adjoining parts or functions are apt to be reduced; but this principle is probably of much less importance than the more general one of the economy of growth. Through mere mechanical pressure hard parts occasionally affect soft adjoining parts. With plants the position of the flowers on the axis, and of the seeds in the capsule, sometimes leads, through a freer flow of sap, to changes of structure; but these changes are often due to reversion. Modifications, in whatever manner caused, will be to a certain extent regulated by that co-ordinating power or *nisus formativus*, which is in fact a remnant of one of the forms of reproduction, displayed by many lowly organised beings in their power of fissiparous generation and budding. Finally, the effects of the laws, which directly or indirectly govern variability, may be largely influenced by man's selection, and will so far be determined by natural selection that changes advantageous to any race will be favoured and disadvantageous changes checked.

Domestic races descended from the same species, or from two or more allied species, are liable to revert to characters derived from their common progenitor, and, as they have much in common in their constitutions, they are also liable under changed conditions to vary in the same manner; from these

two causes analogous varieties often arise. When we reflect
on the several foregoing laws, imperfectly as we understand
them, and when we bear in mind how much remains to be discovered, we need not be surprised at the extremely intricate
manner in which our domestic productions have varied, and
still go on varying.

CHAPTER XXVII.

PROVISIONAL HYPOTHESIS OF PANGENESIS.

PRELIMINARY REMARKS. — FIRST PART :— THE FACTS TO BE CONNECTED UNDER A
SINGLE POINT OF VIEW, NAMELY, THE VARIOUS KINDS OF REPRODUCTION — THE
DIRECT ACTION OF THE MALE ELEMENT ON THE FEMALE — DEVELOPMENT —
THE FUNCTIONAL INDEPENDENCE OF THE ELEMENTS OR UNITS OF THE BODY —
VARIABILITY — INHERITANCE — REVERSION.
SECOND PART :— STATEMENT OF THE HYPOTHESIS — HOW FAR THE NECESSARY ASSUMP-
TIONS ARE IMPROBABLE — EXPLANATION BY AID OF THE HYPOTHESIS OF THE
SEVERAL CLASSES OF FACTS SPECIFIED IN THE FIRST PART — CONCLUSION.

In the previous chapters large classes of facts, such as those
bearing on bud-variation, the various forms of inheritance, the
causes and laws of variation, have been discussed; and it is
obvious that these subjects, as well as the several modes of reproduction, stand in some sort of relation to each other. I have been
led, or rather forced, to form a view which to a certain extent
connects these facts by a tangible method. Every one would
wish to explain to himself, even in an imperfect manner, how it
is possible for a character possessed by some remote ancestor
suddenly to reappear in the offspring; how the effects of
increased or decreased use of a limb can be transmitted to the
child; how the male sexual element can act not solely on
the ovule, but occasionally on the mother-form; how a limb can
be reproduced on the exact line of amputation, with neither too
much nor too little added; how the various modes of reproduction are connected, and so forth. I am aware that my view is
merely a provisional hypothesis or speculation; but until a
better one be advanced, it may be serviceable by bringing
together a multitude of facts which are at present left dis-
connected by any efficient cause. As Whewell, the historian of
the inductive sciences, remarks :—" Hypotheses may often be
" of service to science, when they involve a certain portion of
" incompleteness, and even of error." Under this point of view
I venture to advance the hypothesis of Pangenesis, which im-

plies that the whole organisation, in the sense of every separate atom or unit, reproduces itself. Hence ovules and pollengrains,—the fertilised seed or egg, as well as buds,—include and consist of a multitude of germs thrown off from each separate atom of the organism.

In the First Part I will enumerate as briefly as I can the groups of facts which seem to demand connection; but certain subjects, not hitherto discussed, must be treated at disproportionate length. In the Second Part the hypothesis will be given; and we shall see, after considering how far the necessary assumptions are in themselves improbable, whether it serves to bring under a single point of view the various facts.

PART I.

Reproduction may be divided into two main classes, namely, sexual and asexual. The latter is effected in many ways—by gemmation, that is by the formation of buds of various kinds, and by fissiparous generation, that is by spontaneous or artificial division. It is notorious that some of the lower animals, when cut into many pieces, reproduce so many perfect individuals: Lyonnet cut a Nais or freshwater worm into nearly forty pieces, and these all reproduced perfect animals.[1] It is probable that segmentation could be carried much further in some of the protozoa, and with some of the lowest plants each cell will reproduce the parent-form. Johannes Müller thought that there was an important distinction between gemmation and fission; for in the latter case the divided portion, however small, is more perfectly organised; but most physiologists are now convinced that the two processes are essentially alike.[2] Prof. Huxley remarks, " fission is little more than a peculiar

[1] Quoted by Paget, ' Lectures on Pathology,' 1853, p. 159.

[2] Dr. Lachmann, also, observes (' Annals and Mag. of Nat. History,' 2nd series, vol. xix., 1857, p. 231) with respect to infusoria, that " fissa-" tion and gemmation pass into each " other almost imperceptibly." Again, Mr. W. C. Minor (' Annals and Mag. of Nat. Hist.,' 3rd series, vol. xi. p. 328) shows that with Annelids the distinction that has been made between fission and budding is not a fundamental one. See Bonnet, ' Œuvres d'Hist. Nat.,' tom. v., 1781, p. 339, for remarks on the budding-out of the amputated limbs of Salamanders. See, also, Professor Clark's work ' Mind in Nature,' New York, 1865, pp. 62, 94.

" mode of budding," and Prof. H. J. Clark, who has especially attended to this subject, shows in detail that there is sometimes "a compromise between self-division and budding." When a limb is amputated, or when the whole body is bisected, the cut extremities are said to bud forth; and as the papilla, which is first formed, consists of undeveloped cellular tissue like that forming an ordinary bud, the expression is apparently correct. We see the connection of the two processes in another way; for Trembley observed that with the hydra the reproduction of the head after amputation was checked as soon as the animal began to bud.[3]

Between the production, by fissiparous generation, of two or more complete individuals, and the repair of even a very slight injury, we have, as remarked in a former chapter, so perfect and insensible a gradation, that it is impossible to doubt that they are connected processes. Between the power which repairs a trifling injury in any part, and the power which previously "was occupied in its maintenance by the continued " mutation of its particles," there cannot be any great difference ; and we may follow Mr. Paget in believing them to be the selfsame power. As at each stage of growth an amputated part is replaced by one in the same state of development, we must likewise follow Mr. Paget in admitting " that the powers " of development from the embryo are identical with those " exercised for the restoration from injuries : in other words, " that the powers are the same by which perfection is first " achieved, and by which, when lost, it is recovered."[4] Finally, we may conclude that the several forms of gemmation, and of fissiparous generation, the repair of injuries, the maintenance of each part in its proper state, and the growth or progressive development of the whole structure of the embryo, are all essentially the results of one and the same great power.

Sexual Generation.—The union of the two sexual elements seems to make a broad distinction between sexual and asexual reproduction. But the well-ascertained cases of Parthenogenesis prove that the distinction is not really so great as it at first appears ; for ovules occasionally, and even in some cases fre-

[3] Paget, 'Lectures on Pathology,' 1853, p. 158.   [4] Idem. pp. 152, 164.

quently, become developed into perfect beings, without the concourse of the male element.   J. Müller and others admit that
ovules and buds have the same essential nature.   Certain bodies,
which during their early development cannot be distinguished
by any external character from true ovules, nevertheless must
be classed as buds, for though formed within the ovarium they
are incapable of fertilisation.   This is the case with the germ-
balls of the Cecidomyide larvæ, as described by Leuckart.[5]
Ovules and the male element, before they become united, have,
like buds, an independent existence.[6]   Both have the power of
transmitting every single character possessed by the parent-
form.   We see this clearly when hybrids are paired *inter se*,
for the characters of either grandparent often reappear, either
perfectly or by segments, in the progeny.   It is an error to
suppose that the male transmits certain characters and the
female other characters; though no doubt, from unknown
causes, one sex sometimes has a stronger power of transmission
than the other.

It has been maintained by some authors that a bud differs
essentially from a fertilised germ, by always reproducing the
perfect character of the parent-stock; whilst fertilised germs
become developed into beings which differ, in a greater or less
degree, from each other and from their parents.   But there is
no such broad distinction as this.   In the eleventh chapter,
numerous cases were given showing that buds occasionally
grow into plants having new and strongly marked characters;
and varieties thus produced can be propagated for a length
of time by buds, and occasionally by seed.   Nevertheless, it
must be admitted that beings produced sexually are much more
liable to vary than those produced asexually; and of this fact a
partial explanation will hereafter be attempted.   The variability
in both cases is determined by the same general causes, and is
governed by the same laws.   Hence new varieties arising from
buds cannot be distinguished from those arising from seed.
Although bud-varieties usually retain their character during

---

[5] On the Asexual Reproduction of
Cecydomyide Larvæ, translated in
'Annals and Mag. of Nat. Hist.,' March
1866, pp. 167, 171.

[6] *See* some excellent remarks on this
head by Quatrefages, in 'Annales des
Sc. Nat.,' Zoolog., 3rd series, 1850, p.
138.

successive bud-generations, yet they occasionally revert, even after a long series of bud-generations, to their former character. This tendency to reversion in buds is one of the most remarkable of the several points of agreement between the offspring from bud and seminal reproduction.

There is, however, one difference between beings produced sexually and asexually, which is very general. The former usually pass in the course of their development from a lower to a higher grade, as we see in the metamorphoses of insects and in the concealed metamorphoses of the vertebrata; but this passage from a lower to a higher grade cannot be considered as a necessary accompaniment of sexual reproduction, for hardly anything of the kind occurs in the development of Aphis amongst insects, or with certain crustaceans, cephalopods, or with any of the higher vascular plants. Animals propagated asexually by buds or fission are on the other hand never known to undergo a retrogressive metamorphosis; that is, they do not first sink to a lower, before passing on to their higher and final stage of development. But during the act of asexual production or subsequently to it, they often advance in organisation, as we see in the many cases of " alternate generation." In thus speaking of alternate generation, I follow those naturalists who look at the process as essentially one of internal budding or of fissiparous generation. Some of the lower plants, however, such as mosses and certain algæ, according to Dr. L. Radlkofer,[7] when propagated asexually, do undergo a retrogressive metamorphosis. We can to a certain extent understand, as far as the final cause is concerned, why beings propagated by buds should so rarely retrogress during development; for with each organism the structure acquired at each stage of development must be adapted to its peculiar habits. Now, with beings produced by gemmation, —and this, differently from sexual reproduction, may occur at any period of growth,—if there were places for the support of many individuals at some one stage of development, the simplest plan would be that they should be multiplied by gemmation at that stage, and not that they should first retrograde in their development to an earlier or simpler structure, which might not be fitted for the surrounding conditions.

[7] ' Annals and Mag. of Nat. Hist.,' 2nd series, vol. xx., 1857, pp. 153-455.

From the several foregoing considerations we may conclude that the difference between sexual and asexual generation is not nearly so great as it at first appears; and we have already seen that there is the closest agreement between gemmation, fissiparous generation, the repair of injuries, and ordinary growth or development. The capacity of fertilisation by the male element seems to be the chief distinction between an ovule and a bud; and this capacity is not invariably brought into action, as in the cases of parthenogenetic reproduction. We are here naturally led to inquire what the final cause can be of the necessity in ordinary generation for the concourse of the two sexual elements.

Seeds and ova are often highly serviceable as the means of disseminating plants and animals, and of preserving them during one or more seasons in a dormant state; but unimpregnated seeds or ova, and detached buds, would be equally serviceable for both purposes. We can, however, indicate two important advantages gained by the concourse of the two sexes, or rather of two individuals belonging to opposite sexes; for, as I have shown in a former chapter, the structure of every organism appears to be especially adapted for the concurrence, at least occasionally, of two individuals. In nearly the same manner as it is admitted by naturalists that hybridism, from inducing sterility, is of service in keeping the forms of life distinct and fitted for their proper places; so, when species are rendered highly variable by changed conditions of life, the free inter-crossing of the varying individuals will tend to keep each form fitted for its proper place in nature; and crossing can be effected only by sexual generation, but whether the end thus gained is of sufficient importance to account for the first origin of sexual intercourse is very doubtful. Secondly, I have shown, from the consideration of a large body of facts, that, as a slight change in the conditions of life is beneficial to each creature, so, in an analogous manner, is the change effected in the germ by sexual union with a distinct individual; and I have been led, from observing the many widely-extended provisions throughout nature for this purpose, and from the greater vigour of crossed organisms of all kinds, as proved by direct experiments, as well as from the evil effects of close interbreeding when long continued, to believe that the advantage thus gained is very great. Besides these two important ends, there may, of course, be others, as yet unknown to us, gained by the concourse of the two sexes.

Why the germ, which before impregnation undergoes a certain amount of development, ceases to progress and perishes, unless it be acted on by the male element; and why conversely the male element, which is enabled to keep alive for even four or five years within the spermatheca of a female insect, likewise perishes, unless it acts on or unites with the germ, are questions which cannot be answered with any certainty. It is, however, possible that both sexual elements perish, unless brought into union, simply from including too little formative matter for independent existence and development; for certainly they do not in ordinary cases differ in their power of giving character to the embryo. This view of the importance of the quantity of formative matter seems probable from the following considerations. There is no reason to suspect that the spermatozoa or pollen-grains of the same individual animal or plant differ from each other; yet Quatrefages has shown in the case of the Teredo,[8] as did formerly Prevost and Dumas with other animals, that more than one spermatozoon is requisite to fertilise an ovule. This has likewise been clearly proved by Newport,[9] who adds the important fact, established by numerous experiments, that, when a very small number of spermatozoa are applied to the ova of Batrachians, they are only partially impregnated and the embryo is never fully developed: the first step, however, towards development, namely, the partial segmentation of the yelk, does occur to a greater or less extent, but is never completed up to granulation. The rate of the segmentation is likewise determined by the number of the spermatozoa. With respect to plants, nearly the same results were obtained by Kölreuter and Gärtner. This last careful observer found,[10] after making successive trials on a Malva with more and more pollen-grains, that even thirty grains did not fertilise a single seed; but when forty grains were applied to the

---

[8] 'Annales des Sc. Nat.,' 3rd series, 1850, tom. xiii.
[9] 'Transact. Phil. Soc.,' 1851, pp. 196, 208, 210 ; 1853, p. 245, 247.
[10] 'Beitrage zur Kenntniss,' &c., 1844, s. 345.

stigma, a few seeds of small size were formed. The pollen-grains of Mirabilis are extraordinarily large, and the ovarium contains only a single ovule; and these circumstances led Naudin[11] to make the following interesting experiments: a flower was fertilised by three grains and succeeded perfectly; twelve flowers were fertilised by two grains, and seventeen flowers by a single grain, and of these one flower alone in each lot perfected its seed; and it deserves especial notice that the plants produced by these two seeds never attained their proper dimensions, and bore flowers of remarkably small size. From these facts we clearly see that the quantity of the peculiar formative matter which is contained within the spermatozoa and pollen-grains is an all-important element in the act of fertilisation, not only in the full development of the seed, but in the vigour of the plant produced from such seed. We see something of the same kind in certain cases of parthenogenesis, that is, when the male element is wholly excluded; for M. Jourdan[12] found that, out of about 58,000 eggs laid by unimpregnated silk-moths, many passed through their early embryonic stages, showing that they were capable of self-development, but only twenty-nine out of the whole number produced caterpillars. Therefore it is not an improbable view that deficient bulk or quantity in the formative matter, contained within the sexual elements, is the main cause of their not having the capacity of prolonged separate existence and development. The belief that it is the function of the spermatozoa to communicate life to the ovule seems a strange one, seeing that the unimpregnated ovule is already alive and continues for a considerable time alive. We shall hereafter see that it is probable that the sexual elements, or possibly only the female element, include certain primordial cells, that is, such as have undergone no differentiation, and which are not present in an active state in buds.

Graft-hybrids.—When discussing in the eleventh chapter the curious case of the *Cytisus adami*, facts were given which render it to a certain degree probable, in accordance with the belief of some distinguished botanists, that, when the tissues of two plants

[11] 'Nouvelles Archives du Muséum,' tom. i. p. 27.
[10] As quoted by Sir J. Lubbock in 'Nat. Hist. Review,' 1862, p. 345.

belonging to distinct species or varieties are intimately united, buds are afterwards occasionally produced which, like hybrids, combine the characters of the two united forms. It is certain that when trees with variegated leaves are grafted or budded on a common stock, the latter sometimes produces buds bearing variegated leaves; but this may perhaps be looked at as a case of inoculated disease. Should it ever be proved that hybridised buds can be formed by the union of two distinct vegetative tissues, the essential identity of sexual and asexual reproduction would be shown in the most interesting manner; for the power of combining in the offspring the characters of both parents, is the most striking of all the functions of sexual generation.

Direct Action of the Male Element on the Female.—In the chapter just referred to, I have given abundant proofs that foreign pollen occasionally affects the mother-plant in a direct manner. Thus, when Gallesio fertilised an orange-flower with pollen from the lemon, the fruit bore stripes of perfectly characterised lemon-peel: with peas, several observers have seen the colour of the seed-coats and even of the pod directly affected by the pollen of a distinct variety; so it has been with the fruit of the apple, which consists of the modified calyx and upper part of the flower-stalk. These parts in ordinary cases are wholly formed by the mother-plant. We here see the male element affecting and hybridising not that part which it is properly adapted to affect, namely the ovule, but the partially developed tissues of a distinct individual. We are thus brought half-way towards a graft-hybrid, in which the cellular tissue of one form, instead of its pollen, is believed to hybridise the tissues of a distinct form. I formerly assigned reasons for rejecting the belief that the mother-plant is affected through the intervention of the hybridised embryo; but even if this view were admitted, the case would become one of graft-hybridism, for the fertilised embryo and the mother-plant must be looked at as distinct individuals.

With animals which do not breed until nearly mature, and of which all the parts are then fully developed, it is hardly possible that the male element should directly affect the female. But we have the analogous and perfectly well-ascertained case of the male element of a distinct form, as with the quagga and Lord Morton's mare, affecting the ovarium of the female, so that the ovules and offspring subsequently produced by her when impregnated by other males are plainly affected and hybridised by the first male.

Development.—The fertilised germ reaches maturity by a vast number of changes: these are either slight and slowly effected, as when the child grows into the man, or are great and sudden, as with the metamorphoses of most insects. Between these extremes we have, even within the same class, every gradation: thus, as Sir J. Lubbock has shown,[13] there is an Ephemerous insect which moults above twenty times, undergoing each time a slight but decided change of structure; and these changes, as he further remarks, probably reveal to us the normal stages of development which are concealed and hurried through, or suppressed, in most other insects. In ordinary metamorphoses, the parts and organs appear to become changed into the corresponding parts in the next stage of development; but there is another form of development, which has been called by Professor Owen metagenesis. In this case " the new parts are " not moulded upon the inner surface of the old ones. The " plastic force has changed its course of operation. The outer " case, and all that gave form and character to the precedent " individual, perish and are cast off; they are not changed into " the corresponding parts of the new individual. These are due " to a new and distinct developmental process," &c.[14] Metamorphosis, however, graduates so insensibly into metagenesis, that the two processes cannot be distinctly separated. For instance, in the last change which Cirripedes undergo, the alimentary canal and some other organs are moulded on pre-existing parts; but the eyes of the old and the young animal are developed in entirely different parts of the body; the tips of the mature limbs are formed within the larval limbs, and may be said to be metamorphosed from them; but their basal portions and the whole thorax are developed in a plane actually at right angles to the limbs and thorax of the larva; and this

[13] ' Transact. Linn. Soc.,' vol. xxiv., 1863, p. 62.

[14] ' Parthenogenesis,' 1849, pp. 25–26. Prof. Huxley has some excellent remarks (' Medical Times,' 1850, p. 637) on this subject, in reference to the development of star-fishes, and shows how curiously metamorphosis graduates into gemmation or zoid-formation, which is in fact the same as metagenesis.

may be called metagenesis. The metagenetic process is carried to an extreme degree in the development of some Echinoderms, for the animal in the second stage of development is formed almost like a bud within the animal of the first stage, the latter being then cast off like an old vestment, yet sometimes still maintaining for a short period an independent vitality.[15]

If, instead of a single individual, several were to be thus developed metagenetically within a pre-existing form, the process would be called one of alternate generation. The young thus developed may either closely resemble the encasing parent-form, as with the larvæ of Cecidomyia, or may differ to an astonishing degree, as with many parasitic worms and with jelly-fishes; but this does not make any essential difference in the process, any more than the greatness or abruptness of the change in the metamorphoses of insects.

The whole question of development is of great importance for our present subject. When an organ, the eye for instance, is metagenetically formed in a part of the body where during the previous stage of development no eye existed, we must look at it as a new and independent growth. The absolute independence of new and old structures, which correspond in structure and function, is still more obvious when several individuals are formed within a previous encasing form, as in the cases of alternate generation. The same important principle probably comes largely into play even in the case of continuous growth, as we shall see when we consider the inheritance of modifications at corresponding ages.

We are led to the same conclusion, namely, the independence of parts successively developed, by another and quite distinct group of facts. It is well known that many animals belonging to the same class, and therefore not differing widely from each other, pass through an extremely different course of development. Thus certain beetles, not in any way remarkably different from others of the same order, undergo what has been called a hyper-metamorphosis—that is, they pass through an early stage wholly different from the ordinary grub-like larva. In the same sub-order of crabs, namely, the Macroura, as Fritz

[15] Prof. J. Reay Greene, in Günther's ' Record of Zoolog. Lit.,' 1865, p. 625.

Müller remarks, the river cray-fish is hatched under the same form which it ever afterwards retains; the young lobster has divided legs, like a Mysis; the Palæmon appears under the form of a Zoea, and Peneus under the Nauplius-form; and how wonderfully these larval forms differ from each other, is known to every naturalist.[16] Some other crustaceans, as the same author observes, start from the same point and arrive at nearly the same end, but in the middle of their development are widely different from each other. Still more striking cases could be given with respect to the Echinodermata. With the Medusæ or jelly-fishes Professor Allman observes, "the classi-" fication of the Hydroida would be a comparatively simple " task if, as has been erroneously asserted, generically-identical " medusoids always arose from generically-identical polypoids; " and on the other hand, that generically-identical polypoids " always gave origin to generically-identical medusoids." So, again, Dr. Strethill Wright remarks, " in the life-history of the " Hydroidæ any phase, planuloid, polypoid, or medusoid, may " be absent." [17]

According to the belief now generally accepted by our best naturalists, all the members of the same order or class, the Macrourous crustaceans for instance, are descended from a common progenitor. During their descent they have diverged much in structure, but have retained much in common; and this divergence and retention of character has been effected, though they have passed and still pass through marvellously different metamorphoses. This fact well illustrates how independent each structure must be from that which precedes and follows it in the course of development.

The Functional Independence of the Elements or Units of the Body.—Physiologists agree that the whole organism consists of a multitude of elemental parts, which are to a great extent independent of each other. Each organ, says Claude Bernard,[18]

---

[16] Fritz Müller's 'Für Darwin,' 1864, s. 65, 71. The highest authority on crustaceans, Prof. Milne Edwards, insists ('Annal. des Sci. Nat.,' 2nd series, Zoolog., tom. iii. p. 322) on their metamorphoses differing even in closely allied genera.

[17] Prof. Allman, in 'Annals and Mag. of Nat. Hist.,' 3rd series, vol. xiii., 1864, p. 348; Dr. S. Wright, idem, vol. viii., 1861, p. 127. See also p. 358 for analogous statements by Sars.

[18] 'Tissus Vivants,' 1866, p. 22.

has its proper life, its autonomy; it can develop and reproduce itself independently of the adjoining tissues. The great German authority, Virchow,[19] asserts still more emphatically that each system, as the nervous or osseous system, or the blood, consists of an "enormous mass of minute centres of action. . . . . Every " element has its own special action, and even though it derive " its stimulus to activity from other parts, yet alone effects the " actual performance of its duties. . . . . Every single epithelial " and muscular fibre-cell leads a sort of parasitical existence " in relation to the rest of the body. . . . . Every single bone- " corpuscle really possesses conditions of nutrition peculiar to " itself." Each element, as Mr. Paget remarks, lives its appointed time, and then dies, and, after being cast off or absorbed, is replaced.[20] I presume that no physiologist doubts that, for instance, each bone-corpuscle of the finger differs from the corresponding corpuscle in the corresponding joint of the toe; and there can hardly be a doubt that even those on the corresponding sides of the body differ, though almost identical in nature. This near approach to identity is curiously shown in many diseases in which the same exact points on the right and left sides of the body are similarly affected; thus Mr. Paget[21] gives a drawing of a diseased pelvis, in which the bone has grown into a most complicated pattern, but "there is not one " spot or line on one side which is not represented, as exactly as " it would be in a mirror, on the other."

Many facts support this view of the independent life of each minute element of the body. Virchow insists that a single bone-corpuscle or a single cell in the skin may become diseased. The spur of a cock, after being inserted into the eye of an ox, lived for eight years, and acquired a weight of 306 grammes, or nearly fourteen ounces.[22] The tail of a pig has been grafted into the middle of its back, and reacquired sensibility. Dr. Ollier[23] inserted a piece of periosteum from the bone of a young dog under the skin of a rabbit, and true bone was developed. A multitude of similar facts could be given The

---

[19] 'Cellular Pathology,' translat. by Dr. Chance, 1860, pp. 14, 18, 83, 460.

[20] Paget, 'Surgical Pathology,' vol. i., 1853, pp. 12-14.

[21] Idem, p. 19.

[22] Mantegazza, quoted in 'Popular Science Review,' July 1865, p. 522.

[23] 'De la Production Artificielle des Os,' p. 8.

frequent presence of hairs and of perfectly developed teeth, even teeth of the second dentition, in ovarian tumours,[24] are facts leading to the same conclusion.

Whether each of the innumerable autonomous elements of the body is a cell or the modified product of a cell, is a more doubtful question, even if so wide a definition be given to the term, as to include cell-like bodies without walls and without nuclei.[25]   Professor Lionel Beale uses the term "germinal " matter" for the contents of cells, taken in this wide accepta- tion, and he draws a broad distinction between germinal matter and "formed material" or the various products of cells.[26]   But the doctrine of *omnis cellula e cellulâ* is admitted for plants, and is a widely prevalent belief with respect to animals.[27]   Thus Virchow, the great supporter of the cellular theory, whilst allowing that difficulties exist, maintains that every atom of tissue is derived from cells, and these from pre-existing cells, and these primarily from the egg, which he regards as a great cell.   That cells, still retaining the same nature, increase by self-division or proliferation, is admitted by almost every one. But when an organism undergoes a great change of structure during development, the cells, which at each stage are supposed to be directly derived from previously-existing cells, must likewise be greatly changed in nature; this change is apparently attri- buted by the supporters of the cellular doctrine to some inherent power which the cells possess, and not to any external agency.

Another school maintains that cells and tissues of all kinds may be formed, independently of pre-existing cells, from plastic lymph or blastema; and this it is thought is well exhibited in the repair of wounds.   As I have not especially attended to histology, it would be presumptuous in me to express an opi- nion on the two opposed doctrines.   But every one appears to admit that the body consists of a multitude of "organic units,"[28]

[24] Isidore Geoffroy St. Hilaire, ' Hist. des Anomalies,' tom. ii. pp. 549, 560, 562 ; Virchow, idem, p. 484.

[25] For the most recent classification of cells, *see* Ernst Häckel's ' Generelle Morpholog.,' Band ii., 1866, s. 275.

[26] ' The Structure and Growth of Tissues,' 1865, p. 21, &c.

[27] Dr. W. Turnor, ' The present

Aspect of Cellular Pathology,' ' Edin- burgh Medical Journal,' April, 1863.

[28] This term is used by Dr. E. Mont- gomery (' On the Formation of so-called Cells in Animal Bodies,' 1867, p. 42), who denies that cells are derived from other cells by a process of growth, but believes that they originate through certain chemical changes.

each of which possesses its own proper attributes, and is to a certain extent independent of all others. Hence it will be convenient to use indifferently the terms cells or organic units or simply units.

Variability and Inheritance.—We have seen in the twenty-second chapter that variability is not a principle co-ordinate with life or reproduction, but results from special causes, generally from changed conditions acting during successive generations. Part of the fluctuating variability thus induced is apparently due to the sexual system being easily affected by changed conditions, so that it is often rendered impotent; and when not so seriously affected, it often fails in its proper function of transmitting truly the characters of the parents to the offspring. But variability is not necessarily connected with the sexual system, as we see from the cases of bud-variation; and although we may not be able to trace the nature of the connexion, it is probable that many deviations of structure which appear in sexual offspring result from changed conditions acting directly on the organisation, independently of the reproductive organs. In some instances we may feel sure of this, when all, or nearly all the individuals which have been similarly exposed are similarly and definitely affected—as in the dwarfed and otherwise changed maize brought from hot countries when cultivated in Germany; in the change of the fleece in sheep within the tropics; to a certain extent in the increased size and early maturity of our highly-improved domesticated animals; in inherited gout from intemperance; and in many other such cases. Now, as such changed conditions do not especially affect the reproductive organs, it seems mysterious on any ordinary view why their product, the new organic being, should be similarly affected.

How, again, can we explain to ourselves the inherited effects of the use or disuse of particular organs? The domesticated duck flies less and walks more than the wild duck, and its limb-bones have become in a corresponding manner diminished and increased in comparison with those of the wild duck. A horse is trained to certain paces, and the colt inherits similar consensual movements. The domesticated rabbit becomes tame from close confinement; the dog intelligent from associating with man; the retriever is taught to fetch and carry: and these

2 B 2

mental endowments and bodily powers are all inherited. Nothing in the whole circuit of physiology is more wonderful. How can the use or disuse of a particular limb or of the brain affect a small aggregate of reproductive cells, seated in a distant part of the body, in such a manner that the being developed from these cells inherits the characters of either one or both parents? Even an imperfect answer to this question would be satisfactory.

Sexual reproduction does not essentially differ, as we have seen, from budding or self-division, and these processes graduate through the repair of injuries into ordinary development and growth; it might therefore be expected that every character would be as regularly transmitted by all the methods of reproduction as by continued growth. In the chapters devoted to inheritance it was shown that a multitude of newly-acquired characters, whether injurious or beneficial, whether of the lowest or highest vital importance, are often faithfully transmitted —frequently even when one parent alone possesses some new peculiarity. It deserves especial attention that characters appearing at any age tend to reappear at a corresponding age. We may on the whole conclude that in all cases inheritance is the rule, and non-inheritance the anomaly. In some instances a character is not inherited, from the conditions of life being directly opposed to its development; in many instances, from the conditions incessantly inducing fresh variability, as with grafted fruit-trees and highly cultivated flowers. In the remaining cases the failure may be attributed to reversion, by which the child resembles its grandparents or more remote progenitors, instead of its parents.

This principle of Reversion is the most wonderful of all the attributes of Inheritance. It proves to us that the transmission of a character and its development, which ordinarily go together and thus escape discrimination, are distinct powers; and these powers in some cases are even antagonistic, for each acts alternately in successive generations. Reversion is not a rare event, depending on some unusual or favourable combination of circumstances, but occurs so regularly with crossed animals and plants, and so frequently with uncrossed breeds, that it is evidently an essential part of the principle of inheritance. We know that changed conditions have the power of evoking long-lost characters, as in the case of some feral animals. The act of crossing in itself possesses this power in a high degree. What can be more wonderful than that characters, which have disappeared during scores, or hundreds, or even thousands of generations, should suddenly reappear perfectly developed, as in the case of pigeons and fowls when purely bred, and especially when crossed ; or as with the zebrine stripes on dun-coloured horses, and other such cases? Many monstrosities come under this same head, as when rudimentary organs are redeveloped, or when an organ which we must believe was possessed by an early progenitor, but of which not even a rudiment is left, suddenly reappears, as with the fifth stamen in some Scrophulariaceæ. We have already seen that reversion acts in bud-reproduction ; and we know that it occasionally acts during the growth of the same individual animal, especially, but not exclusively, when of crossed parentage,—as in the rare cases described of individual fowls, pigeons, cattle, and rabbits, which have reverted as they advanced in years to the colours of one of their parents or ancestors.

We are led to believe, as formerly explained, that every character which occasionally reappears is present in a latent form in each generation, in nearly the same manner as in male and female animals secondary characters of the opposite sex lie latent, ready to be evolved when the reproductive organs are injured. This comparison of the secondary sexual characters which are latent in both sexes, with other latent characters, is the more appropriate from the case recorded of the Hen, which assumed some of the masculine characters, not of her own race, but of an early progenitor; she thus exhibited at the same time the redevelopment of latent characters of both kinds and connected both classes. In every living creature we may feel assured that a host of lost characters lie ready to be evolved under proper conditions. How can we make intelligible, and connect with other facts, this wonderful and common capacity of reversion,—this power of calling back to life long-lost characters ?

Part II.

I have now enumerated the chief facts which every one would desire to connect by some intelligible bond. This can be done, as it seems to me, if we make the following assumptions; if the first and chief one be not rejected, the others, from being supported by various physiological considerations, will not appear very improbable. It is almost universally admitted that cells, or the units of the body, propagate themselves by self-division or proliferation, retaining the same nature, and ultimately becoming converted into the various tissues and substances of the body. But besides this means of increase I assume that cells, before their conversion into completely passive or "formed material," throw off minute granules or atoms, which circulate freely throughout the system, and when supplied with proper nutriment multiply by self-division, subsequently becoming developed into cells like those from which they were derived. These granules for the sake of distinctness may be called cell-gemmules, or, as the cellular theory is not fully established, simply gem-mules. They are supposed to be transmitted from the parents to the offspring, and are generally developed in the genera-tion which immediately succeeds, but are often transmitted in a dormant state during many generations and are then developed. Their development is supposed to depend on their union with other partially developed cells or gemmules which precede them in the regular course of growth. Why I use the term union, will be seen when we discuss the direct action of pollen on the tissues of the mother-plant. Gemmules are supposed to be thrown off by every cell or unit, not only during the adult state, but during all the stages of develop-ment. Lastly, I assume that the gemmules in their dormant state have a mutual affinity for each other, leading to their aggregation either into buds or into the sexual elements. Hence, speaking strictly, it is not the reproductive elements, nor the buds, which generate new organisms, but the cells themselves throughout the body. These assumptions constitute the pro-visional hypothesis which I have called Pangenesis. Nearly similar views have been propounded, as I find, by other authors, more especially by Mr. Herbert Spencer;[29] but they are here modified and amplified.

[29] Prof. Huxley has called my attention to the views of Buffon and Bonnet. The former ('Hist. Nat. Gén.,' edit. of 1749, tom. ii. pp. 54, 62, 329, 333, 420, 425) supposes that organic molecules exist in the food consumed by every living creature; and that these molecules are analogous in nature with the various organs by which they are absorbed. When the organs thus become fully developed, the molecules being no longer required collect and form buds or the sexual elements. If Buffon had assumed that his organic molecules had been formed by each separate unit throughout the body, his view and mine would have been closely similar.

Bonnet ('Œuvres d'Hist. Nat.,' tom. v., part i., 1781, 4to edit., p. 334) speaks of the limbs having germs adapted for the reparation of all possible losses; but whether these germs are supposed to be the same with those within the buds and sexual organs is not clear. His famous but now exploded theory of *emboîtement* implies that perfect germs are included within germs in endless succession, pre-formed and ready for all succeeding generations. According to my view, the germs or gemmules of each separate part were not originally pre-formed, but are continually produced at all ages during each generation, with some handed down from preceding generations.

Prof. Owen remarks ('Parthenogenesis,' 1849, pp. 5-8), "Not all the " progeny of the primary impregnated " germ-cell are required for the formation " of the body in all animals: certain of " the derivative germ-cells may remain " unchanged and become included in " that body which has been composed " of their metamorphosed and diversely " combined or confluent brethren: so " included, any derivative germ-cell, or " the nucleus of such, may commence and " repeat the same processes of growth by " imbibition, and of propagation by spon-

" taneous fission, as those to which itself " owed its origin;" &c. By the agency of these germ-cells Prof. Owen accounts for parthenogenesis, for propagation by self-division during successive generations, and for the repairs of injuries. His view agrees with mine in the assumed transmission and multiplication of his germ-cells, but differs fundamentally from mine in the belief that the primary germ-cell was formed within the ovarium of the female and was fertilised by the male. My gemmules are supposed to be formed, quite independently of sexual concourse, by each separate cell or unit throughout the body, and to be merely aggregated within the reproductive organs.

Lastly, Mr. Herbert Spencer ('Principles of Biology, vol. i., 1863-4, chaps. iv. and viii.) has discussed at considerable length what he designates as physiological units. These agree with my gemmules in being supposed to multiply and to be transmitted from parent to child; the sexual elements are supposed to serve merely as their vehicles; they are the efficient agents in all the forms of reproduction and in the repairs of injuries; they account for inheritance, but they are not brought to bear on reversion or atavism, and this is unintelligible to me; they are supposed to possess polarity, or, as I call it, affinity; and apparently they are believed to be derived from each separate part of the whole body. But gemmules differ from Mr. Spencer's physiological units, inasmuch as a certain number, or mass of them, are, as we shall see, requisite for the development of each cell or part. Nevertheless I should have concluded that Mr. Spencer's views were fundamentally the same with mine, had it not been for several passages which, as far as I understand them, indicate something quite different. I will quote some of these passages from pp. 254-256. " In the fertilised germ

Before proceeding to show, firstly, how far these assumptions are in themselves probable, and secondly, how far they connect and explain the various groups of facts with which we are concerned, it may be useful to give an illustration of the hypothesis. If one of the simplest Protozoa be formed, as appears under the microscope, of a small mass of homogeneous gelatinous matter, a minute atom thrown off from any part and nourished under favourable circumstances would naturally reproduce the whole; but if the upper and lower surfaces were to differ in texture from the central portion, then all three parts would have to throw off atoms or gemmules, which when aggregated by mutual affinity would form either buds or the sexual elements. Precisely the same view may be extended to one of the higher animals; although in this case many thousand gemmules must be thrown off from the various parts of the body. Now, when the leg, for instance, of a salamander is cut off, a slight crust forms over the wound, and beneath this crust the uninjured cells or units of bone, muscle, nerves, &c., are supposed to unite with the diffused gemmules of those cells which in the perfect leg come next in order; and these as they become slightly developed unite with others, and so on until a papilla of soft cellular tissue, the "budding leg," is formed, and in time a perfect leg.[30] Thus, that portion of the leg which had

"we have two groups of physiological "units, slightly different in their struc- "tures." . . . "It is not obvious that "change in the form of the part, caused "by changed action, involves such change "in the physiological units throughout "the organism, that these, when groups "of them are thrown off in the shape of "reproductive centres, will unfold into "organisms that have this part similarly "changed in form. Indeed, when treat- "ing of Adaptation, we saw that an "organ modified by increase or decrease "of function can but slowly so react on "the system at large as to bring about "those correlative changes required to "produce a new equilibrium; and yet "only when such new equilibrium has "been established, can we expect it to be "*fully* expressed in the modified physio- "logical units of which the organism is

"built—only then can we count on a "complete transfer of the modification "to descendants." . . . . . "That the "change in the offspring must, other "things equal, be in the same direction "as the change in the parent, we may "dimly see is implied by the fact, that "the change propagated throughout the "parental system is a change towards "a new state of equilibrium—a change "tending to bring the actions of all "organs, reproductive included, into "harmony with these new actions."

[30] M. Philipeaux (' Comptes Rendus,' Oct. 1, 1866, p. 576, and June, 1867) has lately shown that when the entire fore-limb, including the scapula, is ex- tirpated, the power of regrowth is lost. From this he concludes that it is neces- sary for regrowth that a small portion of the limb should be left. But as in

been cut off, neither more nor less, would be reproduced. If the tail or leg of a young animal had been cut off, a young tail or leg would have been reproduced, as actually occurs with the amputated tail of the tadpole; for gemmules of all the units which compose the tail are diffused throughout the body at all ages. But during the adult state the gemmules of the larval tail would remain dormant, for they would not meet with pre-existing cells in a proper state of development with which to unite. If from changed conditions or any other cause any part of the body should become permanently modified, the gem-mules, which are merely minute portions of the contents of the cells forming the part, would naturally reproduce the same modification. But gemmules previously derived from the same part before it had undergone any change, would still be diffused throughout the organisation, and would be transmitted from generation to generation, so that under favourable circumstances they might be redeveloped, and then the new modification would be for a time or for ever lost. The aggregation of gemmules derived from every part of the body, through their mutual affinity, would form buds, and their aggregation in some special manner, apparently in small quantity, together probably with the presence of gemmules of certain primordial cells, would constitute the sexual elements. By means of these illustrations the hypothesis of pangenesis has, I hope, been rendered intelligible.

Physiologists maintain, as we have seen, that each cell, though to a large extent dependent on others, is likewise, to a certain extent, independent or autonomous. I go one small step further, and assume that each cell casts off a free gemmule, which is capable of reproducing a similar cell. There is some analogy between this view and what we see in compound animals and in the flower-buds on the same tree; for these are distinct individuals capable of true or seminal reproduction, yet have parts in common and are dependent on each other; thus

the lower animals the whole body may be bisected and both halves be reproduced, this belief does not seem probable. May not the early closing of a deep wound, as in the case of the extirpation of the scapula, prevent the formation or protrusion of the nascent limb?

the tree has its bark and trunk, and. certain corals, as the Virgularia, have not only parts, but movements in common.

The existence of free gemmules is a gratuitous assumption, yet can hardly be considered as very improbable, seeing that cells have the power of multiplication through the self-division of their contents. Gemmules differ from true ovules or buds inasmuch as they are supposed to be capable of multiplication in their undeveloped state. No one probably will object to this capacity as improbable. The blastema within the egg has been known to divide and give birth to two embryos; and Thuret[31] has seen the zoospore of an alga divide itself, and both halves germinate. An atom of small-pox matter, so minute as to be borne by the wind, must multiply itself many thousand-fold in a person thus inoculated.[32] It has recently been ascer-tained[33] that a minute portion of the mucous discharge from an animal affected with rinderpest, if placed in the blood of a healthy ox, increases so fast that in a short space of time "the " whole mass of blood, weighing many pounds, is infected, and " every small particle of that blood contains enough poison to " give, within less than forty-eight hours, the disease to another " animal."

The retention of free and undeveloped gemmules in the same body from early youth to old age may appear improbable, but we should remember how long seeds lie dormant in the earth and buds in the bark of a tree. Their transmission from genera-tion to generation may appear still more improbable; but here again we should remember that many rudimentary and useless organs are transmitted and have been transmitted during an indefinite number of generations. We shall presently see how well the long-continued transmission of undeveloped gemmules explains many facts.

As each unit, or group of similar units throughout the body, casts off its gemmules, and as all are contained within the smallest egg or seed, and within each spermatozoon or pollen-grain, their number and minuteness must be something incon-

---

[31] 'Annal. des Sc. Nat.,' 3rd series, Bot., tom. xiv., 1850, p. 244.

[32] *See* some very interesting papers on this subject by Prof. Lionel Beale, in 'Medical Times and Gazette,' Sept.

9th, 1865, pp. 273, 330.

[33] Third Report of the R. Comm. on the Cattle Plague, as quoted in 'Gard. Chronicle,' 1866, p. 446.

ceivable. I shall hereafter recur to this objection, which at
first appears so formidable; but it may here be remarked that
a cod-fish has been found to produce 4,872,000 eggs, a single
Ascaris about 64,000,000 eggs, and a single Orchidaceous plant
probably as many million seeds.[34] In these several cases, the
spermatozoa and pollen-grains must exist in considerably larger
numbers. Now, when we have to deal with numbers such as
these, which the human intellect cannot grasp, there is no
good reason for rejecting our present hypothesis on account of
the assumed existence of cell-gemmules a few thousand times
more numerous.

The gemmules in each organism must be thoroughly diffused;
nor does this seem improbable considering their minuteness, and
the steady circulation of fluids throughout the body. So it
must be with the gemmules of plants, for with certain kinds
even a minute fragment of a leaf will reproduce the whole.
But a difficulty here occurs; it would appear that with plants,
and probably with compound animals, such as corals, the gemmules do not spread from bud to bud, but only through the
tissues developed from each separate bud. We are led to this
conclusion from the stock being rarely affected by the insertion
of a bud or graft from a distinct variety. This non-diffusion of
the gemmules is still more plainly shown in the case of ferns;
for Mr. Bridgman[35] has proved that, when spores (which it
should be remembered are of the nature of buds) are taken from
a monstrous part of a frond, and others from an ordinary part,

[34] In a cod-fish, weighing 20 lb., Mr.
F. Buckland ('Land and Water,' 1867,
p. 57) calculated the above number of
eggs. In another instance, Harmer
('Phil. Transact.,' 1767, p. 280) found
3,681,760 eggs. For the Ascaris, see Carpenter's 'Comp. Phys.,' 1854, p. 590. Mr.
J. Scott, of the Royal Botanic Garden of
Edinburgh, calculated, in the same
manner as I have done for some British
orchids ('Fertilisation of Orchids,' p.
344), the number of seeds in a capsule of
an Acropera, and found the number to
be 371,250. Now this plant produces
several flowers on a raceme and many
racemes during a season. In an allied
genus, Gongora, Mr. Scott has seen

twenty capsules produced on a single
raceme : ten such racemes on the
Acropera would yield above seventy-four
millions of seed. I may add that Fritz
Müller informs me that he found in
a capsule of a Maxillaria, in South
Brazil, that the seed weighed 42½
grains : he then arranged half a grain
of seed in a narrow line, and by counting
a measured length found the number in
the half-grain to be 20,667, so that in
the capsule there must have been
1,756,440 seeds ! The same plant sometimes produces half-a-dozen capsules.

[35] 'Annals and Mag. of Nat. Hist.'
3rd series, vol. viii., 1861, p. 490.

each reproduces the form of the part whence derived. But this non-diffusion of the gemmules from bud to bud may be only apparent, depending, as we shall hereafter see, on the nature of the first-formed cells in the buds.

The assumed elective affinity of each gemmule for that particular cell which precedes it in the order of development is supported by many analogies. In all ordinary cases of sexual reproduction the male and female elements have a mutual affinity for each other: thus, it is believed that about ten thousand species of Compositæ exist, and there can be no doubt that if the pollen of all these species could be, simultaneously or successively, placed on the stigma of any one species, this one would elect with unerring certainty its own pollen. This elective capacity is all the more wonderful, as it must have been acquired since the many species of this great group of plants branched off from a common progenitor. On any view of the nature of sexual reproduction, the protoplasm contained within the ovules and within the sperm-cells (or the " spermatic force" of the latter, if so vague a term be preferred) must act on each other by some law of special affinity, either during or subsequently to impregnation, so that corresponding parts alone affect each other; thus, a calf produced from a short-horned cow by a long-horned bull has its horns and not its horny hoofs affected by the union of the two forms, and the offspring from two birds with differently coloured tails have their tails and not their whole plumage affected.

The various tissues of the body plainly show, as many physiologists have insisted,[36] an affinity for special organic substances, whether natural or foreign to the body. We see this in the cells of the kidneys attracting urea from the blood; in the worrara poison affecting the nerves; upas and digitalis the muscles; the Lytta vesicatoria the kidneys; and in the poisonous matter of many diseases, as small-pox, scarlet-fever, hooping-cough, glanders, cancer, and hydrophobia, affecting certain definite parts of the body or certain tissues or glands.

The affinity of various parts of the body for each other during

---

[36] Paget, 'Lectures on Pathology,' p. 27; Virchow, 'Cellular Pathology,' translat. by Dr. Chance, pp. 123, 126, 294; Claude Bernard, 'Des Tissus Vivants,' pp. 177, 210, 337; Müller's 'Physiology,' Eng. translat., p. 290.

their early development was shown in the last chapter,
when discussing the tendency to fusion in homologous parts.
This affinity displays itself in the normal fusion of organs
which are separate at an early embryonic age, and still more
plainly in those marvellous cases of double monsters in which
each bone, muscle, vessel, and nerve in the one embryo, blends
with the corresponding part in the other. The affinity between
homologous organs may come into action with single parts,
or with the entire individual, as in the case of flowers or fruits
which are symmetrically blended together with all their parts
doubled, but without any other trace of fusion.

It has also been assumed that the development of each gemmule depends on its union with another cell or unit which has
just commenced its development, and which, from preceding it in
order of growth, is of a somewhat different nature. Nor is it a
very improbable assumption that the development of a gemmule
is determined by its union with a cell slightly different in nature,
for abundant evidence was given in the seventeenth chapter,
showing that a slight degree of differentiation in the male and
female sexual elements favours in a marked manner their union
and subsequent development. But what determines the development of the gemmules of the first-formed or primordial cell in
the unimpregnated ovule, is beyond conjecture.

It must also be admitted that analogy fails to guide us
towards any determination on several other points : for instance,
whether cells, derived from the same parent-cell, may, in the
regular course of growth, become developed into different structures, from absorbing peculiar kinds of nutriment, independently
of their union with distinct gemmules. We shall appreciate this
difficulty if we call to mind, what complex yet symmetrical
growths the cells of plants yield when they are inoculated by
the poison of a gall-insect. With animals various polypoid
excrescences and tumours are now generally admitted[37] to be
the direct product, through proliferation, of normal cells which
have become abnormal. In the regular growth and repair of
bones, the tissues undergo, as Virchow remarks,[38] a whole series
of permutations and substitutions. "The cartilage-cells may be

[37] Virchow, ' Cellular Pathology,' trans. by Dr. Chance, 1860. pp. 60, 162, 245,
441, 454.   [38] Idem, pp. 412-426.

" converted by a direct transformation into marrow-cells, and
" continue as such ; or they may first be converted into osseous
" and then into medullary tissue ; or lastly, they may first be
" converted into marrow and then into bone. So variable are
" the permutations of these tissues, in themselves so nearly
" allied, and yet in their external appearance so completely
" distinct." But as these tissues thus change their nature at
any age, without any obvious change in their nutrition, we
must suppose in accordance with our hypothesis that gemmules
derived from one kind of tissue combine with the cells of another
kind, and cause the successive modifications.

It is useless to speculate at what period of development each
organic unit casts off its gemmules ; for the whole subject of the
development of the various elemental tissues is as yet involved
in much doubt. Some physiologists, for instance, maintain that
muscle or nerve-fibres are developed from cells, which are
afterwards nourished by their own proper powers of absorption ;
whilst other physiologists deny their cellular origin ; and Beale
maintains that such fibres are renovated exclusively by the
conversion of fresh germinal matter (that is the so-called nuclei)
into " formed material." However this may be, it appears probable that all external agencies, such as changed nutrition,
increased use or disuse, &c., which induced any permanent modification in a structure, would at the same time or previously act
on the cells, nuclei, germinal or formative matter, from which
the structures in question were developed, and consequently
would act on the gemmules or cast-off atoms.

There is another point on which it is useless to speculate,
namely, whether all gemmules are free and separate, or whether
some are from the first united into small aggregates. A feather,
for instance, is a complex structure, and, as each separate part
is liable to inherited variations, I conclude that each feather
certainly generates a large number of gemmules ; but it is
possible that these may be aggregated into a compound gemmule. The same remark applies to the petals of a flower,
which in some cases are highly complex, with each ridge and
hollow contrived for special purposes, so that each part must
have been separately modified, and the modifications transmitted ;
consequently, separate gemmules, according to our hypothesis,

must have been thrown off from each cell or part. But, as we sometimes see half an anther or a small portion of a filament becoming petaliform, or parts or mere stripes of the calyx assuming the colour and texture of the corolla, it is probable that with petals the gemmules of each cell are not aggregated together into a compound gemmule, but are freely and separately diffused.

Having now endeavoured to show that the several foregoing assumptions are to a certain extent supported by analogous facts, and having discussed some of the most doubtful points, we will consider how far the hypothesis brings under a single point of view the various cases enumerated in the First Part. All the forms of reproduction graduate into each other and agree in their product; for it is impossible to distinguish between organisms produced from buds, from self-division, or from fertilised germs; such organisms are liable to variations of the same nature and to reversion of character; and as we now see that all the forms of reproduction depend on the aggregation of gemmules derived from the whole body, we can understand this general agreement. It is satisfactory to find that sexual and asexual generation, by both of which widely different processes the same living creature is habitually produced, are fundamentally the same. Parthenogenesis is no longer wonderful; in fact, the wonder is that it should not oftener occur. We see that the reproductive organs do not actually create the sexual elements; they merely determine or permit the aggregation of the gemmules in a special manner. These organs, together with their accessory parts, have, however, high functions to perform; they give to both elements a special affinity for each other, independently of the contents of the male and female cells, as is shown in the case of plants by the mutual reaction of the stigma and pollen-grains; they adapt one or both elements for independent temporary existence, and for mutual union. The contrivances for these purposes are sometimes wonderfully complex, as with the spermatophores of the Cephalopoda. The male element sometimes possesses attributes which, if observed in an independent animal, would be put down to instinct guided by sense-organs, as when the spermato-

zoon of an insect finds its way into the minute micropyle of the
egg, or as when the antherozoids of certain algæ swim by
the aid of their ciliæ to the female plant, and force themselves
a minute orifice. In these latter cases, however, we must into
believe that the male element has acquired its powers, on the
same principle with the larvæ of animals, namely by successive
modifications developed at corresponding periods of life: we
can hardly avoid in these cases looking at the male element
as a sort of premature larva, which unites, or, like one of the
lower algæ, conjugates, with the female element. What determines the aggregation of the gemmules within the sexual organs
we do not in the least know; nor do we know why buds are
formed in certain definite places, leading to the symmetrical
growth of trees and corals, nor why adventitious buds may be
formed almost anywhere, even on a petal, and frequently upon
healed wounds.[39] As soon as the gemmules have aggregated
themselves, development apparently commences, but in the
case of buds is often afterwards suspended, and in the case of
the sexual elements soon ceases, unless the elements of the
opposite sexes combine; even after this has occurred, the fertilised germ, as with seeds buried in the ground, may remain
during a lengthened period in a dormant state.

The antagonism which has long been observed,[40] though exceptions occur,[41] between active growth and the power of sexual
reproduction—between the repair of injuries and gemmation—
and with plants, between rapid increase by buds, rhizomes, &c.,
and the production of seed, is partly explained by the gemmules not existing in sufficient numbers for both processes.

[39] See Rev. J. M. Berkeley, in 'Gard.
Chron.,' April 28th, 1866, on a bud
developed on the petal of the Clarkia.
See also H. Schacht, 'Lehrbuch der
Anat.,' &c., 1859, Theile ii. s. 12, on
adventitious buds.
[40] Mr. Herbert Spencer ('Principles
of Biology,' vol. ii. p. 430) has fully
discussed the antagonism between
growth and reproduction.
[41] The male salmon is known to breed
at a very early age. The Triton and
Siredon, whilst retaining their larval
branchiæ, according to Filippi and
Duméril ('Annals and Mag. of Nat

Hist.,' 3rd series, 1866, p. 157), are
capable of reproduction. Ernst Häckel
has recently (' Monatsbericht Akad.
Wiss. Berlin,' Feb. 2nd, 1865) observed the surprising case of a medusa,
with its reproductive organs active,
which produces by budding a widely
different form of medusa; and this
latter also has the power of sexual reproduction. Krohn has shown ('Annals
and Mag. of Nat. Hist.,' 3rd series, vol.
xix., 1862, p. 6) that certain other
medusæ, whilst sexually mature, propagate by gemmæ.

But this explanation hardly applies to those plants which naturally produce a multitude of seeds, but which, through a comparatively small increase in the number of the buds on their rhizomes or offsets, yield few or no seed. As, however, we shall presently see that buds probably include tissue which has already been to a certain extent developed or differentiated, some additional organised matter will thus have been expended.

From one of the forms of Reproduction, namely, spontaneous self-division, we are led by insensible steps to the repair of the slightest injury; and the existence of gemmules, derived from every cell or unit throughout the body and everywhere diffused, explains all such cases,—even the wonderful fact that, when the limbs of the salamander were cut off many times successively by Spallanzani and Bonnet, they were exactly and completely reproduced. I have heard this process compared with the re-crystallisation which occurs when the angles of a broken crystal are repaired; and the two processes have this much in common, that in the one case the polarity of the molecules is the efficient cause, and in the other the affinity of the gemmules for particular nascent cells.

Pangenesis does not throw much light on Hybridism, but agrees well with most of the ascertained facts. We may conclude from the fact of a single spermatozoon or pollen-grain being insufficient for impregnation, that a certain number of gemmules derived from each cell or unit are required for the development of each part. From the occurrence of parthenogenesis, more especially in the case of the silk-moth, in which the embryo is often partially formed, we may also infer that the female element includes nearly sufficient gemmules of all kinds for independent development, so that when united with the male element the gemmules must be superabundant. Now, as a general rule, when two species or races are crossed reciprocally, the offspring do not differ, and this shows that both sexual elements agree in power, in accordance with the view that they include the same gemmules. Hybrids and mongrels are generally intermediate in character between the two parent-forms, yet occasionally they closely resemble one parent in one part and the other parent in another part, or even in their whole structure: nor is this difficult to understand on the admission that the gemmules in the fertilised germ are superabundant in number, and that those derived from one parent have some advantage in number, affinity, or vigour over those derived from the other parent. Crossed forms sometimes exhibit the colour or other characters of either parent in stripes or blotches; and this may occur in the first generation, or through reversion in succeeding bud and seminal generations, as in the several instances given in the eleventh chapter. In these cases we must follow Naudin,[42] and admit that the "essence" or "element" of the two species, which terms I should translate into the gemmules, have an affinity for their own kind, and thus separate themselves into distinct stripes or blotches; and reasons were given, when discussing in the fifteenth chapter the incompatibility of certain characters to unite, for believing in such mutual affinity. When two forms are crossed, one is not rarely found to be prepotent in the transmission of character over the other; and this we can explain only by again assuming that the one form has some advantage in the number, vigour, or affinity of its gemmules, except in those cases, where certain characters are present in the one form and latent in the other. For instance, there is a latent tendency in all pigeons to become blue, and, when a blue pigeon is crossed with one of any other colour, the blue tint is generally prepotent. When we consider latent characters, the explanation of this form of prepotency will be obvious.

When one species is crossed with another it is notorious that they do not yield the full or proper number of offspring; and we can only say on this head that, as the development of each organism depends on such nicely-balanced affinities between a host of gemmules and developing cells or units, we need not feel at all surprised that the commixture of gemmules derived from two distinct species should lead to a partial or complete failure of development. With respect to the sterility of hybrids produced from the union of two distinct species, it was shown in the nineteenth chapter that this depends exclusively on the reproductive organs being specially affected; but why these organs should be thus affected we do not know, any more than

---

[42] *See* his excellent discussion on this subject in ' Nouvelles Archives du Muséum.' tom. i. p. 151.

why unnatural conditions of life, though compatible with health, should cause sterility; or why continued close interbreeding, or the illegitimate unions of dimorphic and trimorphic plants, induce the same result. The conclusion that the reproductive organs alone are affected, and not the whole organisation, agrees perfectly with the unimpaired or even increased capacity in hybrid plants for propagation by buds; for this implies, according to our hypothesis, that the cells of the hybrids throw off hybridised cell-gemmules, which become aggregated into buds, but fail to become aggregated within the reproductive organs, so as to form the sexual elements. In a similar manner many plants, when placed under unnatural conditions, fail to produce seed, but can readily be propagated by buds. We shall presently see that pangenesis agrees well with the strong tendency to reversion exhibited by all crossed animals and plants.

It was shown in the discussion on graft-hybrids that there is some reason to believe that portions of cellular tissue taken from distinct plants become so intimately united, as afterwards occasionally to produce crossed or hybridised buds. If this fact were fully established, it would, by the aid of our hypothesis, connect gemmation and sexual reproduction in the closest manner.

Abundant evidence has been advanced proving that pollen taken from one species or variety and applied to the stigma of another sometimes directly affects the tissues of the mother-plant. It is probable that this occurs with many plants during fertilisation, but can only be detected when distinct forms are crossed. On any ordinary theory of reproduction this is a most anomalous circumstance, for the pollen-grains are manifestly adapted to act on the ovule, but in these cases they act on the colour, texture, and form of the coats of the seeds, on the ovarium itself, which is a modified leaf, and even on the calyx and upper part of the flower-peduncle. In accordance with the hypothesis of pangenesis pollen includes gemmules, derived from every part of the organisation, which diffuse themselves and multiply by self-division; hence it is not surprising that gemmules within the pollen, which are derived from the parts near the reproductive organs, should sometimes be able to affect the same parts, whilst still undergoing development, in the mother-plant.

As, during all the stages of development, the tissues of plants consist of cells, and as new cells are not known to be formed between, or independently of, pre-existing cells, we must conclude that the gemmules derived from the foreign pollen do not become developed merely in contact with pre-existing cells, but actually penetrate the nascent cells of the mother-plant. This process may be compared with the ordinary act of fertilisation, during which the contents of the pollen-tubes penetrate the closed embryonic sack within the ovule, and determine the development of the embryo. According to this view, the cells of the mother-plant may almost literally be said to be fertilised by the gemmules derived from the foreign pollen. With all organisms, as we shall presently see, the cells or organic units of the embryo during the successive stages of development may in like manner be said to be fertilised by the gemmules of the cells, which come next in the order of formation.

Animals, when capable of sexual reproduction, are fully developed, and it is scarcely possible that the male element should affect the tissues of the mother in the same direct manner as with plants; nevertheless it is certain that her ovaria are sometimes affected by a previous impregnation, so that the ovules subsequently fertilised by a distinct male are plainly influenced in character; and this, as in the case of foreign pollen, is intelligible through the diffusion, retention, and action of the gemmules included within the spermatozoa of the previous male.

Each organism reaches maturity through a longer or shorter course of development. The changes may be small and insensibly slow, as when a child grows into a man, or many, abrupt, and slight, as in the metamorphoses of certain ephemerous insects, or again few and strongly marked, as with most other insects. Each part may be moulded within a previously existing and corresponding part, and in this case it will appear, falsely as I believe, to be formed from the old part; or it may be developed within a wholly distinct part of the body, as in the extreme cases of metagenesis. An eye, for instance, may be developed at a spot where no eye previously existed. We have also seen that allied organic beings in the course of their metamorphoses sometimes attain nearly the same structure after passing through widely different forms; or conversely, after passing through nearly the same early forms, arrive at a widely different termination. In these cases it is very difficult to believe that the early cells or units possess the inherent power, independently of any external agent, of producing new structures wholly different in form, position, and function. But these cases become plain on the hypothesis of pangenesis. The organic units, during each stage of development, throw off gemmules, which, multiplying, are transmitted to the offspring. In the offspring, as soon as any particular cell or unit in the proper order of development becomes partially developed, it unites with (or to speak metaphorically is fertilised by) the gemmule of the next succeeding cell, and so onwards. Now, supposing that at any stage of development, certain cells or aggregates of cells had been slightly modified by the action of some disturbing cause, the cast-off gemmules or atoms of the cell-contents could hardly fail to be similarly affected, and consequently would reproduce the same modification. This process might be repeated until the structure of the part at this particular stage of development became greatly changed, but this would not necessarily affect other parts whether previously or subsequently developed. In this manner we can understand the remarkable independence of structure in the successive metamorphoses, and especially in the successive metageneses of many animals.

The term growth ought strictly to be confined to mere increase of size, and development to change of structure.[43] Now, a child is said to grow into a man, and a foal into a horse, but, as in these cases there is much change of structure, the process properly belongs to the order of development. We have indirect evidence of this in many variations and diseases supervening during so-called growth at a particular period, and being inherited at a corresponding period. In the case, however, of diseases which supervene during old age, subsequently to the ordinary period of procreation, and which nevertheless are sometimes inherited, as occurs with brain and heart complaints, we

---

[43] Various physiologists have insisted on this distinction between growth and development. Prof. Marshall ('Phil. Transact.,' 1864, p. 544) gives a good instance in microcephalous idiots, in which the brain continues to grow after having been arrested in its development.

must suppose that the organs were in fact affected at an earlier age and threw off at this period affected gemmules; but that the affection became visible or injurious only after the prolonged growth of the part in the strict sense of the word. In all the changes of structure which regularly supervene during old age, we see the effects of deteriorated growth, and not of true development.

In the so-called process of *alternate generation* many individuals are generated asexually during very early or later stages of development. These individuals may closely resemble the preceding larval form, but generally are wonderfully dissimilar. To understand this process we must suppose that at a certain stage of development the gemmules are multiplied at an unusual rate, and become aggregated by mutual affinity at many centres of attraction, or buds. These buds, it may be remarked, must include gemmules not only of all the succeeding but likewise of all the preceding stages of development; for when mature they have the power of transmitting by sexual generation gemmules of all the stages, however numerous these may be. It was shown in the First Part, at least in regard to animals, that the new beings which are thus at any period asexually generated do not retrograde in development—that is, they do not pass through those earlier stages, through which the fertilised germ of the same animal has to pass; and an explanation of this fact was attempted as far as the final or teleological cause is concerned. We can likewise understand the proximate cause, if we assume, and the assumption is far from improbable, that buds, like chopped-up fragments of a hydra, are formed of tissue which has already passed through several of the earlier stages of development; for in this case their component cells or units would not unite with the gemmules derived from the earlier-formed cells, but only with those which came next in the order of development. On the other hand, we must believe that, in the sexual elements, or probably in the female alone, gemmules of certain primordial cells are present; and these, as soon as their development commences, unite in due succession with the gemmules of every part of the body, from the first to the last period of life.

The principle of the independent formation of each part, in

so far as its development depends on the union of the proper gemmules with certain nascent cells, together with the super-abundance of the gemmules derived from both parents and self-multiplied, throws light on a widely different group of facts, which on any ordinary view of development appears very strange. I allude to organs which are abnormally multiplied or transposed. Thus gold-fish often have supernumerary fins placed on various parts of their bodies. We have seen that, when the tail of a lizard is broken off, a double tail is sometimes reproduced, and when the foot of the salamander is divided longitudinally, additional digits are occasionally formed. When frogs, toads, &c., are born with their limbs doubled, as sometimes occurs, the doubling, as Gervais remarks,[44] cannot be due to the complete fusion of two embryos, with the exception of the limbs, for the larvæ are limbless. The same argument is applicable[45] to certain insects produced with multiple legs or antennæ, for these are metamorphosed from apodal or antennæless larvæ. Alphonse Milne-Edwards[46] has described the curious case of a crustacean in which one eye-peduncle supported, instead of a complete eye, only an imperfect cornea, out of the centre of which a portion of an antenna was developed. A case has been recorded[47] of a man who had during both dentitions a double tooth in place of the left second incisor, and he inherited this peculiarity from his paternal grandfather. Several cases are known[48] of additional teeth having been developed in the palate, more especially with horses, and in the orbit of the eye. Certain breeds of sheep bear a whole crowd of horns on their foreheads. Hairs occasionally appear in strange situations, as within the ears of the Siamese hairy family; and hairs "quite natural in structure" have been observed "within the substance of the brain."[49] As many as five spurs have been seen on both legs in certain Game-fowls. In the Polish fowl the male is ornamented with a topknot of hackles

[44] 'Compte Rendu,' Nov. 14, 1864, p. 800.

[45] As previously remarked by Quatrefages, in his 'Metamorphoses de l'Homme,' &c., 1862, p. 129.

[46] Günther's 'Zoological Record,' 1864, p. 279.

[47] Sedgwick, in 'Medico-Chirurg. Review,' April 1863, p. 454.

[48] Isid. Geoffroy St. Hilaire, 'Hist. des Anomalies.' tom. i., 1832, pp. 435, 657; and tom. ii. p. 560.

[49] Virchow, 'Cellular Pathology,' 1860, p. 66.

like those on his neck, whilst the female has one of common feathers. In feather-footed pigeons and fowls, feathers like those on the wing arise from the outer side of the legs and toes. Even the elemental parts of the same feather may be transposed ; for in the Sebastopol goose, barbules are developed on the divided filaments of the shaft.

Analogous cases are of such frequent occurrence with plants that they do not strike us with sufficient surprise. Supernumerary petals, stamens, and pistils, are often produced. I have seen a leaflet low down in the compound leaf of *Vicia sativa* converted into a tendril, and a tendril possesses many peculiar properties, such as spontaneous movement and irritability. The calyx sometimes assumes, either wholly or by stripes, the colour and texture of the corolla. Stamens are so frequently converted, more or less completely, into petals, that such cases are passed over as not deserving notice ; but as petals have special functions to perform, namely, to protect the included organs, to attract insects, and in not a few cases to guide their entrance by well-adapted contrivances, we can hardly account for the conversion of stamens into petals merely by unnatural or superfluous nourishment. Again, the edge of a petal may occasionally be found including one of the highest products of the plant, namely the pollen; for instance, I have seen in an Ophrys a pollen-mass with its curious structure of little packets, united together and to the caudicle by elastic threads, formed between the edges of an upper petal. The segments of the calyx of the common pea have been observed partially converted into carpels, including ovules, and with their tips converted into stigmas. Numerous analogous facts could be given.[50]

I do not know how physiologists look at such facts as the foregoing. According to the doctrine of pangenesis, the free and superabundant gemmules of the transposed organs are developed in the wrong place, from uniting with wrong cells or aggregates of cells during their nascent state; and this would follow from a slight modification in the elective affinity of such cells, or possibly of certain gemmules. Nor ought we to feel much surprise at the affinities of cells and gemmules varying

---

[50] Moquin-Tandon, ' Tératologie Vég.,' 1841, pp. 218, 220, 353.   For the case of the pea, see ' Gardener's Chron.,' 1866, p. 897.

under domestication, when we remember the many curious cases given, in the seventeenth chapter, of cultivated plants which absolutely refuse to be fertilised by their own pollen or by that of the same species, but are abundantly fertile with pollen of a distinct species; for this implies that their sexual elective affinities—and this is the term used by Gärtner —have been modified. As the cells of adjoining or homologous parts will have nearly the same nature, they will be liable to acquire by variation each other's elective affinities; and we can thus to a certain extent understand such cases as a crowd of horns on the heads in certain sheep, of several spurs on the leg, and of hackles on the head of the fowl, and with the pigeon the occurrence of wing-feathers on their legs and of membrane between their toes; for the leg is the homologue of the wing. As all the organs of plants are homologous and spring from a common axis, it is natural that they should be eminently liable to transposition. It ought to be observed that when any compound part, such as an additional limb or an antenna, springs from a false position, it is only necessary that the few first gemmules should be wrongly attached; for these whilst developing would attract others in due succession, as in the regrowth of an amputated limb. When parts which are homologous and similar in structure, as the vertebræ in snakes or the stamens in polyandrous flowers, &c., are repeated many times in the same organism, closely allied gemmules must be extremely numerous, as well as the points to which they ought to become united; and, in accordance with the foregoing views, we can to a certain extent understand Isid. Geoffroy St. Hilaire's law, namely, that parts, which are already multiple, are extremely liable to vary in number.

The same general principles apply to the fusion of homologous parts; and with respect to mere cohesion there is probably always some degree of fusion, at least near the surface. When two embryos during their early development come into close contact, as both include corresponding gemmules, which must be in all respects almost identical in nature, it is not surprising that some derived from one embryo and some from the other should unite at the point of contact with a single nascent cell or aggregate of cells, and thus give rise to a single part or organ. For instance, two embryos might thus come to have on their

adjoining sides a single symmetrical arm, which in one sense
will have been formed by the fusion of the bones, muscles, &c.,
belonging to the arms of both embryos.   In the case of the fish
described by Lereboullet, in which a double head was seen
gradually to fuse into a single one, the same process must have
taken place, together with the absorption of all the parts which
had been already formed. These cases are exactly the reverse
of those in which a part is doubled either spontaneously or after
an injury; for in the case of doubling, the superabundant gemmules of the same part are separately developed in union with
adjoining points; whilst in the case of fusion the gemmules
derived from two homologous parts become mingled and form a
single part; or it may be that the gemmules from one of two
adjoining embryos alone become developed.

Variability often depends, as I have attempted to show,
on the reproductive organs being injuriously affected by
changed conditions; and in this case the gemmules derived
from the various parts of the body are probably aggregated in
an irregular manner, some superfluous and others deficient.
Whether a superabundance of gemmules, together with fusion
during development, would lead to the increased size of any
part cannot be told; but we can see that their partial deficiency,
without necessarily leading to the entire abortion of the part,
might cause considerable modifications; for in the same manner
as a plant, if its own pollen be excluded, is easily hybridised, so,
in the case of a cell, if the properly succeeding gemmules were
absent, it would probably combine easily with other and allied
gemmules.   We see this in the case of imperfect nails growing
on the stumps of amputated fingers,[51] for the gemmules of the
nails have manifestly been developed at the nearest point.

In variations caused by the direct action of changed conditions,
whether of a definite or indefinite nature, as with the fleeces of
sheep in hot countries, with maize grown in cold countries, with
inherited gout, &c., the tissues of the body, according to the
doctrine of pangenesis, are directly affected by the new conditions, and consequently throw off modified gemmules, which
are transmitted with their newly acquired peculiarities to the
offspring. On any ordinary view it is unintelligible how changed

<hr>

[51] Müller's 'Physiology,' Eng. translat., vol. i. p. 407.

conditions, whether acting on the embryo, the young or adult animal, can cause inherited modifications. It is equally or even more unintelligible on any ordinary view, how the effects of the long-continued use or disuse of any part, or of changed habits of body or mind, can be inherited. A more perplexing problem can hardly be proposed; but on our view we have only to suppose that certain cells become at last not only functionally but structurally modified; and that these throw off similarly modified gemmules. This may occur at any period of development, and the modification will be inherited at a corresponding period; for the modified gemmules will unite in all ordinary cases with the proper preceding cells, and they will consequently be developed at the same period at which the modification first arose. With respect to mental habits or instincts, we are so profoundly ignorant on the relation between the brain and the power of thought that we do not know whether an inveterate habit or trick induces any change in the nervous system; but when any habit or other mental attribute, or insanity, is inherited, we must believe that some actual modification is transmitted;[52] and this implies, according to our hypothesis, that gemmules derived from modified nerve-cells are transmitted to the offspring.

It is generally, perhaps always, necessary that an organism should be exposed during several generations to changed conditions or habits, in order that any modification in the structure of the offspring should ensue. This may be partly due to the changes not being at first marked enough to catch the attention, but this explanation is insufficient; and I can account for the fact, only by the assumption, which we shall see under the head of reversion is strongly supported, that gemmules derived from each cell before it had undergone the least modification are transmitted in large numbers to successive generations, but that the gemmules derived from the same cells after modification, naturally go on increasing under the same favouring conditions, until at last they become sufficiently numerous to overpower and supplant the old gemmules.

Another difficulty may be here noticed; we have seen that

[52] *See* some remarks to this effect by Sir H. Holland in his 'Medical Notes,' 1839, p. 32.

there is an important difference in the frequency, though not
in the nature, of the variations in plants propagated by sexual
and asexual generation.   As far as variability depends on the
imperfect action of the reproductive organs under changed conditions, we can at once see why seedlings should be far more
variable than plants propagated by buds.   We know that
extremely slight causes,—for instance, whether a tree has been
grafted or grows on its own stock, the position of the seeds
within the capsule, and of the flowers on the spike,—sometimes
suffice to determine the variation of a plant, when raised from
seed.   Now, it is probable, as explained when discussing alternate generation, that a bud is formed of a portion of already
differentiated tissue; consequently an organism thus formed
does not pass through the earlier phases of development, and
cannot be so freely exposed, at the age when its structure
would be most readily modified, to the various causes inducing
variability; but it is very doubtful whether this is a sufficient
explanation of the difficulty.

With respect to the tendency to reversion, there is a similar
difference between plants propagated from buds and seed.
Many varieties, whether originally produced from seed or buds,
can be securely propagated by buds, but generally or invariably
revert by seed.   So, also, hybridised plants can be multiplied
to any extent by buds, but are continually liable to reversion
by seed,—that is, to the loss of their hybrid or intermediate
character.   I can offer no satisfactory explanation of this fact.
Here is a still more perplexing case : certain plants with variegated leaves, phloxes with striped flowers, barberries with
seedless fruit, can all be securely propagated by the buds on
cuttings ; but the buds developed from the roots of these cuttings
almost invariably lose their character and revert to their former
condition.

Finally, we can see on the hypothesis of pangenesis that
variability depends on at least two distinct groups of causes.
Firstly, on the deficiency, superabundance, fusion, and transposition of gemmules, and on the redevelopment of those
which have long been dormant.   In these cases the gemmules
themselves have undergone no modification ; but the mutations
in the above respects will amply account for much fluctuating

variability. Secondly, in the cases in which the organisation has been modified by changed conditions, the increased use or disuse of parts, or any other cause, the gemmules cast off from the modified units of the body will be themselves modified, and, when sufficiently multiplied, will be developed into new and changed structures.

Turning now to Inheritance: if we suppose a homogeneous gelatinous protozoon to vary and assume a reddish colour, a minute separated atom would naturally, as it grew to full size, retain the same colour; and we should have the simplest form of inheritance.[53] Precisely the same view may be extended to the infinitely numerous and diversified units of which the whole body in one of the higher animals is composed; and the separated atoms are our gemmules. We have already sufficiently discussed the inheritance of the direct effects of changed conditions, and of increased use or disuse of parts, and, by implication, the important principle of inheritance at corresponding ages. These groups of facts are to a large extent intelligible on the hypothesis of pangenesis, and on no other hypothesis as yet advanced.

A few words must be added on the complete abortion or suppression of organs. When a part becomes diminished by disuse prolonged during many generations, the principle of economy of growth, as previously explained, will tend to reduce it still further; but this will not account for the complete or almost complete obliteration of, for instance, a minute papilla of cellular tissue representing a pistil, or of a microscopically minute nodule of bone representing a tooth. In certain cases of suppression not yet completed, in which a rudiment occasionally reappears through reversion, diffused gemmules derived from this part must, according to our view, still exist; hence we must suppose that the cells, in union with which the rudiment was formerly developed, in these cases fail in their affinity for such gemmules. But in the cases of complete and final abortion the gemmules themselves no doubt have perished; nor is this

[53] This is the view taken by Prof. Häckel, in his ' Generelle Morphologie ' (B. ii. s. 171), who says: "Lediglich die partielle Identität der specifisch- constituirten Materie im elterlichen und im kindlichen Organismus, die Theilung dieser Materie bei der Fortpflanzung, ist die Ursache der Erblichkeit."

in any way improbable, for, though a vast number of active and
long-dormant gemmules are diffused and nourished in each living
creature, yet there must be some limit to their number; and it
appears natural that gemmules derived from an enfeebled and useless rudiment would be more liable to perish than those derived
from other parts which are still in full functional activity.

With respect to mutilations, it is certain that a part may
be removed or injured during many generations, and no inherited result follow; and this is an apparent objection to the
hypothesis which will occur to every one. But, in the first
place, a being can hardly be intentionally mutilated during its
early stages of growth whilst in the womb or egg; and such
mutilations, when naturally caused, would appear like congenital deficiencies, which are occasionally inherited. In the
second place, according to our hypothesis, gemmules multiply
by self-division and are transmitted from generation to generation; so that during a long period they would be present and
ready to reproduce a part which was repeatedly amputated.
Nevertheless it appears, from the facts given in the twelfth
chapter, that in some rare cases mutilations have been inherited, but in most of these the mutilated surface became diseased.
In this case it may be conjectured that the gemmules of the lost
part were gradually all attracted by the partially diseased surface,
and thus perished. Although this would occur in the injured
individual alone, and therefore in only one parent, yet this might
suffice for the inheritance of a mutilation, on the same principle
that a hornless animal of either sex, when crossed with a perfect
animal of the opposite sex, often transmits its deficiency.

The last subject that need here be discussed, namely Reversion, rests on the principle that transmission and development,
though generally acting in conjunction, are distinct powers; and
the transmission of gemmules and their subsequent development
show us how the existence of these two distinct powers is
possible. We plainly see this distinction in the many cases in
which a grandfather transmits to his grandson, through his
daughter, characters which she does not, or cannot, possess.
Why the development of certain characters, not necessarily in
any way connected with the reproductive organs, should be confined to one sex alone—that is, why certain cells in one sex

should unite with and cause the development of certain gemmules—we do not in the least know; but it is the common
attribute of most organic beings in which the sexes are separate.

The distinction between transmission and development is
likewise seen in all ordinary cases of Reversion; but before
discussing this subject it may be advisable to say a few words
on those characters which I have called latent, and which
would not be classed under Reversion in its usual sense. Most,
or perhaps all, the secondary characters, which appertain to
one sex, lie dormant in the other sex; that is, gemmules
capable of development into the secondary male sexual characters
are included within the female; and conversely female characters in the male. Why in the female, when her ovaria
become diseased or fail to act, certain masculine gemmules
become developed, we do not clearly know, any more than why
when a young bull is castrated his horns continue growing
until they almost resemble those of a cow; or why, when a stag
is castrated, the gemmules derived from the antlers of his progenitors quite fail to be developed. But in many cases, with
variable organic beings, the mutual affinities of the cells and
gemmules become modified, so that parts are transposed or
multiplied; and it would appear that a slight change in the
constitution of an animal, in connection with the state of
the reproductive organs, leads to changed affinities in the
tissues of various parts of the body. Thus, when male animals
first arrive at puberty, and subsequently during each recurrent season, certain cells or parts acquire an affinity for certain gemmules, which become developed into the secondary
masculine characters; but if the reproductive organs be destroyed, or even temporarily disturbed by changed conditions,
these affinities are not excited. Nevertheless, the male, before
he arrives at puberty, and during the season when the species
does not breed, must include the proper gemmules in a latent
state. The curious case formerly given of a Hen which assumed
the masculine characters, not of her own breed but of a remote
progenitor, illustrates the connexion between latent sexual characters and ordinary reversion. With those animals and plants
which habitually produce several forms, as with certain butterflies described by Mr. Wallace, in which three female forms and

the male exist, or as with the trimorphic species of Lythrum and Oxalis, gemmules capable of reproducing several widely-different forms must be latent in each individual.

The same principle of the latency of certain characters, combined with the transposition of organs, may be applied to those singular cases of butterflies and other insects, in which exactly one half or one quarter of the body resembles the male, and the other half or three quarters the female; and when this occurs the opposite sides of the body, separated from each other by a distinct line, sometimes differ in the most conspicuous manner. Again, these same principles apply to the cases given in the thirteenth chapter, in which the right and left sides of the body differ to an extraordinary degree, as in the spiral winding of certain shells, and as in the genus Verruca among cirripedes; for in these cases it is known that either side indifferently may undergo the same remarkable change of development.

Reversion, in the ordinary sense of the word, comes into action so incessantly, that it evidently forms an essential part of the general law of inheritance. It occurs with beings, however propagated, whether by buds or seminal generation, and sometimes may even be observed in the same individual as it advances in age. The tendency to reversion is often induced by a change of conditions, and in the plainest manner by the act of crossing. Crossed forms are generally at first nearly intermediate in character between their two parents; but in the next generation the offspring generally revert to one or both of their grand-parents, and occasionally to more remote ancestors. How can we account for these facts? Each organic unit in a hybrid must throw off, according to the doctrine of pangenesis, an abundance of hybridised gemmules, for crossed plants can be readily and largely propagated by buds; but by the same hypothesis there will likewise be present dormant gemmules derived from both pure parent-forms; and as these latter retain their normal condition, they would, it is probable, be enabled to multiply largely during the lifetime of each hybrid. Consequently the sexual elements of a hybrid will include both pure and hybridised gemmules; and when two hybrids pair, the combination of pure gemmules derived from the one hybrid with the pure gemmules of the same parts derived from the other would necessarily lead to complete reversion of character; and it is, perhaps, not too bold a supposition that unmodified and undeteriorated gemmules of the same nature would be especially apt to combine. Pure gemmules in combination with hybridised gemmules would lead to partial reversion. And lastly, hybridised gemmules derived from both parent-hybrids would simply reproduce the original hybrid form.[54] All these cases and degrees of reversion incessantly occur.

It was shown in the fifteenth chapter that certain characters are antagonistic to each other or do not readily blend together; hence, when two animals with antagonistic characters are crossed, it might well happen that a sufficiency of gemmules in the male alone for the reproduction of his peculiar characters, and in the female alone for the reproduction of her peculiar characters, would not be present; and in this case dormant gemmules derived from some remote progenitor might easily gain the ascendency, and cause the reappearance of long-lost characters. For instance, when black and white pigeons, or black and white fowls, are crossed,—colours which do not readily blend,—blue plumage in the one case, evidently derived from the rock-pigeon, and red plumage in the other case, derived from the wild jungle-cock, occasionally reappear. With uncrossed breeds the same result would follow, under conditions which favoured the multiplication and development of certain dormant gemmules, as when animals become feral and revert to their pristine character. A certain number of gemmules being requisite for the development of each character, as is known to be the case from several spermatozoa or pollen-grains being necessary for fertilisation, and time favouring their multiplication, will together account for the curious cases, insisted on by Mr. Sedgwick, of certain diseases regularly appearing in alternate generations. This likewise holds good, more or less strictly, with other weakly inherited modifications. Hence, as I have heard it remarked, certain diseases appear actually to gain strength by the intermission of a generation. The transmission of dormant gemmules during many successive generations is hardly in itself more improbable, as

---

[54] In these remarks I, in fact, follow Naudin, who speaks of the elements or essences of the two species which are crossed. See his excellent memoir in the 'Nouvelles Archives du Muséum,' tom. i. p. 151.

previously remarked, than the retention during many ages of rudimentary organs, or even only of a tendency to the production of a rudiment; but there is no reason to suppose that all dormant gemmules would be transmitted and propagated for ever. Excessively minute and numerous as they are believed to be, an infinite number derived, during a long course of modification and descent, from each cell of each progenitor, could not be supported or nourished by the organism. On the other hand, it does not seem improbable that certain gemmules, under favourable conditions, should be retained and go on multiplying for a longer period than others. Finally, on the views here given, we certainly gain some clear insight into the wonderful fact that the child may depart from the type of both its parents, and resemble its grandparents, or ancestors removed by many generations.

Conclusion.

The hypothesis of Pangenesis, as applied to the several great classes of facts just discussed, no doubt is extremely complex; but so assuredly are the facts. The assumptions, however, on which the hypothesis rests cannot be considered as complex in any extreme degree—namely, that all organic units, besides having the power, as is generally admitted, of growing by self-division, throw off free and minute atoms of their contents, that is gemmules. These multiply and aggregate themselves into buds and the sexual elements; their development depends on their union with other nascent cells or units; and they are capable of transmission in a dormant state to successive generations.

In a highly organised and complex animal, the gemmules thrown off from each different cell or unit throughout the body must be inconceivably numerous and minute. Each unit of each part, as it changes during development, and we know that some insects undergo at least twenty metamorphoses, must throw off its gemmules. All organic beings, moreover, include many dormant gemmules derived from their grandparents and more remote progenitors, but not from all their progenitors. These almost infinitely numerous and minute gemmules must be included in each bud, ovule, spermatozoon, and pollen-grain. Such an admission will be declared impossible; but, as previously remarked, number and size are only relative difficulties, and the eggs or seeds produced by certain animals or plants are so numerous that they cannot be grasped by the intellect.

The organic particles with which the wind is tainted over miles of space by certain offensive animals must be infinitely minute and numerous; yet they strongly affect the olfactory nerves. An analogy more appropriate is afforded by the contagious particles of certain diseases, which are so minute that they float in the atmosphere and adhere to smooth paper; yet we know how largely they increase within the human body, and how powerfully they act. Independent organisms exist which are barely visible under the highest powers of our recently-improved microscopes, and which probably are fully as large as the cells or units in one of the higher animals; yet these organisms no doubt reproduce themselves by germs of extreme minuteness, relatively to their own minute size. Hence the difficulty, which at first appears insurmountable, of believing in the existence of gemmules so numerous and so small as they must be according to our hypothesis, has really little weight.

The cells or units of the body are generally admitted by physiologists to be autonomous, like the buds on a tree, but in a less degree. I go one step further and assume that they throw off reproductive gemmules. Thus an animal does not, as a whole, generate its kind through the sole agency of the reproductive system, but each separate cell generates its kind. It has often been said by naturalists that each cell of a plant has the actual or potential capacity of reproducing the whole plant; but it has this power only in virtue of containing gemmules derived from every part. If our hypothesis be provisionally accepted, we must look at all the forms of asexual reproduction, whether occurring at maturity or as in the case of alternate generation during youth, as fundamentally the same, and dependent on the mutual aggregation and multiplication of the gemmules. The regrowth of an amputated limb or the healing of a wound is the same process partially carried out. Sexual generation differs in some important respects, chiefly, as it would appear, in an insufficient number of gemmules being aggregated within the separate sexual elements, and probably in the presence of certain primordial cells. The development of each being, including all the

forms of metamorphosis and metagenesis, as well as the so-
called growth of the higher animals, in which structure changes
though not in a striking manner, depends on the presence of
gemmules thrown off at each period of life, and on their development, at a corresponding period, in union with preceding cells.
Such cells may be said to be fertilised by the gemmules which
come next in the order of development.   Thus the ordinary act
of impregnation and the development of each being are closely
analogous processes.   The child, strictly speaking, does not grow
into the man, but includes germs which slowly and successively
become developed and form the man.   In the child, as well as
in the adult, each part generates the same part for the next
generation.   Inheritance must be looked at as merely a form
of growth, like the self-division of a lowly-organised unicellular
plant.   Reversion depends on the transmission from the fore-
father to his descendants of dormant gemmules, which occasionally become developed under certain known or unknown
conditions.   Each animal and plant may be compared to a bed
of mould full of seeds, most of which soon germinate, some lie
for a period dormant, whilst others perish.   When we hear it
said that a man carries in his constitution the seeds of an inherited disease, there is much literal truth in the expression.
Finally, the power of propagation possessed by each separate
cell, using the term in its largest sense, determines the reproduction, the variability, the development and renovation of each
living organism.   No other attempt, as far as I am aware, has
been made, imperfect as this confessedly is, to connect under
one point of view these several grand classes of facts.   We
cannot fathom the marvellous complexity of an organic being;
but on the hypothesis here advanced this complexity is much
increased.   Each living creature must be looked at as a mi-
crocosm—a little universe, formed of a host of self-propagating
organisms, inconceivably minute and as numerous as the stars in
heaven.

CHAPTER XXVIII.

CONCLUDING REMARKS.

DOMESTICATION — NATURE AND CAUSES OF VARIABILITY — SELECTION — DIVER-
GENCE AND DISTINCTNESS OF CHARACTER — EXTINCTION OF RACES — CIRCUM-
STANCES FAVOURABLE TO SELECTION BY MAN — ANTIQUITY OF CERTAIN RACES —
THE QUESTION WHETHER EACH PARTICULAR VARIATION HAS BEEN SPECIALLY PRE-
ORDAINED.

As summaries have been added to nearly all the chapters, and
as, in the chapter on pangenesis, various subjects, such as the
forms of reproduction, inheritance, reversion, the causes and
laws of variability, &c., have been recently discussed, I will
here only make a few general remarks on the more important
conclusions which may be deduced from the multifarious details
given throughout this work.

Savages in all parts of the world easily succeed in taming
wild animals; and those inhabiting any country or island, when
first invaded by man, would probably have been still more
easily tamed. Complete subjugation generally depends on an
animal being social in its habits, and on receiving man as the
chief of the herd or family. Domestication implies almost
complete fertility under new and changed conditions of life, and
this is far from being invariably the case. An animal would not
have been worth the labour of domestication, at least during early
times, unless of service to man. From these circumstances the
number of domesticated animals has never been large. With
respect to plants, I have shown in the ninth chapter how their
varied uses were probably first discovered, and the early steps in
their cultivation. Man could not have known, when he first
domesticated an animal or plant, whether it would flourish and
multiply when transported to other countries, therefore he could
not have been thus influenced in his choice. We see that the
close adaptation of the reindeer and camel to extremely cold and
hot countries has not prevented their domestication. Still less

could man have foreseen whether his animals and plants would vary in succeeding generations and thus give birth to new races; and the small capacity of variability in the goose and ass has not prevented their domestication from the remotest epoch.

With extremely few exceptions, all animals and plants which have been long domesticated, have varied greatly. It matters not under what climate, or for what purpose, they are kept, whether as food for man or beast, for draught or hunting, for clothing or mere pleasure,—under all these circumstances domesticated animals and plants have varied to a much greater extent than the forms which in a state of nature are ranked as one species. Why certain animals and plants have varied more under domestication than others we do not know, any more than why some are rendered more sterile than others under changed conditions of life. But we frequently judge of the amount of variation by the production of numerous and diversified races, and we can clearly see why in many cases this has not occurred, namely, because slight successive variations have not been steadily accumulated; and such variations will never be accumulated when an animal or plant is not closely observed, or much valued, or kept in large numbers.

The fluctuating, and, as far as we can judge, never-ending variability of our domesticated productions,—the plasticity of their whole organisation,—is one of the most important facts which we learn from the numerous details given in the earlier chapters of this work. Yet domesticated animals and plants can hardly have been exposed to greater changes in their conditions than have many natural species during the incessant geological, geographical, and climatal changes of the whole world. The former will, however, commonly have been exposed to more sudden changes and to less continuously uniform conditions. As man has domesticated so many animals and plants belonging to widely different classes, and as he certainly did not with prophetic instinct choose those species which would vary most, we may infer that all natural species, if subjected to analogous conditions, would, on an average, vary to the same degree. Few men at the present day will maintain that animals and plants were created with a tendency to vary, which long remained dormant, in order that fanciers in after ages might rear, for instance, curious breeds of the fowl, pigeon, or canary-bird.

From several causes it is difficult to judge of the amount of modification which our domestic productions have under-gone. In some cases the primitive parent-stock has become extinct, or cannot be recognised with certainty owing to its supposed descendants having been so much modified. In other cases two or more closely allied forms, after being do-mesticated, have crossed; and then it is difficult to estimate how much of the change ought to be attributed to variation. But the degree to which our domestic breeds have been modified by the crossing of distinct natural forms has probably been exaggerated by some authors. A few individuals of one form would seldom permanently affect another form existing in much greater numbers; for, without careful selection, the stain of the foreign blood would soon be obliterated, and during early and barbarous times, when our animals were first domesticated, such care would seldom have been taken.

There is good reason to believe that several of the breeds of the dog, ox, pig, and of some other animals, are respectively descended from distinct wild prototypes; nevertheless the belief in the multiple origin of our domesticated animals has been extended by some few naturalists and by many breeders to an unauthorised extent. Breeders refuse to look at the whole subject under a single point of view; I have heard one, who maintained that our fowls were the descendants of at least half-a-dozen aboriginal species, protest that he was in no way concerned with the origin of pigeons, ducks, rabbits, horses, or any other animal. They overlook the improbability of many species having been domesticated at an early and barbarous period. They do not consider the improbability of species having existed in a state of nature which, if like our present domestic breeds, would have been highly abnormal in comparison with all their congeners. They maintain that certain species, which formerly existed, have become extinct or unknown, although the world is now so much better known. The assump-tion of so much recent extinction is no difficulty in their eyes; for they do not judge of its probability by the facility or difficulty of the extinction of other closely allied wild forms. Lastly, they often ignore the whole subject of geographical distribution as completely as if its laws were the result of chance.

Although from the reasons just assigned it is often difficult to judge accurately of the amount of change which our domesticated productions have undergone, yet this can be ascertained in the cases in which we know that all the breeds are descended from a single species, as with the pigeon, duck, rabbit, and almost certainly with the fowl; and by the aid of analogy this is to a certain extent possible in the case of animals descended from several wild stocks. It is impossible to read the details given in the earlier chapters, and in many published works, or to visit our various exhibitions, without being deeply impressed with the extreme variability of our domesticated animals and cultivated plants. I have in many instances purposely given details on new and strange peculiarities which have arisen. No part of the organisation escapes the tendency to vary. The variations generally affect parts of small vital or physiological importance, but so it is with the differences which exist between closely allied species. In these unimportant characters there is often a greater difference between the breeds of the same species than between the natural species of the same genus, as Isidore Geoffroy has shown to be the case with size, and as is often the case with the colour, texture, form, &c., of the hair, feathers, horns, and other dermal appendages.

It has often been asserted that important parts never vary under domestication, but this is a complete error. Look at the skull of the pig in any one of the highly improved breeds, with the occipital condyles and other parts greatly modified; or look at that of the niata ox. Or again, in the several breeds of the rabbit, observe the elongated skull, with the differently shaped occipital foramen, atlas, and other cervical vertebræ. The whole shape of the brain, together with the skull, has been modified in Polish fowls; in other breeds of the fowl the number of the vertebræ and the forms of the cervical vertebræ have been changed. In certain pigeons the shape of the lower jaw, the relative length of the tongue, the size of the nostrils and eyelids, the number and shape of the ribs, the form and size of the œsophagus, have all varied. In certain quadrupeds the length of the intestines has been much increased or diminished. With plants we see wonderful differences in the stones of various fruits. In the Cucurbitaceæ several highly important characters have varied, such as the sessile position of the stigmas on the ovarium, the position of the carpels within the ovarium, and its projection out of the receptacle. But it would be useless to run through the many facts given in the earlier chapters.

It is notorious how greatly the mental disposition, tastes, habits, consensual movements, loquacity or silence, and the tone of voice have varied and been inherited with our domesticated animals. The dog offers the most striking instance of changed mental attributes, and these differences cannot be accounted for by descent from distinct wild types. New mental characters have certainly often been acquired, and natural ones lost, under domestication.

New characters may appear and disappear at any stage of growth, and be inherited at a corresponding period. We see this in the difference between the eggs of various breeds of the fowl, and in the down on chickens; and still more plainly in the differences between the caterpillars and cocoons of various breeds of the silk-moth. These facts, simple as they appear, throw light on the characters which distinguish the larval and adult states of natural species, and on the whole great subject of embryology. New characters are liable to become attached exclusively to that sex in which they first appeared, or they may be developed in a much higher degree in the one than the other sex; or again, after having become attached to one sex, they may be partially transferred to the opposite sex. These facts, and more especially the circumstance that new characters seem to be particularly liable, from some unknown cause, to become attached to the male sex, have an important bearing on the acquirement by animals in a state of nature of secondary sexual characters.

It has sometimes been said that our domestic productions do not differ in constitutional peculiarities, but this cannot be maintained. In our improved cattle, pigs, &c., the period of maturity, including that of the second dentition, has been much hastened. The period of gestation varies much, but has been modified in a fixed manner in only one or two cases. In our poultry and pigeons the acquirement of down and of the first plumage by the young, and of the secondary sexual characters by the males, differ. The number of moults through which the larvæ of silk-moths pass, varies. The tendency to fatten, to yield much milk, to produce many young or eggs at a birth or during life, differs in different breeds. We find different degrees of adaptation to climate, and different tendencies to certain diseases, to the attacks of parasites, and to the action of certain vegetable poisons. With plants, adaptation to certain soils, as with some kinds of plums, the power of resisting frost, the period of flowering and fruiting, the duration of life, the period of shedding the leaves and of retaining them throughout the winter, the proportion and nature of certain chemical compounds in the tissues or seeds, all vary.

There is, however, one important constitutional difference between domestic races and species; I refer to the sterility which almost invariably follows, in a greater or less degree, when species are crossed, and to the perfect fertility of the most distinct domestic races, with the exception of a very few plants, when similarly crossed. It certainly appears a remarkable fact that many closely allied species which in appearance differ extremely little should yield when united only a few, more or less sterile offspring, or none at all; whilst domestic races which differ conspicuously from each other, are when united remarkably fertile, and yield perfectly fertile offspring. But this fact is not in reality so inexplicable as it at first appears. In the first place, it was clearly shown in the nineteenth chapter that the sterility of crossed species does not closely depend on differences in their external structure or general constitution, but results exclusively from differences in the reproductive system, analogous with those which cause the lessened fertility of the illegitimate unions and illegitimate offspring of dimorphic and trimorphic plants. In the second place, the Pallasian doctrine, that species after having been long domesticated lose their natural tendency to sterility when crossed, has been shown to be highly probable; we can scarcely avoid this conclusion when we reflect on the parentage and present fertility of the several breeds of the dog, of Indian and European cattle, sheep, and pigs. Hence it would be unreasonable to expect that races formed under domestication

should acquire sterility when crossed, whilst at the same time
we admit that domestication eliminates the normal sterility of
crossed species. Why with closely allied species their reproductive systems should almost invariably have been modified in so
peculiar a manner as to be mutually incapable of acting on each
other—though in unequal degrees in the two sexes, as shown by
the difference in fertility between reciprocal crosses in the same
species—we do not know, but may with much probability infer
the cause to be as follows. Most natural species have been
habituated to nearly uniform conditions of life for an incom-
parably longer period of time than have domestic races; and
we positively know that changed conditions exert an especial
and powerful influence on the reproductive system. Hence this
difference in habituation may well account for the different
action of the reproductive organs when domestic races and when
species are crossed. It is a nearly analogous fact, that most
domestic races may be suddenly transported from one climate
to another, or be placed under widely different conditions, and
yet retain their fertility unimpaired; whilst a multitude of
species subjected to lesser changes are rendered incapable of
breeding.

With the exception of fertility, domestic varieties resemble
species when crossed in transmitting their characters in the
same unequal manner to their offspring, in being subject to the
prepotency of one form over the other, and in their liability
to reversion. By repeated crosses a variety or a species may
be made completely to absorb another. Varieties, as we shall
see when we treat of their antiquity, sometimes inherit their
new characters almost, or even quite, as firmly as species.
With both, the conditions leading to variability and the
laws governing its nature appear to be the same. Domestic
varieties can be classed in groups under groups, like species
under genera, and these under families and orders; and the
classification may be either artificial,—that is, founded on any
arbitrary character,—or natural. With varieties a natural classification is certainly founded, and with species is apparently
founded, on community of descent, together with the amount of
modification which the forms have undergone. The characters
by which domestic varieties differ from each other are more

variable than those distinguishing species, though hardly more
so than with certain protean species; but this greater degree
of variability is not surprising, as varieties have generally been
exposed within recent times to fluctuating conditions of life,
are much more liable to have been crossed, and are still in
many cases undergoing, or have recently undergone, modification by man's methodical or unconscious selection.

Domestic varieties as a general rule certainly differ from each
other in less important parts of their organisation than do
species; and when important differences occur, they are seldom
firmly fixed; but this fact is intelligible if we consider man's
method of selection. In the living animal or plant he cannot
observe internal modifications in the more important organs;
nor does he regard them as long as they are compatible with
health and life. What does the breeder care about any slight
change in the molar teeth of his pigs, or for an additional
molar tooth in the dog; or for any change in the intestinal
canal or other internal organ? The breeder cares for the
flesh of his cattle being well marbled with fat, and for an
accumulation of fat within the abdomen of his sheep, and this
he has effected. What would the floriculturist care for any
change in the structure of the ovarium or of the ovules? As
important internal organs are certainly liable to numerous
slight variations, and as these would probably be inherited, for
many strange monstrosities are transmitted, man could undoubtedly effect a certain amount of change in these organs.
When he has produced any modification in an important part,
it has generally been unintentionally in correlation with some
other conspicuous part, as when he has given ridges and protuberances to the skulls of fowls, by attending to the form of the
comb, and in the case of the Polish fowl to the plume of feathers
on the head. By attending to the external form of the pouter-
pigeon, he has enormously increased the size of the œsophagus,
and has added to the number of the ribs, and given them
greater breadth. With the carrier-pigeon, by increasing, through
steady selection, the wattles on the upper mandible, he has
greatly modified the form of the lower mandible; and so in many
other cases. Natural species, on the other hand, have been modified exclusively for their own good, to fit them for infinitely

diversified conditions of life, to avoid enemies of all kinds, and to struggle against a host of competitors. Hence, under such complex conditions, it would often happen that modifications of the most varied kinds, in important as well as in unimportant parts, would be advantageous or even necessary; and they would slowly but surely be acquired through the survival of the fittest. Various indirect modifications would likewise arise through the law of correlated variation.

Domestic breeds often have an abnormal or semi-monstrous character, as the Italian greyhound, bulldog, Blenheim spaniel, and bloodhound amongst dogs,—some breeds of cattle and pigs, several breeds of the fowl, and the chief breeds of the pigeon. The differences between such abnormal breeds occur in parts which in closely-allied natural species differ but slightly or not at all. This may be accounted for by man's often selecting, especially at first, conspicuous and semi-monstrous deviations of structure. We should, however, be cautious in deciding what deviations ought to be called monstrous: there can hardly be a doubt that, if the brush of horse-like hair on the breast of the turkey-cock had first appeared on the domesticated bird, it would have been considered a monstrosity; the great plume of feathers on the head of the Polish cock has been thus designated, though plumes are common with many kinds of birds; we might call the wattle or corrugated skin round the base of the beak of the English carrier-pigeon a monstrosity, but we do not thus speak of the globular fleshy excrescence at the base of the beak of the male *Carpophaga oceanica*.

Some authors have drawn a wide distinction between artificial and natural breeds; although in extreme cases the distinction is plain, in many other cases an arbitrary line has to be drawn. The difference depends chiefly on the kind of selection which has been applied. Artificial breeds are those which have been intentionally improved by man; they frequently have an unnatural appearance, and are especially liable to loss of excellence through reversion and continued variability. The so-called natural breeds, on the other hand, are those which are now found in semi-civilised countries, and which formerly inhabited separate districts in nearly all the European kingdoms. They have been rarely acted on by man's intentional selection; more frequently, it is probable, by un-conscious selection, and partly by natural selection, for animals kept in semi-civilised countries have to provide largely for their own wants. Such natural breeds will also, it may be presumed, have been directly acted on to some extent by the differences, though slight, in the surrounding physical conditions.

It is a much more important distinction that some breeds have been from their first origin modified in so slow and insensible a manner, that if we could see their early progenitors we should hardly be able to say when or how the breed first arose; whilst other breeds have originated from a strongly-marked or semi-monstrous deviation of structure, which, however, may subse-quently have been augmented by selection. From what we know of the history of the racehorse, greyhound, gamecock, &c., and from their general appearance, we may feel nearly con-fident that they were formed by a slow process of improvement: and with the carrier-pigeon, as well as with some other pigeons, we know that this has been the case. On the other hand, it is certain that the ancon and mauchamp breeds of sheep, and almost certain that the niata cattle, turnspit and pug-dogs, jumper and frizzled fowls, short-faced tumbler pigeons, hook-billed ducks, &c., and with plants a multitude of varieties, sud-denly appeared in nearly the same state as we now see them. The frequency of these cases is likely to lead to the false belief that natural species have often originated in the same abrupt manner. But we have no evidence of the appearance, or at least of the continued procreation, under nature, of abrupt modifications of structure; and various general reasons could be assigned against such a belief: for instance, without separation a single monstrous variation would almost certainly be soon obliterated by crossing.

On the other hand, we have abundant evidence of the con-stant occurrence under nature of slight individual differences of the most diversified kinds; and thus we are led to conclude that species have generally originated by the natural selection, not of abrupt modifications, but of extremely slight differences. This process may be strictly compared with the slow and gradual im-provement of the racehorse, greyhound, and gamecock. As every detail of structure in each species is closely adapted to its general habits of life, it will rarely happen that one part alone will be modified; but the co-adapted modifications, as formerly shown, need not be absolutely simultaneous. Many variations, however, are from the first connected by the law of correlation. Hence it follows that even closely-allied species rarely or never differ from each other by some one character alone; and this same remark applies to a certain extent to domestic races; for these, if they differ much, generally differ in many respects.

Some naturalists boldly insist[1] that species are absolutely distinct productions, never passing by intermediate links into each other; whilst they maintain that domestic varieties can always be connected either with each other or with their parent-forms. But if we could always find the links between the several breeds of the dog, horse, cattle, sheep, pigs, &c., the incessant doubts whether they are descended from one or several species would not have arisen. The greyhound genus, if such a term may be used, cannot be closely connected with any other breed, unless, perhaps, we go back to the ancient Egyptian monuments. Our English bulldog also forms a very distinct breed. In all these cases crossed breeds must of course be excluded, for the most distinct natural species can thus be connected. By what links can the Cochin fowl be closely united with others? By searching for breeds still preserved in distant lands, and by going back to historical records, tumbler-pigeons, carriers, and barbs can be closely connected with the parent rock-pigeon; but we cannot thus connect the turbit or the pouter. The degree of distinctness between the various domestic breeds depends on the amount of modification which they have undergone, and especially on the neglect and final extinction of the linking, intermediate, and less valued forms.

It has often been argued that no light is thrown, from the admitted changes of domestic races, on the changes which natural species are believed to undergo, as the former are said to be mere temporary productions, always reverting, as soon as they become feral, to their pristine form. This argument has been well combated by Mr. Wallace;[2] and full details were given in the thirteenth chapter, showing that the tendency to reversion in feral

<hr>

[1] Godron, 'De l'Espèce,' 1859, tom. ii. p. 44, &c.

[2] Journal Proc. Linn. Soc., 1858, vol. iii. p. 60.

animals and plants has been greatly exaggerated, though no doubt to a certain extent it exists. It would be opposed to all the principles inculcated in this work, if domestic animals, when exposed to new conditions and compelled to struggle for their own wants against a host of foreign competitors, were not in the course of time in some manner modified. It should also be remembered that many characters lie latent in all organic beings ready to be evolved under fitting conditions; and in breeds modified within recent times the tendency to reversion is particularly strong. But the antiquity of various breeds clearly proves that they remain nearly constant as long as their conditions of life remain the same.

It has been boldly maintained by some authors that the amount of variation to which our domestic productions are liable is strictly limited; but this is an assertion resting on little evidence. Whether or not the amount in any particular direction is fixed, the tendency to general variability seems unlimited. Cattle, sheep, and pigs have been domesticated and have varied from the remotest period, as shown by the researches of Rüti- meyer and others, yet these animals have, within quite recent times, been improved in an unparalleled degree; and this im- plies continued variability of structure. Wheat, as we know from the remains found in the Swiss lake-habitations, is one of the most anciently cultivated plants, yet at the present day new and better varieties occasionally arise. It may be that an ox will never be produced of larger size or finer proportions than our present animals, or a race-horse fleeter than Eclipse, or a gooseberry larger than the London variety; but he would be a bold man who would assert that the extreme limit in these respects has been finally attained. With flowers and fruit it has repeatedly been asserted that perfection has been reached, but the standard has soon been excelled. A breed of pigeons may never be produced with a beak shorter than that of the present short-faced tumbler, or with one longer than that of the English carrier, for these birds have weak constitutions and are bad breeders; but the shortness and length of the beak are the points which have been steadily improved during at least the last 150 years; and some of the best judges deny that the goal has yet been reached. We may, also, reasonably suspect, from what we see in natural species of the variability of extremely modified parts, that any structure, after remaining constant during a long series of generations, would, under new and changed conditions of life, recommence its course of variability, and might again be acted on by selection. Nevertheless, as Mr. Wallace[3] has recently remarked with much force and truth, there must be both with natural and domestic productions a limit to change in certain directions; for instance, there must be a limit to the fleetness of any terrestrial animal, as this will be determined by the friction to be overcome, the weight to be carried, and the power of contraction in the muscular fibres. The English race-horse may have reached this limit; but it already surpasses in fleetness its own wild progenitor, and all other equine species.

It is not surprising, seeing the great difference between many domestic breeds, that some few naturalists have concluded that all are descended from distinct aboriginal stocks, more especially as the principle of selection has been ignored, and the high antiquity of man, as a breeder of animals, has only recently become known. Most naturalists, however, freely admit that various extremely dissimilar breeds are descended from a single stock, although they do not know much about the art of breeding, cannot show the connecting links, nor say where and when the breeds arose. Yet these same naturalists will declare, with an air of philosophical caution, that they can never admit that one natural species has given birth to another until they behold all the transitional steps. But fanciers have used exactly the same language with respect to domestic breeds; thus an author of an excellent treatise says he will never allow that carrier and fantail pigeons are the descendants of the wild rock-pigeon, until the transitions have "actually been observed, and can be repeated "whenever man chooses to set about the task." No doubt it is difficult to realise that slight changes added up during long centuries can produce such results; but he who wishes to understand the origin of domestic breeds or natural species must overcome this difficulty.

The causes inducing and the laws governing variability have been so lately discussed, that I need here only enumerate the leading points. As domesticated organisms are much more

[3] 'The Quarterly Journal of Science,' Oct. 1867, p. 486.

liable to slight deviations of structure and to monstrosities, than species living under their natural conditions, and as widely-ranging species vary more than those which inhabit restricted areas, we may infer that variability mainly depends on changed conditions of life. We must not overlook the effects of the unequal combination of the characters derived from both parents, nor reversion to former progenitors. Changed conditions have an especial tendency to render the reproductive organs more or less impotent, as shown in the chapter devoted to this subject; and these organs consequently often fail to transmit faithfully the parental characters. Changed conditions also act directly and definitely on the organisation, so that all or nearly all the individuals of the same species thus exposed become modified in the same manner; but why this or that part is especially affected we can seldom or never say. In most cases, however, of the direct action of changed conditions, independently of the indirect variability caused by the reproductive organs being affected, indefinite modifications are the result; in nearly the same manner as exposure to cold or the absorption of the same poison affects different individuals in various ways. We have reason to suspect that an habitual excess of highly nutritious food, or an excess relatively to the wear and tear of the organisation from exercise, is a powerful exciting cause of variability. When we see the symmetrical and complex outgrowths, caused by a minute atom of the poison of a gall-insect, we may believe that slight changes in the chemical nature of the sap or blood would lead to extraordinary modifications of structure.

The increased use of a muscle with its various attached parts, and the increased activity of a gland or other organ, lead to their increased development. Disuse has a contrary effect. With domesticated productions organs sometimes become rudimentary through abortion; but we have no reason to suppose that this has ever followed from mere disuse. With natural species, on the contrary, many organs appear to have been rendered rudimentary through disuse, aided by the principle of the economy of growth, and by the hypothetical principle discussed in the last chapter, namely, the final destruction of the germs or gemmules of such useless parts. This difference may be partly accounted for by disuse having acted on domestic forms for an insufficient length of time, and partly from their exemption from any severe struggle for existence, entailing rigid economy in the development of each part, to which all species under nature are subjected. Nevertheless the law of compensation or balancement apparently affects, to a certain extent, our domesticated productions.

We must not exaggerate the importance of the definite action of changed conditions in modifying all the individuals of the same species in the same manner, or of use and disuse. As every part of the organisation is highly variable, and as variations are so easily selected, both consciously and unconsciously, it is very difficult to distinguish between the effects of the selection of indefinite variations, and the direct action of the conditions of life. For instance, it is possible that the feet of our water-dogs, and of the American dogs which have to travel much over the snow, may have become partially webbed from the stimulus of widely extending their toes; but it is far more probable that the webbing, like the membrane between the toes of certain pigeons, spontaneously appeared and was afterwards increased by the best swimmers and the best snow-travellers being preserved during many generations. A fancier who wished to decrease the size of his bantams or tumbler-pigeons would never think of starving them, but would select the smallest individuals which spontaneously appeared. Quadrupeds are sometimes- born destitute of hair, and hairless breeds have been formed, but there is no reason to believe that this is caused by a hot climate. Within the tropics heat often causes sheep to lose their fleeces, and on the other hand wet and cold act as a direct stimulus to the growth of hair; it is, however, possible that these changes may merely be an exaggeration of the regular yearly change of coat; and who will pretend to decide how far this yearly change, or the thick fur of arctic animals, or as I may add their white colour, is due to the direct action of a severe climate, and how far to the preservation of the best protected individuals during a long succession of generations?

Of all the laws governing variability, that of correlation is the most important. In many cases of slight deviations of structure as well as of grave monstrosities, we cannot even conjecture what is the nature of the bond of connexion. But between homologous parts—between the fore and hind limbs—between the hair, hoofs, horns, and teeth—we can see that parts which are closely similar during their early development, and which are exposed to similar conditions, would be liable to be modified in the same manner. Homologous parts, from having the same nature, are apt to blend together, and, when many exist, to vary in number.

Although every variation is either directly or indirectly caused by some change in the surrounding conditions, we must never forget that the nature of the organisation which is acted on essentially governs the result. Distinct organisms, when placed under similar conditions, vary in different manners, whilst closely-allied organisms under dissimilar conditions often vary in nearly the same manner. We see this in the same modification frequently reappearing at long intervals of time in the same variety, and likewise in the several striking cases given of analogous or parallel varieties. Although some of these latter cases are simply due to reversion, others cannot thus be accounted for.

From the indirect action of changed conditions on the organisation, through the impaired state of the reproductive organs—from the direct action of such conditions (and this will cause the individuals of the same species either to vary in the same manner, or differently in accordance with slight differences in their constitution)—from the effects of the increased or decreased use of parts,—and from correlation,—the variability of our domesticated productions is complicated in an extreme degree. The whole organisation becomes slightly plastic. Although each modification must have its proper exciting cause, and though each is subjected to law, yet we can so rarely trace the precise relation between cause and effect, that we are tempted to speak of variations as if they spontaneously arose. We may even call them accidental, but this must be only in the sense in which we say that a fragment of rock dropped from a height owes its shape to accident.

It may be worth while briefly to consider the results of the exposure to unnatural conditions of a large number of animals of the same species, allowed to cross freely, with no selection of any kind; and afterwards to consider the results when selection is brought into play. Let us suppose that 500 wild rock-pigeons were confined in their native land in an aviary, and fed in the same manner as pigeons usually are; and that they were not allowed to increase in number. As pigeons propagate so rapidly, I suppose that a thousand or fifteen hundred birds would have to be annually killed by mere chance. After several generations had been thus reared, we may feel sure that some of the young birds would vary, and the variations would tend to be inherited; for at the present day slight deviations of structure often occur, but, as most breeds are already well established, these modifications are rejected as blemishes. It would be tedious even to enumerate the multitude of points which still go on varying or have recently varied. Many variations would occur in correlation, as the length of the wing and tail feathers—the number of the primary wing-feathers, as well as the number and breadth of the ribs, in correlation with the size and form of the body—the number of the scutellæ, with the size of the feet— the length of the tongue, with the length of the beak—the size of the nostrils and eyelids and the form of lower jaw in correlation with the development of wattle—the nakedness of the young with the future colour of the plumage—the size of the feet and beak, and other such points. Lastly, as our birds are supposed to be confined in an aviary, they would use their wings and legs but little, and certain parts of the skeleton, such as the sternum and scapulæ and the feet, would in consequence become slightly reduced in size.

As in our assumed case many birds have to be indiscriminately killed every year, the chances are against any new variety surviving long enough to breed. And as the variations which arise are of an extremely diversified nature, the chances are very great against two birds pairing which have varied in the same manner; nevertheless, a varying bird even when not thus paired would occasionally transmit its character to its young; and these would not only be exposed to the same conditions which first caused the variation in question to appear, but would in addition inherit from their one modified parent a tendency again to vary in the same manner. So that, if the conditions decidedly tended to induce some particular variation, all the birds might in the course of time become similarly modified. But a far commoner result would be, that one bird would vary in one way and another bird in another way; one would be born with a little longer beak, and another with a shorter beak; one would gain some black feathers, another some white or red feathers. And as these birds would be continually intercrossing, the final result would be a body of individuals differing from each other slightly in many ways, yet far more than did the original rock-pigeons. But there would not be the least tendency to the formation of distinct breeds.

If two separate lots of pigeons were to be treated in the manner just described, one in England and the other in a tropical country, the two lots being supplied with different food, would they, after many generations had passed, differ? When we reflect on the cases given in the twenty-third chapter, and on such facts as the difference in former times between the breeds of cattle, sheep, &c., in almost every district of Europe, we are strongly inclined to admit that the two lots would be differently modified through the influence of climate and food. But the evidence on the definite action of changed conditions is in most cases insufficient; and, with respect to pigeons, I have had the opportunity of examining a large collection of domesticated birds, sent to me by Sir W. Elliot from India, and they varied in a remarkably similar manner with our European birds.

If two distinct breeds were to be confined together in equal numbers, there is reason to suspect that they would to a certain extent prefer pairing with their own kind; but they would like-wise intercross. From the greater vigour and fertility of the crossed offspring, the whole body would by this means become interblended sooner than would otherwise have occurred. From certain breeds being prepotent over others, it does not follow that the interblended progeny would be strictly intermediate in character. I have, also, proved that the act of crossing in itself gives a strong tendency to reversion, so that the crossed offspring would tend to revert to the state of the aboriginal rock-pigeon. In the course of time they would probably be not much more heterogeneous in character than in our first case, when birds of the same breed were confined together.

I have just said that the crossed offspring would gain in vigour and fertility. From the facts given in the seventeenth chapter there can be no doubt of this; and there can be little doubt, though the evidence on this head is not so easily acquired, that long-continued close interbreeding leads to evil results. With hermaphrodites of all kinds, if the sexual elements of the same individual habitually acted on each other, the closest possible interbreeding would be perpetual. Therefore we should bear in mind that with all hermaphrodite animals, as far as I can learn, their structure permits and frequently necessitates a cross with a distinct individual. With hermaphrodite plants we incessantly meet with elaborate and perfect contrivances for this same end. It is no exaggeration to assert that, if the use of the talons and tusks of a carnivorous animal, or the use of the viscid threads of a spider's web, or of the plumes and hooks on a seed may be safely inferred from their structure, we may with equal safety infer that many flowers are constructed for the express purpose of ensuring a cross with a distinct plant. From these various considerations, the conclusion arrived at in the chapter just referred to—namely, that great good of some kind is derived from the sexual concourse of distinct individuals—must be admitted.

To return to our illustration: we have hitherto assumed that the birds were kept down to the same number by indiscriminate slaughter; but if the least choice be permitted in their preservation and slaughter, the whole result will be changed. Should the owner observe any slight variation in one of his birds, and wish to obtain a breed thus characterised, he would succeed in a surprisingly short time by carefully selecting and pairing the young. As any part which has once varied generally goes on varying in the same direction, it is easy, by continually preserving the most strongly marked individuals, to increase the amount of difference up to a high, predetermined standard of excellence. This is methodical selection.

If the owner of the aviary, without any thought of making a new breed, simply admired, for instance, short-beaked more than long-beaked birds, he would, when he had to reduce the number, generally kill the latter; and there can be no doubt that he would thus in the course of time sensibly modify his stock. It is improbable, if two men were to keep pigeons and act in this manner, that they would prefer exactly the same characters; they would, as we know, often prefer directly opposite characters, and the two lots would ultimately come to differ. This has actually occurred with strains or families of cattle, sheep, and pigeons, which have been long kept and carefully attended to by different breeders without any wish on their part to form new and distinct sub-breeds. This unconscious kind of selection will more especially come into action with animals which are highly serviceable to man; for every one tries to get the best dog, horse, cow, or sheep, and these animals will transmit more or less surely their good qualities to their offspring. Hardly any one is so careless as to breed from his worst animals. Even savages, when compelled from extreme want to kill some of their animals, would destroy the worst and preserve the best. With animals kept for use and not for mere amusement, different fashions prevail in different districts, leading to the preservation, and consequently to the transmission, of all sorts of trifling peculiarities of character. The same process will have been pursued with our fruit-trees and vegetables, for the best will always have been the most largely cultivated, and will occasionally have yielded seedlings better than their parents.

The different strains, just alluded to, which have been raised by different breeders without any wish for such a result, and the unintentional modification of foreign breeds in their new homes, both afford excellent evidence of the power of unconscious selection. This form of selection has probably led to far more important results than methodical selection, and is likewise more important under a theoretical point of view from closely resembling natural selection. For during this process the best or most valued individuals are not separated and prevented crossing with others of the same breed, but are simply preferred and preserved; but this inevitably leads during a long succession of generations to their increase in number and to their gradual improvement; so that finally they prevail to the exclusion of the old parent-form.

With our domesticated animals natural selection checks the production of races with any injurious deviation of structure. In the case of animals kept by savages and semi-civilised people, which have to provide largely for their own wants under different circumstances, natural selection will probably play a more important part. Hence such animals often closely resemble natural species.

As there is no limit to man's desire to possess animals and plants more and more useful in any respect, and as the fancier always wishes, from fashion running into extremes, to produce each character more and more strongly pronounced, there is a constant tendency in every breed, through the prolonged action of methodical and unconscious selection, to become more and more different from its parent-stock; and when several breeds have been produced and are valued for different qualities, to differ more and more from each other. This leads to Divergence of Character. As improved sub-varieties and races are slowly formed, the older and less improved breeds are neglected and decrease in number. When few individuals of any breed exist within the same locality, close interbreeding, by lessening their vigour and fertility, aids in their final extinction. Thus the intermediate links are lost, and breeds which have already diverged gain Distinctness of Character.

In the chapters on the Pigeon, it was proved by historical details and by the existence of connecting sub-varieties in distant lands that several breeds have steadily diverged in character, and that many old and intermediate sub-breeds have become extinct. Other cases could be adduced of the extinction of domestic breeds, as of the Irish wolf-dog, the old English hound, and of two breeds in France, one of which was formerly highly valued.[4] Mr. Pickering remarks[5] that "the sheep figured on "the most ancient Egyptian monuments is unknown at the pre- "sent day; and at least one variety of the bullock, formerly "known in Egypt, has in like manner become extinct." So it has been with some animals, and with several plants cultivated by the ancient inhabitants of Europe during the neolithic period. In Peru, Von Tschudi[6] found in certain tombs, apparently prior to the dynasty of the Incas, two kinds of maize not now known in the country. With our flowers and culinary vegetables,

---

[4] M. Rufz de Lavison, in ' Bull. Soc. Imp. d'Acclimat.,' Dec. 1862. p. 1009.
[5] ' Races of Man,' 1850, p. 315.   [6] ' Travels in Peru,' Eng. translat., p. 177.

the production of new varieties and their extinction has incessantly recurred. At the present time improved breeds sometimes displace at an extraordinarily rapid rate older breeds; as has recently occurred throughout England with pigs. The Long-horn cattle in their native home were "suddenly swept " away as if by some murderous pestilence," by the introduction of Short-horns.[7]

What grand results have followed from the long-continued action of methodical and unconscious selection, checked and regulated to a certain extent by natural selection, is seen on every side of us. Compare the many animals and plants which are displayed at our exhibitions with their parent-forms when these are known, or consult old historical records with respect to their former state. Almost all our domesticated animals have given rise to numerous and distinct races, excepting those which cannot be easily subjected to selection—such as cats, the cochineal insect, and the hive-bee,—and excepting those animals which are not much valued. In accordance with what we know of the process of selection, the formation of our many races has been slow and gradual. The man who first observed and preserved a pigeon with its œsophagus a little enlarged, its beak a little longer, or its tail a little more expanded than usual, never dreamed that he had made the first step in the creation of the pouter, carrier, and fantail-pigeon. Man can create not only anomalous breeds, but others with their whole structure admirably co-ordinated for certain purposes, such as the race-horse and dray-horse, or the greyhound. It is by no means necessary that each small change of structure throughout the body, leading towards excellence, should simultaneously arise and be selected. Although man seldom attends to differences in organs which are important under a physiological point of view, yet he has so profoundly modified some breeds, that assuredly, if found wild, they would be ranked under distinct genera.

The best proof of what selection has effected is perhaps afforded by the fact that whatever part or quality in any animal, and more especially in any plant, is most valued by man, that part or quality differs most in the several races. This result is well seen by comparing the amount of difference

---

[7] Youatt on Cattle, 1834, p 200 : on Pigs, see ' Gard. Chronicle,' 1854, p. 410.

between the fruits produced by the varieties of the same fruit-tree, between the flowers of the varieties in our flower-garden, between the seeds, roots, or leaves of our culinary and agricultural plants, in comparison with the other and not valued parts of the same plants. Striking evidence of a different kind is afforded by the fact ascertained by Oswald Heer,[8] namely, that the seeds of a large number of plants,—wheat, barley, oats, peas, beans, lentils, poppies,—cultivated for their seed by the ancient Lake-inhabitants of Switzerland, were all smaller than the seeds of our existing varieties. Rütimeyer has shown that the sheep and cattle which were kept by the earlier Lake-inhabitants were likewise smaller than our present breeds. In the middens of Denmark, the earliest dog of which the remains have been found was the weakest; this was succeeded during the Bronze age by a stronger kind, and this again during the Iron age by one still stronger. The sheep of Denmark during the Bronze period had extraordinarily slender limbs, and the horse was smaller than our present animal.[9] No doubt in these cases the new and larger breeds were generally introduced from foreign lands by the immigration of new hordes of men. But it is not probable that each larger breed, which in the course of time supplanted a previous and smaller breed, was the descendant of a distinct and larger species; it is far more probable that the domestic races of our various animals were gradually improved in different parts of the great Europæo-Asiatic continent, and thence spread to other countries. This fact of the gradual increase in size of our domestic animals is all the more striking as certain wild or half-wild animals, such as red-deer, aurochs, park-cattle, and boars,[10] have within nearly the same period decreased in size.

The conditions favourable to selection by man are,—the closest attention being paid to every character,—long-continued perseverance,—facility in matching or separating animals,—and especially a large number being kept, so that the inferior individuals may be freely rejected or destroyed, and the better ones preserved. When many are kept there will also be a

[8] 'Die Pflanzen der Pfahlbauten,' 1855.
[9] Morlot, 'So:. Vau l. des Scien. Nat,'
[10] Rütimeyer, 'Die Fauna der Pfahlbauten,' 1861, s. 30.
Mars 1860, p. 298.

greater chance of the occurrence of well-marked deviations of structure.   Length of time is all-important; for as each character, in order to become strongly pronounced, has to be augmented by the selection of successive variations of the same nature, this can only be effected during a long series of generations.   Length of time will, also, allow any new feature to become fixed by the continued rejection of those individuals which revert or vary, and the preservation of those which inherit the new character.   Hence, although some few animals have varied rapidly in certain respects under new conditions of life, as dogs in India and sheep in the West Indies, yet all the animals and plants which have produced strongly marked races were domesticated at an extremely remote epoch, often before the dawn of history.   As a consequence of this, no record has been preserved of the origin of our chief domestic breeds. Even at the present day new strains or sub-breeds are formed so slowly that their first appearance passes unnoticed.   A man attends to some particular character, or merely matches his animals with unusual care, and after a time a slight difference is perceived by his neighbours;—the difference goes on being augmented by unconscious and methodical selection, until at last a new sub-breed is formed, receives a local name, and spreads; but, by this time, its history is almost forgotten.   When the new breed has spread widely, it gives rise to new strains and sub-breeds, and the best of these succeed and spread, supplanting other and older breeds; and so always onwards in the march of improvement.

When a well-marked breed has once been established, if not supplanted by still improving sub-breeds, and if not exposed to greatly changed conditions of life, inducing further variability. or reversion to long-lost characters, it may apparently last for an enormous period.   We may infer that this is the case from the high antiquity of certain races; but some caution is necessary on this head, for the same variation may appear independently after long intervals of time, or in distant places.   We may safely assume that this has occurred with the turnspit-dog which is figured on the ancient Egyptian monuments, with the solid-hoofed swine [11] mentioned by Aristotle, with five-toed fowls de-

[11] Godron, 'De l'Espèce,' tom. i., 1859, p. 368.

scribed by Columella, and certainly with the nectarine. The dogs represented on the Egyptian monuments, about 2000 B.C., show us that some of the chief breeds then existed, but it is extremely doubtful whether any are identically the same with our present breeds. A great mastiff sculptured on an Assyrian tomb, 640 B.C., is said to be the same with the dog still imported into the same region from Thibet. The true greyhound existed during the Roman classical period. Coming down to a later period, we have seen that, though most of the chief breeds of the pigeon existed between two and three centuries ago, they have not all retained to the present day exactly the same character; but this has occurred in certain cases in which improvement was not desired, for instance in the case of the Spot or the Indian ground-tumbler.

De Candolle [12] has fully discussed the antiquity of various races of plants; he states that the black-seeded poppy was known in the time of Homer, the white-seeded sesamum by the ancient Egyptians, and almonds with sweet and bitter kernels by the Hebrews; but it does not seem improbable that some of these varieties may have been lost and reappeared. One variety of barley and apparently one of wheat, both of which were cultivated at an immensely remote period by the Lake-inhabitants of Switzerland, still exist. It is said [13] that " specimens of a " small variety of gourd which is still common in the market " of Lima were exhumed from an ancient cemetery in Peru." De Candolle remarks that, in the books and drawings of the sixteenth century, the principal races of the cabbage, turnip, and gourd can be recognised; this might have been expected at so late a period, but whether any of these plants are absolutely identical with our present sub-varieties is not certain. It is, however, said that the Brussels sprout, a variety which in some places is liable to degeneration, has remained genuine for more than four centuries in the district where it is believed to have originated. [14]

In accordance with the views maintained by me in this work and elsewhere, not only the various domestic races, but the

[12] 'Géographie Botan.,' 1855, p. 989.

[13] Pickering, ' Races of Man,' 1850, p. 318.

[14] ' Journal of a Horticultural Tour,' by a Deputation of the Caledonian Hist. Soc., 1823, p. 293.

most distinct genera and orders within the same great class,—
for instance, whales, mice, birds, and fishes—are all the descendants of one common progenitor, and we must admit that the
whole vast amount of difference between these forms of life has
primarily arisen from simple variability. To consider the subject under this point of view is enough to strike one dumb
with amazement. But our amazement ought to be lessened
when we reflect that beings, almost infinite in number, during
an almost infinite lapse of time, have often had their whole
organisation rendered in some degree plastic, and that each
slight modification of structure which was in any way beneficial under excessively complex conditions of life, will have
been preserved, whilst each which was in any way injurious
will have been rigorously destroyed. And the long-continued
accumulation of beneficial variations will infallibly lead to
structures as diversified, as beautifully adapted for various
purposes, and as excellently co-ordinated, as we see in the
animals and plants all around us. Hence I have spoken of
selection as the paramount power, whether applied by man to
the formation of domestic breeds, or by nature to the production of species. I may recur to the metaphor given in a former
chapter: if an architect were to rear a noble and commodious
edifice, without the use of cut stone, by selecting from the
fragments at the base of a precipice wedge-formed stones
for his arches, elongated stones for his lintels, and flat stones for
his roof, we should admire his skill and regard him as the
paramount power. Now, the fragments of stone, though indispensable to the architect, bear to the edifice built by him the
same relation which the fluctuating variations of each organic
being bear to the varied and admirable structures ultimately
acquired by its modified descendants.

Some authors have declared that natural selection explains
nothing, unless the precise cause of each slight individual difference be made clear. Now, if it were explained to a savage
utterly ignorant of the art of building, how the edifice had been
raised stone upon stone, and why wedge-formed fragments were
used for the arches, flat stones for the roof, &c. ; and if the use
of each part and of the whole building were pointed out, it
would be unreasonable if he declared that nothing had been

made clear to him, because the precise cause of the shape of each fragment could not be given. But this is a nearly parallel case with the objection that selection explains nothing, because we know not the cause of each individual difference in the structure of each being.

The shape of the fragments of stone at the base of our precipice may be called accidental, but this is not strictly correct; for the shape of each depends on a long sequence of events, all obeying natural laws; on the nature of the rock, on the lines of deposition or cleavage, on the form of the mountain which depends on its upheaval and subsequent denudation, and lastly on the storm or earthquake which threw down the fragments. But in regard to the use to which the fragments may be put, their shape may be strictly said to be accidental. And here we are led to face a great difficulty, in alluding to which I am aware that I am travelling beyond my proper province. An omniscient Creator must have foreseen every consequence which results from the laws imposed by Him. But can it be reasonably maintained that the Creator intentionally ordered, if we use the words in any ordinary sense, that certain fragments of rock should assume certain shapes so that the builder might erect his edifice? If the various laws which have determined the shape of each fragment were not predetermined for the builder's sake, can it with any greater probability be maintained that He specially ordained for the sake of the breeder each of the innumerable variations in our domestic animals and plants;— many of these variations being of no service to man, and not beneficial, far more often injurious, to the creatures themselves? Did He ordain that the crop and tail-feathers of the pigeon should vary in order that the fancier might make his grotesque pouter and fantail breeds? Did He cause the frame and mental qualities of the dog to vary in order that a breed might be formed of indomitable ferocity, with jaws fitted to pin down the bull for man's brutal sport? But if we give up the principle in one case,—if we do not admit that the variations of the primeval dog were intentionally guided in order that the greyhound, for instance, that perfect image of symmetry and vigour, might be formed,—no shadow of reason can be assigned for the belief that variations, alike in nature and the result

of the same general laws, which have been the groundwork
through natural selection of the formation of the most perfectly adapted animals in the world, man included, were intentionally and specially guided. However much we may wish
it, we can hardly follow Professor Asa Gray in his belief " that
" variation has been led along certain beneficial lines," like a
stream " along definite and useful lines of irrigation." If we
assume that each particular variation was from the beginning
of all time preordained, the plasticity of organisation, which
leads to many injurious deviations of structure, as well as that
redundant power of reproduction which inevitably leads to a
struggle for existence, and, as a consequence, to the natural
selection or survival of the fittest, must appear to us superfluous
laws of nature. On the other hand, an omnipotent and omniscient Creator ordains everything and foresees everything.
Thus we are brought face to face with a difficulty as insoluble
as is that of free will and predestination.

INDEX.

THE END.

Printed in the United States
By Bookmasters